Leno Mascia

Polymers in Industry from A—Z

Related Titles

Schlüter, Dieter A. / Hawker, Craig / Sakamoto, Junji (eds.)

Synthesis of Polymers

New Structures and Methods

2012
ISBN: 978-3-527-32757-7

Mathers, Robert T. / Meier, Michael A. R. (eds.)

Green Polymerization Methods

Renewable Starting Materials, Catalysis and Waste Reduction

2011
ISBN: 978-3-527-32625-9

Cosnier, S., Karyakin, A. (eds.)

Electropolymerization

Concepts, Materials and Applications

2010
ISBN: 978-3-527-32414-9

Mittal, V. (ed.)

Polymer Nanotube Nanocomposites

Synthesis, Properties, and Applications

Hardcover
ISBN: 978-0-470-62592-7

Eftekhari, A. (ed.)

Nanostructured Conductive Polymers

ISBN: 978-0-470-74585-4

Gujrati, P. D., Leonov, A. I. (eds.)

Modeling and Simulation in Polymers

2010
ISBN: 978-3-527-32415-6

Leclerc, M., Morin, J.-F. (eds.)

Design and Synthesis of Conjugated Polymers

2010
ISBN: 978-3-527-32474-3

Xanthos, M. (ed.)

Functional Fillers for Plastics

2010
ISBN: 978-3-527-32361-6

Pascault, J.-P., Williams, R. J. J. (eds.)

Epoxy Polymers

New Materials and Innovations

2010
ISBN: 978-3-527-32480-4

Mittal, V. (ed.)

Optimization of Polymer Nanocomposite Properties

2010
ISBN: 978-3-527-32521-4

Geckeler, K. E., Nishide, H. (eds.)

Advanced Nanomaterials

2010
ISBN: 978-3-527-31794-3

Leno Mascia

Polymers in Industry from A–Z

A Concise Encyclopedia

WILEY-VCH

WILEY-VCH Verlag GmbH & Co. KGaA

The Author

Dr. Leno Mascia
Department of Materials
Loughborough University
Loughborough LE11 3TU
United Kingdom

All books published by **Wiley-VCH** are carefully produced. Nevertheless, authors, editors, and publisher do not warrant the information contained in these books, including this book, to be free of errors. Readers are advised to keep in mind that statements, data, illustrations, procedural details or other items may inadvertently be inaccurate.

Library of Congress Card No.: applied for

British Library Cataloguing-in-Publication Data
A catalogue record for this book is available from the British Library.

Bibliographic information published by the Deutsche Nationalbibliothek
The Deutsche Nationalbibliothek lists this publication in the Deutsche Nationalbibliografie; detailed bibliographic data are available on the Internet at http://dnb.d-nb.de.

© 2012 WILEY-VCH Verlag GmbH & Co. KGaA, Boschstr. 12, 69469 Weinheim, Germany

All rights reserved (including those of translation into other languages). No part of this book may be reproduced in any form – by photoprinting, microfilm, or any other means – nor transmitted or translated into a machine language without written permission from the publishers. Registered names, trademarks, etc. used in this book, even when not specifically marked as such, are not to be considered unprotected by law.

Typesetting   Thomson Digital, Noida, India
Printing and Binding   Markono Print Media Pte Ltd, Singapore
Cover Design   Adam-Design, Weinheim

Printed in Singapore
Printed on acid-free paper

Print ISBN: 978-3-527-32964-9
ePDF ISBN: 978-3-527-64405-6
oBook ISBN: 978-3-527-64403-2
ePub ISBN: 978-3-527-64404-9
Mobi ISBN: 978-3-527-64406-3

Dedication

To my grandchildren and their future

Contents

Preface *IX*
Acknowledgements *XI*
Overview Guide *XIII*
List of Acronyms *XXIII*

Chapter	*Entries*	*Pages*
A	Abrasion Resistance — Azeotropic Copolymerization	*1*
B	Back-Flow — Butyl Rubber	*24*
C	Calcium Carbonate ($CaCO_3$) — Cyanate Ester	*37*
D	DABCO — Dynamic Vulcanization	*80*
E	Effective Modulus — Eyring Equation	*99*
F	Fabrication — Fusion Promoter	*126*
G	Gate — Gutta Percha	*153*
H	Halogenated Fire Retardant — Hyperbranched Polymer	*157*
I	Impact Modifier — Izod Impact Test	*163*
J	*J* integral — Joint	*175*
K	*K* Value — Kneading	*176*
L	Lamella — Lüder Lines	*178*
M	M_{100} and M_{300} — Mylar	*191*
N	Nafion — Nylon Screw	*215*
O	Oil Absorption — Ozone	*225*
P	Paint — Pyrolysis	*229*
Q	Q Meter — Quinone Structure	*267*
R	Rabinowitsch Equation — Rutile	*268*
S	Sag — Syntactic Foam	*277*
T	Tack — Tyre Construction	*301*
U	Ubbelohde Viscometer — UV Stabilizer	*318*
V	Vacuum Forming — Vulcanization	*326*

W		Wall Slip — Work of Adhesion	335
X		X-Ray Diffraction (XRD) — Xenon Arc Lamp	342
Y		Y Calibration Factor — Young's Modulus	343
Z		Z-Blade Mixer — Zisman Plot	347
		References	349

Preface

Polymers are a well-established class of materials both in the commercial sector and in educational curricula. Polymers are the main component of commercial products known as plastics, composites, rubber (or elastomers), surface coatings, fibres and adhesives.

Although many books have been published over the past 50 years or so to satisfy the demands of those concerned with the scientific and technological aspects of the subject, the author feels that there is a need for a reference text that can provide easy access to brief and concise information on the terminology, concepts, principles and industrial practice related to the constitution, manufacture and properties of polymer materials.

A compact encyclopaedia provides the easiest and most rapid route for retrieving both specific and general information about the subject of interest. This is particularly valuable when the reader is interested primarily in the basics of the subject.

Although the central focus of this book is on aspects concerned with the constitution, properties and processing of polymer-based materials, the treatment extends into related areas, including synthesis and characterization. The amount of information and details provided for each entry, therefore, varies according to the anticipated needs and interests of the potential reader within the core areas.

The contents covered by the text have been derived with the view that the field spans various disciplines and branches of industry and, therefore, the needs of the potential reader beyond the boundaries of these areas are served by complementary texts related to other sectors, such as the petrochemical industry and specific manufacturing concerns.

The information is presented in two sections. The main part of the book consists of an "A to Z encyclopaedic outline", which enables the reader to search directly for a particular topic or item of interest. This is complemented by the preliminary section "Overview" and "Search Guide", which should assist the reader to identify the specific topic or term for the search. All the terms that appear in the "Overview" and "Search Guide" are identified as individual entries in the main part of the book that follows, so that the reader can scan the entire field and select the topics and aspects of the subject that are of interest.

Acknowledgements

A book that covers such a wide field could not have been written without the contribution of very many authors over a large number of years. Although credit has been given to authors and publishers, there have been a few occasions where the original source could not be ascertained. Some of the diagrams used in the text have been taken from personal lecture notes produced as early as 1970. Many of these were extracted from the immensely rich literature provided by the industrial sector, whose identity was not recorded at the time. The author wishes to thank these anonymous contributors and hopes to be able to overcome any such deficiency in any future editions, if the required information is brought to his attention or is otherwise obtained.

Overview

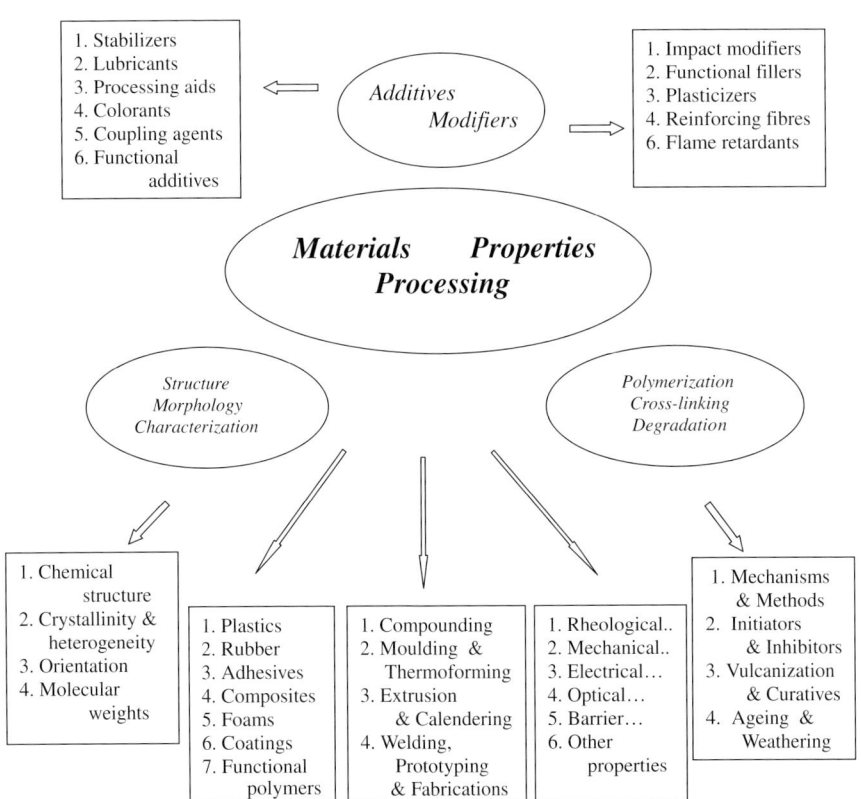

Supporting fundamental principles

Search Guide

1. Materials

1.1 Plastics

Thermoplastics

Acetals: Oxymethylene polymer, Polyoxymethylene, Poly(methylene oxide) (PMO)

Acrylics: Poly(methyl methacrylate) (PMMA), Methyl methacrylate–butadiene–styrene (MBS)

Barrier polymers: Phenoxy, Poly(vinylidene chloride) (PVDC), Ethylene–vinyl alcohol copolymer (EVOH)

Cellulosics: Cellulose acetate, Cellulose acetate butyrate, Cellulose nitrate, Cellulose propionate

Fluoropolymers: Perfluoropolymer (PFA), Polychlorotrifluoroethylene (PCTFE), Polytetrafluoroethylene (PTFE), Poly(tetrafluoroethylene–ethylene) copolymer (PETFE), Poly(tetrafluoroethylene–hexafluoropropylene) copolymer (FEP), Poly(vinylidene fluoride) (PVDF or PVF_2)

High-temperature polymers: Polybenzimidazole (PBI), Poly(ether ketone) (PEK), Poly(ether ether ketone) (PEEK), Polyketone, Poly(phenylene oxide) (PPO), Polysulfone (PSU), Poly(ether sulfone) (PES), Poly(phenylene sulfide) (PPS), Poly(ether imide) (PEI)

Polyamides: Nylon 6 (PA 6), Nylon 4,6 (PA 4,6), Nylon 6,6 (PA 6,6), Nylon 11 (PA 11), Nylon 12 (PA 12)

Polyesters: Caprolactone polymer (PLC), Poly(butylene terephthalate) (PBT), Polycarbonate (PC), Poly(ethylene terephthalate) (PET), Aromatic polyester

Polyolefins: Chlorinated polyethylene, Chlorosulfonated polyethylene, Coupled polypropylene, Ethylene–carbon monoxide copolymer (ECO), Ethylene–ethyl acrylate copolymer (EEA), Ethylene–methyl methacrylate copolymer (EMA), Ethylene–vinyl acetate copolymer (EVA), Ethenoid polymer, Polyethylene (LDPE, VLDPE, LLDPE, MDPE, HDPE, UHMWPE), Polypropylene (PP), Polybutene, Poly(4-methyl pent-1-ene)

Styrene polymers: Acrylonitrile–butadiene–styrene terpolymer (ABS), Acrylonitrile–acrylate–styrene terpolymer (ASA), High-impact polystyrene (HIPS), Polystyrene (PS), Styrene–acrylonitrile copolymer (SAN), Styrene–maleic anhydride copolymer (SMA)

Vinyl polymers: Chlorinated PVC, Poly(vinyl acetate) (PVAc), Poly(vinyl alcohol) (PVA), Poly(vinyl alkyl ether) (PVME), Poly(vinyl butyral), Poly(vinyl carbazole), Poly(vinyl chloride) (PVC, PVC-U), Poly(vinyl fluoride) (PVF), Polyvinylpyridine, Polyvinylpyrrolidone

Thermosets

Amino resins: Melamine formaldehyde (MF), Urea formaldehyde (UF), Casein

Epoxy resins: Bisphenol-A (DGEBA), Cycloaliphatic, Epoxidized novolac resin, Tetraglycidoxylmethylene p,p'-diphenylene diamine (TGDM)

High-temperature resins: Cyanate ester, Diallyl phthalate, Furan resin, PMR, Polyimide, Polybismaleimide

Phenolics: Bakelite, Novolac, Resole

Unsaturated polyesters: Resins for composites, Bulk moulding compound (BMC), Sheet moulding compound (SMC), Vinyl ester

Other Aspects

Gel, Recycling, Polymer blend, Thermoreversible gel, Vitrification

1.2 Rubber

Vulcanized/Cross-Linked Elastomers

Acrylates: Ethylene–ethyl acrylate terpolymer elastomer (EAM)

Diene elastomers: Acrylonitrile–butadiene (NBR), Gutta percha, Natural rubber (NR), Polybutadiene (BR), Polychloroprene (CR), Polyisoprene (IR), Styrene–butadiene (SBR)

Fluoroelastomers: Hexafluoropropylene copolymer and terpolymer

Hydrocarbon (polyolefin) elastomers: Butyl rubber, Ethylene–propylene rubber (EPR), Ethylene–propylene–diene monomer rubber (EPDM), Chlorosulfonated polyethylene (CSM)

Silicone rubber: Polydimethylsiloxane (PDMS, VMQ), Fluorosilicone (FVMQ)

Other elastomers: Epichlorohydrin (ECO), Phosphazene elastomer, Polycarborane–siloxane, Polyurethanes (PUR)

Thermoplastic Elastomers (TPE)
Block copolymers: Styrene–butadiene–styrene (SBS), Styrene–ethylene/butylene–styrene (S-EB-S), Styrene–isoprene–styrene (SIB), Poly(butylene terephthalate-*b*-tetramethylene oxide) (TPE-E), Poly(amide-g-alkylene oxide) (TPE-A), Poly(urethane-g-ester/alkylene oxide) (TPU)

Dynamic vulcanizates: Thermoplastic polyolefin (TPO), Plasticized PVC/NBR blend, NR/EPR blend

Other Aspects
Coagulum, Crepe, Fluidized bed, Latex, Mastication, Microwaving, Vulcanizate

1.3 Adhesives

Hot Melts
Ethylene–vinyl acetate copolymer (EVA), Ethylene–ethyl acrylate copolymer (EEA), Polyamide, Polyester, Phenoxy

Curable Adhesives
Acrylic, Cyanoacrylate, Epoxy, Phenolic

Water-Borne
Poly(vinyl acetate), Styrene–butadiene copolymer (SBR), Poly(ethylene–vinyl acetate) copolymer (VAE)

Pressure-Sensitive Adhesive
Elastomer, Tackifier

Other Aspects
Adherend, Adhesive test, Adhesive wetting, Anodizing, Debonding, Cold plasma, Corona discharge, Critical surface energy, Critical wetting tension, Joint, Lap shear, Pot life, Tack, Work of adhesion

1.4 Composites

Fibre Reinforcement
Continuous fibres: Aramid fibre, Carbon fibre, Glass fibre, Graphite fibre

Short fibres: Asbestos, Carbon fibre, Chopped strand mat laminate (CSM), Glass fibre

Particulate Reinforcement
Fillers: Barite, Bentonite, Calcium carbonate, Carbon black, Channel black, Clay, Fumed silica, Functional filler, Glass flake, Kaolin, Mica, Quartz, Titania

Nanofillers: Carbon nanotube, Exfoliation, Montmorillonite, Nanoclay

Fire retardant and Antitracking: Aluminium trihydrate, Antimony oxide, Magnesium hydroxide, Molybdenum oxide

Other fillers: Bioactive filler, Magnetic filler, Molybdenum disulfide, Metal powder

Matrix
Thermosets: Unsaturated polyester, Epoxy resin, Phenolic, Cyanoacrylate, PMR, Polyimide

Thermoplastics: Coupled polypropylene, Polyamide, Poly(ether ketone), Poly(ether imide), Poly(phenylene sulfide)

Nanocomposites
Ceramer, Exfoliated nanocomposite, Organic modified filler, Organic–inorganic hybrid

Other Aspects
Debonding, BET isotherm, Commingled fibre, Composite manufacture, Composite

property prediction (Reinforcement theory), Composite test, Coupling agent, Critical fibre length, Gel coat, Graphene, Exfoliation, Intercalation, Ion exchange capacity (IEC), Interfacial bonding, Laminate, Low profile additive, Modulus enhancement factor, Oil absorption, Prepreg, Reinforcement factor, Size, Sizing/Finishing, Zeta potential

1.5 Foams

Aspects
Closed cell, Foam density and properties, Foam formation mechanism, Foam manufacture, Open cell, Structural foam, Syntactic foam

1.6 Coatings

Ac Curable Varnishes
Acrylic, Alkyd, Drying resin, Epoxide, Phenolic, Urethane

Water-Borne Dispersions
Microemulsions: Polyurethane, Epoxide
  Polymer emulsions: Poly(vinyl acetate), Acrylic

Powders
Curable thermosetting systems: Epoxide, Poly(ester–epoxide), Polyurethane
  Thermoplastic powders: Polyester, Acrylic

Other Aspects
Anodizing, Blistering, Cold plasma, Contact angle, Critical surface energy, Critical wetting tension, Dip coating, Electrolytic deposition, Electrostatic spraying, Film formation, Metallization, Gel coat, Intumescent coating, Pinhole, Plating, Pot life, Primer, Roughness factor, Sputtering, Surface energy, Work of adhesion, Zisman plot

1.7 Functional Polymers

Biopolymers
Chitin, Chitosan, Ester–amide polymer, Ethylene–carbon monoxide copolymer (ECO), Poly(lactic acid), Polylactide, Oxo-biodegradable polymer, Poly(hydroxy alkanoate), Poly(hydroxy butyrate), Polypeptide, Polysaccharide, Starch

Conductive Polymers
Polyacetylene, Polyaniline (PANI), Polypyrrole

Ionomeric Polymers
Ionomer, Ion exchange resin, Nafion, Polyelectrolyte

Other Aspects
Cross-linked thermoplastic, Dendrimer, Doping, Engineering polymer, Heat-shrinkable product, Light-sensitive polymer, Liquid-crystal polymer, Shape memory polymer, Nonlinear dielectric polymer, Piezoelectric polymer, Photoresist, Poly(amic acid), PTC polymer, Spandex fibre

2. Properties

2.1 Rheological properties

Rheology
Barus effect, Bingham body, Carreau model, Complex viscosity, Consistency index, Converging flow, Couette flow, Deborah number, Deformational behaviour, Die swell, Dilatant fluid, Drag flow, Elongational flow, Elongation rate, Elongational viscosity, Ellis equation, Extension viscosity, G' and G'', Gel time (gel point), Melt elasticity, Mooney equation, Newtonian behaviour, Non-Newtonian behaviour, Normal stress difference, First normal stress coefficient, Plug flow, Power law, Power-law index, Thixotropy, Trouton viscosity, True shear rate, True viscosity, Viscosity, Zero-shear viscosity

Rheometry
Bagley correction, Brookfield viscometer, Capillary rheometer, Cone-and-plate rhe-

ometer, Coaxial cylinder rheometer, Die-exit phenomena, Dynamic (oscillatory) flow, Gelation, Entry effect, Melt flow index (Melt flow rate), Melt fracture, Melt strength, Monsanto rheometer, Mooney viscometer, Parallel-plate rheometer, *PVT* diagram, Rabinowitsch correction, Sharkskin, Slip analysis, Slit rheometer, Torque rheometer, Weissenberg rheogoniometer

Other Aspects
Critical chain length (Z_c), Deformational behaviour, Entanglement, Flow curve, Mark–Huggins equation, Radius of gyration, Reptation, Rubbery state, Viscous state

2.2 Mechanical Properties

Elasticity and Viscoelasticity
Bulk modulus, Compliance, Complex compliance, Complex modulus, Creep, Creep compliance, Creep curves, Creep modulus, Creep period, Damping factor, Deflection under load temperature, Deformational behaviour, Dynamic mechanical spectra, Effective modulus, Elasticity, Elastic behaviour, Elastic memory, Elastic recovery, Flexural modulus, Force–deflection curve, Force–deformation curve, Force–extension curve, Heat distortion temperature (HDT), Hundred-percent (100%) modulus, Kelvin–Voigt model, Load–deflection curve, Load–deformation curve, Load–extension curve, Loss angle, Loss compliance, Loss factor, Loss modulus, Loss tangent, Master curve, Maxwell model, Modulus, M_{100} and M_{300}, Nonlinear viscoelastic behaviour, Phase angle, Recovery, Recoverable strain, Reduced time, Relaxation time, Retardation time, Relaxation modulus, Relaxed modulus, Rubber elasticity, Shear modulus, Shift factor, Standard linear solid model, Tan δ, Tangent modulus, Tensile modulus, Time-dependent modulus, Time–temperature superposition, Viscoelastic behaviour, Voigt model, Young's modulus

Failure and Fracture
a/W ratio, Brittle fracture, Brittle point, Brittle strength, Brittle–tough transition, Cold-flex temperature, Cold flow, Compression set, Crack initiation, Crack length, Crack opening displacement (COD), Crazing, Creep rupture, Critical crazing strain, Critical strain energy release rate (G_c), Critical stress intensity factor (K_c), Ductile, Ductile–brittle transition, Ductile failure, Environmental stress cracking, Fatigue life, Flex cracking, Flex life, Flexural strength, Fracture mechanics, Griffith's equation, Impact strength, Interlaminar shear strength (ILSS), *J*-integral, Lüder lines, Notch sensitivity, Peel strength, Permanent set, Plastic deformation, Tear strength, Tenacity, Tresca criterion, Von Mises criterion, Yield criteria, Yield failure, Yield point, Yield strength, Young's modulus

Test Methods
Ball drop, Bending test, Charpy impact test, Compression test, Creep test, Dilatometer, Dumb-bell specimen, Dynamic mechanical analysis (DMA), Dynamic mechanical thermal analysis (DMTA), Extensometer, Falling-weight impact test, Fatigue test, Fracture test, Impact test, Izod impact test, Mechanical spectroscopy, Microhardness, Notch, Peel test, Pendulum impact test, Rubber elasticity, Tensile test, Viscoelasticity

Other Aspects
Abrasion resistance, Anisotropy, Boltzmann superposition principle, Compression, Compact tension specimen, Damage tolerance, Damping, Deformation, Delamination, Distribution of relaxation times, Distribution of retardation times, Dynamic mechanical spectra, Flexural properties, Friction coefficient, Hardness, Hydrostatic stress, Life prediction, Load, Microvoid, Poisson's ratio, Rockwell hardness, Rubbery state, Shear stress, Shore hardness,

Stiffness, Strain, Strain rate, Stress, Stress concentration, Tack, Transmissibility, Tensor, Torque, Torsion pendulum, Toughness, Vicat softening point, Vickers hardness, Volumetric strain

2.3 Electrical Properties

Low Voltage

Capacitance, Clausius–Mossotti equation, Complex permittivity, Conductance, Conductivity, Corona discharge, Current density, Dielectric, Dielectric constant, Dielectric properties, Dielectric thermal analysis (DETA), Direct current, Electrostatic charge, Loss angle, Loss factor, Loss tangent, Permittivity, Polarization, Resistivity, Specific impedance, Stress concentration, Surface resistivity, Tan δ, Volume resistivity

High Voltage

Breakdown voltage, Comparative tracking index (CTI), Dielectric strength, Dry band, Electrical failure, Electric strength, Tracking

Other Aspects

Dipole, Electronic conductivity, Electrostatic charge, Ionic conductivity, Polarity, Stress grading, Nonlinear dielectric, PTC polymer

2.4 Optical Properties

Aspects

Backscatter, Birefringence, CIE chromaticity diagram, Colour, Colour matching, Direct reflection factor, Extinction index, Forward scatter, Gloss, Haze, Light microscopy, Light scattering, Light transmission factor, Optical brightener, Optical microscopy, Optical path difference, Orientation, Photoelasticity, Reflection factor, Refractive index, See-through clarity, Stress optical coefficient, Transparency

2.5 Barrier Properties

Aspects

Diffusion, Diffusion coefficient, Fickian behaviour, Permeability, Solubility, Time lag

2.6 Other Properties

Types

Acoustic: Attenuation, Loss factor

Fire resistance: Cone calorimeter, Flash point, Limiting oxygen index, Pyrolysis, Self-extinguishing, Thermal gravimetric analysis

Thermal properties: Dilatometer, Dilatation coefficient, Thermal conductivity, Thermal expansion coefficient

3. Processing

3.1 Compounding

Aspects

Banbury mixer, Decompression zone, Devolatilization, Dispersion, Dispersive mixing, Dry blend, Formulation, Master batch, Melting, Miscibility, Mixing, Mixer, Pelletizer, Plastication, Plastograph, Purging, Reciprocating screw mixer, Thermal degradation, Torque rheometer, Twin-screw extruder

3.2 Moulding and Thermoforming

Machine and Moulds

Clamping force, Clamping system, Ejection mechanism, Film gate, Gate, Hot runner mould, Locking mechanism, Mould design, Nozzle, Nylon screw, Reciprocating screw, Runner, Sprue, Stripper plate

Operations

Air-slip forming, Blow moulding, Cavity filling, Cavity packing, Compression moulding, Co-injection moulding, Drape forming, Injection blow moulding, Injection moulding, Plastication, Plug-assisted vacuum forming, Purging, Reaction moulding, Reaction processing, Resin transfer moulding, Rotational moulding, Stretch blow moulding, Thermoforming, Transfer moulding, Vacuum forming

Other Aspects

Back-pressure, Demoulding, Distortion and Warping, Flash, Gel coat, Internal stress, Moulding cycle, Moulding defect, Pre-form, *PVT* diagram, Residual stress, Sink mark, Spiral flow moulding, Void, Weld line

3.3 Extrusion and Calendering

Extruder

Barrel, Barrier flights screw, Co-rotating, Counter-rotating, Decompression zone, Devolatilization, Extruder screw, Gear pump, Nylon screw, Twin-screw extruder

Extrusion Dies

Die gap, Die lip, Fish-tail die, Land length, Sizing die, Spider leg, Mandrel

Other Aspects

Adiabatic heating, Back-flow, Blocking, Blow-up ratio, Blown film, Breaker plate, Calender, Cambering (Roll crowning), Chill-roll casting, Co-extrusion, Crow feet, Extrusion theory, Die analysis, Die drool, Drag flow, Draw ratio, Drawdown ratio, Extrudate, Fish eye, Flow analysis, Flow instability, Foam extrusion, Lamination, Leakage flow, Parison, Plastication, Plate out, Pressure flow, Purging, Screw–die interaction, Surging, Tubular film, Weld line

3.4 Welding, Prototyping and Fabrications

Hot-plate welding, Friction welding, High-frequency welding, Ultrasonic welding, Rapid prototyping, Stereolithography, Casting, Machining, Powder sintering, Tyre construction

4. Structure, Morphology and Characterization

4.1 Chemical Structure

Acid number, Aliphatic, Atactic, Attenuated total reflectance spectroscopy, Block copolymer, Carbonyl index, Chain configuration, Chain regularity, Chain stiffness, Chemical shift, Configuration, Conformation, Conjugated double bond, Coordination, Copolymer, Critical chain length, Cross-link, Degree of cross-linking, Degree of polymerization, Dendrimer, Electron spin resonance, Electron spectroscopy for chemical analysis (ESCA), Free radical, Fluorescence, Functionality, Head-to-head, Head-to-tail, Heterochain, Hydrogen bond, Hydrophilic, Hydrophobic, Hydroxyl equivalent (number), Hygroscopic, Hyperbranched polymer, Infrared spectroscopy (IR, FTIR), Interpenetrating polymer network (IPN), Isotactic, Network, Non-polar polymer, Oligomer, Orientation, Oxirane ring, Polarity, Quinone structure, Radiation, Radical, Radius of gyration, Raman spectroscopy, Random copolymer, Side group, Spectroscopy, Spectrum, Syndiotactic, Tacticity, UV spectroscopy, X-ray, X-ray photoelectron microscopy (XPS)

4.2 Crystallinity and Heterogeneity

Amorphous polymer, Atomic force microscopy (AFM), Birefringence, Chain folding, Cluster, Cold crystallization, Colloidal, Co-continuous domain, Core and shell, Crystalline polymer, Crystallization, Degree of crystallinity, Dielectric thermal analysis (DETA), Differential scanning calorimetry (DSC), Differential thermal analysis (DTA), Dynamic mechanical thermal analysis (DMA, DMTA), Exfoliation, Emulsion, First-order transition, Fractal, Fringe micelle, Intercalation, Lamella, Light microscopy, Light scattering, Melting point, Microscopy, Microcavitation, Microemulsion, Microvoid, Optical microscopy, Scanning electron microscope, Secondary crystallization, Shish-kebab crystal, Small-angle X-ray scattering (SAXS), Spherulite, Transmission electron microscopy (TEM), Unit cell, Wide-angle X-ray diffraction (WAXS), X-ray

4.3 Orientation

Anisotropy, Annealing, Biaxial, Birefringence, Draw ratio, Extension ratio, Fibre, Heat setting

Heat-shrinkable polymer, Monoaxial, Monofilament, Optical path difference, Orientation function, Wide-angle X-ray diffraction (WAXS)

4.4 Molecular weights

Dispersity index, Distribution of molecular weight, Gel permeation chromatography, Huggins' constant, Intrinsic viscosity, Molar mass, Molecular weight, Monodisperse, Measurement of molecular weight, Mark–Houwink equation, Number-average molecular weight, Osmotic pressure, Osmometry, Relative viscosity, Weight-average molecular weight, Size exclusion chromatography, Staudinger, Ubbelohde viscometer, Viscometer, Viscosity

5. Polymerization Cross-linking and Degradation

5.1 Initiators, Inhibitors and Vulcanization

Accelerator, Activator, Cobalt naphthanate and octoate, Co-agent, Cure time, Curing, Efficient vulcanization, Fatty acid, Kicker, Mechanical degradation, Mercaptan, Microwaving, Peroxide, Photoinitiator, Post-curing, Scorch time, Retarder, Thiouram, Zinc oxide

5.2 Polymerization and Cross-Linking

Mechanisms
Addition polymerization, Anionic polymerization, Cationic polymerization, Chain stopper, Condensation reaction, Condensation polymerization, Copolymerization, Depolymerization, DMP-30, Free-radical polymerization, Ionic polymerization, Metallocene catalysis, Propagation reaction, Reactivity ratio, Termination reaction, Ziegler catalyst, Ziegler–Natta catalyst

Methods
Bulk polymerization, Electron beaming, Emulsion polymerization, Heterophase or Heterogeneous polymerization, Mass polymerization, Mechanical degradation, Phase inversion, Photopolymerization, Radiation processing, Reaction processing, Seed polymerization, Solid-state polymerization, Surfactant, Suspension polymerization

5.3 Ageing and Weathering

Depolymerization, Chalking, Chain scission, Composting, Environmental ageing, Fogging, Hydrolysis, Induction time, Life cycle analysis, Life prediction, Ozone, Physical ageing, Propagation reaction, Radiation, Recycling, Secondary crystallization, Solar radiation, Thermal degradation, Ultraviolet (UV) light, UV degradation, Xenon arc lamp, Yellowness index

6. Additives and Modifiers

6.1 Additives

Stabilizers
Antidegradant, Antioxidant, Antiozonant, Chelating agent, Excited state quencher, Hindered amine light stabilizer (HALS), Hindered phenol, Hydrolysis stabilizer, Lead stabilizer, Metal deactivator, Phenolic antioxidant, Primary stabilizer, Processing stabilizer, Radical scavenger, UV stabilizer

Lubricants and Processing Aids
Antiblocking agent, Antifoaming agent, Blocking, Cationic surfactant, External lubricant, Factice, Flow promoter, Fusion promoter, Internal lubricant, Low-profile additive, Lubricant, Peptizer, Stearate, Stearic acid, Tackifier, Thickener, Thixotropic agent

Colorants
Anatase, CIE chromaticity diagram, Dye, Mineral pigment, Pigment, Rutile, Thermochromic pigment

Coupling Agents
Compatibilizing agent, Interfacial bonding, Silane, Size (Sizing), Surfactant

Functional Additives
Antifoaming agent, Antimicrobial (biocidal) agent, Antistatic agent, Antitracking additive, Blowing agent, Flocculant, Fungicide, Nucleating agent

Other Aspects
Additive concentrate, Compounding, Fogging, Law of mixtures, Master batch, Miscibility, Mixing, Solubility parameter, Stabilization, Synergism

6.2 Property Modifiers

Fire (Flame) Retardants
Aluminium trihydrate, Antimony oxide, Brominated compound, Chlorinated fire retardant, Magnesium hydroxide, Molybdenum oxide

Impact Modifiers
Acrylonitrile–butadiene–styrene terpolymer (ABS), Amine-terminated acrylonitrile–butadiene–styrene terpolymer (ATBN), Carboxylic acid-terminated acrylonitrile–butadiene–styrene terpolymer (CTBN), Methacrylate–butadiene–styrene terpolymer (MBS), Liquid rubber, Polymer alloy, Polymer blend, Telechelic oligomer

Functional Fillers
Barite, Bioactive filler, Carbon black, Carbon nanotube, Magnetic filler, Metal powder

Plasticizers
Epoxidized soya-bean oil, Extender, Low-temperature plasticizer, Non-migratory plasticizer, Organosol, Primary plasticizer, Reactive diluent, Secondary plasticizer

Reinforcing Fibres
Glass fibre, Carbon fibre (Acrylic fibre), Aramid fibre

Other Aspects
Antiplasticization, Cold-flex temperature, Compatibility, Glass transition temperature, Internal plasticization, Lower critical solution temperature (LCST), Miscibility, Mixing, Plasticization, Plastisol, Solubility parameter, Upper critical solution temperature (UCST)

7. Supporting Fundamental Principles and Terminology

Absorbance, Activation energy, Activation volume, Alpha transition, Alternating current, Aprotic polar solvent, Arrhenius equation, Beer–Lambert law, Beta transition, Boundary condition, Case II diffusion, Cluster, Cohesive energy density, Cold plasma, Colligative properties, Colloidal, Conformation, Configuration, Conservation law, Continuum mechanics, Coordination, Current density, Deconvolution, Direct current, Elasticity, Electromagnetic radiation, Empirical, Endothermic, Engineering design, Enthalpy, Entropy, Exothermic, Extinction coefficient, Fickian behaviour, First-order transition, Fluorescence, Fractal, Fractography, Free radical, Fundamental property, Gibbs free energy, Hooke's law, Hydrogen bond, Hydrolysis, Hydrophilic, Hydrophobic, Hydrostatic pressure, Hydrostatic stress, Hydrothermal stress, Hygroscopic, Induction time, Insulation, Interaction, Interfacial polarization, Intermolecular, Internal energy, Intramolecular, Isothermal, Isotropic, Law of mixtures, Linear behaviour, Linear elastic, Lower bound, Lubrication approximation, Melting point, Modulus, Momentum, Momentum equation, Morphology, Non-destructive test,

Non-polar, Nuclear magnetic resonance, Ozone, Photoelasticity, Plane strain, Plane stress, Plastic, Poiseuille equation, Poisson's ratio, Polarity, Polarization, Polarized light, Radical, Solar radiation, Spectrum, Stiffness, Stokes equation, Synergism, Tensor, Torque, Transesterification, Van der Waals forces, Vector

List of Acronyms

AAA	acetoxyacetoanilide
ABA	acetoxybenzoic acid
ABS	acrylonitrile–butadiene–styrene (terpolymer)
AC	alternating current
ACM	acrylic rubber, copolymer of ethyl acrylate and butyl acrylate
AED	anionic electrolytic deposition
AFM	atomic force microscopy
AIBN	azo-bis-isobutyronitrile
ANA	acetoxynaphthoic acid
ASA	acrylonitrile–acrylate–styrene (terpolymer)
ATBN	amine-terminated butadiene–acrylonitrile oligomer
ATH	aluminium (or alumina) trihydrate
ATR	attenuated total reflectance (spectroscopy)
a/W ratio	ratio of length of crack (or notch), a, to width of specimen, W, used in evaluation of the toughness of materials
AZBN	azo-bis-dibutyronitrile
AZDN	azo-bis-dibutyronitrile
bis-MPA	bis(hydroxymethyl)propionic acid
BMC	bulk moulding compound
BMI	bismaleimide (resin)
b.p.	boiling point
BP	benzophenone
B-stage	term used to describe an intermediate state of cure of thermosetting resins, particularly phenolic systems
BUR	blow-up ratio
CA	cellulose acetate
CAB	cellulose acetate butyrate
CAD	computer-aided design
CAP	cellulose acetate propionate
CBT	cyclic butylene terephthalate
CEC	cation exchange capacity

CED	cationic electrolytic deposition
CED	cohesive energy density
CFRP	carbon-fibre-reinforced plastics
CIE	Commission International de l'Eclairage
CN	cellulose nitrate
CNT	carbon nanotube
COD	crack-opening displacement
CP	cellulose propionate
CPE	chlorinated polyethylene
CR	chloroprene rubber
CSM	chlorosulfonated polyethylene
CSM	chopped strand mat
CT	compact tension (specimen)
CTBN	carboxyl-terminated butadiene–acrylonitrile oligomer
CTI	comparative tracking index
DABCO	1,4-diazabicyclo[2.2.2]octane (tertiary cyclic diamine)
DAI	diaryliodonium
DAP	diallyl phthalate
DBP	dibutyl phthalate
DBTL	dibutyl tin dilaurate
DC	direct current
DCB	double cantilever beam (method)
DDS	4,4'-diaminodiphenylsulfone
DDSA	dodecylsuccinic anhydride
DEAP	diethoxyacetophenone
DEN	double edge notch
DETA	dielectric thermal analysis
DETA	diethylenetriamine
DGEBA	diglycidyl ether of bisphenol A (resin)
DIAP	diallyl isophthalate
DICY	dicyanodiamide
DIDP	diisododecyl phthalate
DIOP	diisooctyl phthalate
DLTDP	dilaurylthiodipropionoate
DMA	dynamic mechanical analysis
DMC	dough moulding compound
DMF	dimethylformamide
DMP-30	tertiary amine used as curing agent for epoxy resins
DMPA	dimethoxyphenylacetophenone
DMTA	dynamic mechanical thermal analysis
DOA	dioctyl adipate
DOP	dioctyl phthalate
DOS	dioctyl sebacate
DSC	differential scanning calorimetry
DSTDP	distearylthiodipropionoate

DTA	differential thermal analysis
DULT	deflection under load temperature
EAM	ethylene–ethyl acrylate (copolymer)
ECO	epichlorohydrin–ethylene oxide (copolymer)
ECO	ethylene–carbon monoxide (copolymer)
EDS	energy-dispersive scanning (analysis)
EDX	energy-dispersive X-ray (analysis) (same as EDS)
EEA	ethylene–ethyl acrylate (copolymer)
EM	electromagnetic
EMA	ethylene–methyl methacrylate (copolymer)
ENF	end-notched flexure (test/specimen)
EPDM	ethylene–propylene–diene monomer (rubber)
EPR	ethylene–propylene rubber
ESCA	electron spectroscopy for chemical analysis
ESCR	environmental stress cracking resistance
ESR	electron spin resonance
ETER	epichlorohydrin–ethylene oxide–diene terpolymer
ETPV	engineering thermoplastic vulcanizate
EV	efficient vulcanizing (cure)
EVA	ethylene–vinyl acetate (copolymer)
EVOH	ethylene–vinyl alcohol (copolymer)
EW	equivalent weight
FEP	poly(tetrafluoroethylene–hexafluoropropylene) copolymer
FIR	far-infrared
FKM	designation for a fluoroelastomer
FRP	fibre-reinforced polymer
FTIR	Fourier transform infrared
FVMQ	designation for a fluorosilicone
G_c	critical strain energy release rate
GCC	ground calcium carbonate
G_{Ic}, G_{IIc}, G_{IIIc}	critical strain energy release rates for fracture modes I, II and III
GOTMS	γ-glycidyloxypropyltrimethoxysilane
GPC	gel permeation chromatography, known also as size exclusion chromatography (SEC)
GPTMS	γ-glycidyloxypropyltrimethoxysilane
GRP	glass-reinforced plastic
HALS	hindered-amine light stabilizer
HDI	hexamethylene diisocyanate
HDPE	high-density polyethylene (density 0.935–0.955 g/cm^3)
HDT	heat distortion temperature
HHPA	hexahydrophthalic anhydride
HIPS	high-impact polystyrene
HP-PE	high-pressure polyethylene
IEC	ion exchange capacity

IIR	isobutylene–isoprene rubber
ILSS	interlaminar shear strength
IPN	interpenetrating polymer network
IR	infrared spectroscopy
IR	isoprene rubber
IRH	international rubber hardness (scale)
K_c	critical stress intensity factor
$K_{Ic}, K_{IIc}, K_{IIIc}$	critical stress intensity factor for fracture modes I, II and III
K value	empirical parameter used for molecular weight of PVC
LARC	tradename for a variety of PMR products of NASA-Langley
LCP	liquid-crystal polymer
LCST	lower critical solution temperature
LDPE	low-density polyethylene
LEFM	linear elastic fracture mechanics
LLDPE	linear low-density polyethylene
LOI	limiting oxygen index
LP	low-profile (additive)
M_{100}, M_{300}	modulus of a rubbery material at 100% and 300% extension
MBS	methyl methacrylate–butadiene–styrene (terpolymer blend)
MBT	mercaptobenzothiazole
MDI	diphenylmethane-4,4'-diisocyanate [4,4'-methylene diphenylene isocyanate]
MDPE	medium-density polyethylene (density = 0.925–0.935 g/cm^3)
MEK	methyl ethyl ketone
MF	melamine formaldehyde (resin)
MFI	melt flow index, also known as melt index or melt flow rate (MFR)
MFR	melt flow rate, also known as melt flow index (MFI)
MNA	methylnadic anhydride
MPD	m-phenylenediamine
MW	molecular weight
MWD	molecular-weight distribution
MWNT	multiple-walled (carbon) nanotube
N_1	first normal stress difference
NBR	butadiene–acrylonitrile rubber
NBR	nitrile rubber
NIR	near-infrared
NMP	N-methylpyrrolidone
NMR	nuclear magnetic resonance
NR	natural rubber
ODA	oxydianiline

PACM	4,4′-methylene-bis-cyclohexanamine
PAN	polyacrylonitrile
PANI	polyaniline
PB	poly(but-1-ene)
PB	polybutadiene
PBI	polybenzimidazole
PBT	poly(butylene terephthalate)
PC	polycarbonate
PCBT	poly(cyclo(butylene terephthalate))
PCL	polycaprolactone
PCTFE	polychlorotrifluoroethylene
PCTP	pentachlorothiophenol
PDLA	poly(D-lactic acid)
PDLLA	poly(D,L-lactic acid)
PDMS	polydimethylsiloxane
PE	polyethylene
PEEK	poly(aryl ether ether ketone)
PEG	poly(ethylene glycol)
PE-g-MA	polyethylene-*graft*-maleic anhydride
PEI	poly(ether imide)
PEK	poly(aryl ether ketone)
PEM	proton exchange membrane
PEN	poly(ethylene naphthanate)
PEO	poly(ethylene oxide)
PES	poly(ether sulfone)
PET	poly(ethylene terephthalate)
PETFE	poly(tetrafluoroethylene–ethylene) copolymer
PF	phenol formaldehyde (resin)
PFA	perfluoroalkoxy (polymer)
PHA	polyhydroxyalkanoate
PHB	poly(3-hydroxybutyrate)
phr	parts per hundred parts resin
PI	polyimide
PLA	poly(lactic acid)
PLLA	poly(L-lactic acid)
PMDA	pyromellitic dianhydride
PMDI	poly(diphenylmethane-4,4′-diisocyanate)
PMMA	poly(methyl methacrylate)
PMO	poly(methylene oxide)
PMP	poly(4-methylpent-1-ene)
PMPS	polymethylphenylsiloxane
PMR	polymerization of monomer reactants (NASA-Langley process)
POM	polyoxymethylene
PP	polypropylene

PPG	poly(propylene glycol)
PPO	poly(phenylene oxide)
PPS	poly(phenylene sulfide)
PS	polystyrene
PSA	pressure-sensitive adhesive
PTBAEMA	poly(2-*t*-butylaminoethyl methacrylate)
PTC	positive temperature coefficient (polymer)
PTE	thermoplastic elastomer
PTFE	polytetrafluoroethylene
PU	polyurethane
PVA	poly(vinyl alcohol)
PVAc	poly(vinyl acetate)
PVB	poly(vinyl butyral)
PVC	poly(vinyl chloride)
PVDC	poly(vinylidene chloride)
PVDF	poly(vinylidene fluoride)
PVEE	poly(vinyl ethyl ether)
PVF	poly(vinyl fluoride)
PVF_2	poly(vinylidene fluoride)
PVME	poly(vinyl methyl ether)
PVP	polyvinylpyridine
PVP	polyvinylpyrrolidone
PVT	pressure–volume–temperature (diagram)
Q	symbol sometimes used for polydimethylsilicone rubber
Q meter	apparatus used to measure permittivity and loss factor of dielectric
RF	radiofrequency
RIM	reaction injection moulding
RMS	root mean square
RTD	residence time distribution
RTM	resin transfer moulding
RTV	room-temperature vulcanization
SAN	styrene–acrylonitrile (copolymer)
SAXS	small-angle X-ray scattering
SBR	styrene–butadiene rubber
SBS	styrene–butadiene–styrene (thermoplastic elastomer)
SEBS	styrene–ethylene–butylene (thermoplastic elastomer)
S-EB-S	styrene–ethylene/butylene–styrene
SEC	size exclusion chromatography
SEM	scanning electron microscopy
SEN	single edge notch
S-EP-S	styrene–ethylene/propylene–styrene
semi-IPN	semi-interpenetrated polymer network
SIS	styrene–isoprene–styrene (thermoplastic elastomer)
SLS	standard linear solid (model)

SMA	styrene–maleic anhydride (copolymer)
SMS	styrene–α-methylstyrene (copolymer)
SPI	Society of the Plastics Industry
STO	stannous octoate
SWNT	single-walled (carbon) nanotube
synPS	syndiotactic polystyrene
TA	tertiary amine
TAC	triallyl cyanurate
TAIC	triallyl isocyanurate
TAS	triarylsulfonium
TCP	tricresyl phosphate
TDCB	tapered double cantilever beam
TDI	toluene diisocyanate
TEM	transmission electron microscopy
TEOS	tetraethoxysilane
TERT	tracking erosion test
TETA	triethylenetetramine
T_g	glass transition temperature
TGA	thermal gravimetric (thermogravimetric) analysis
TGDM	tetraglycidoxylmethylene p,p'-diphenylene diamine
THF	tetrahydrofuran
TIAC	triallyl isocyanurate
T_m	melting temperature
TMA	thermomechanical analysis
TMP	2-ethyl-2-(hydroxymethyl)-1,3-propanediol
TMPTMA	trimethylolpropane trimethacrylate
TMTD	tetramethylthiouram disulfide
TPE	thermoplastic elastomer
TPE-A	poly(amide-g-alkylene oxide) [g = graft]
TPE-E	poly(butylene terephthalate-b-tetramethylene oxide) [b = block]
TPMK	2-methyl-1-[4-(methylthio)-phenyl]-2-morpholinopropane-1-one
TPO	acronym for thermoplastic polyolefin elastomer
TPP	triphenyl phosphate
TPP	triphenylphosphine
TPS	thermoplastic starch
TPU	thermoplastic polyurethane elastomer [e.g. poly(urethane-g-ester/alkylene oxide)]
TX	thioxanthone
UCST	upper critical solution temperature
UDCB	uniform double cantilever beam
UF	urea formaldehyde (resin)
UHF	ultra-high-frequency
UHMWPE	ultra-high-molecular-weight polyethylene

UL	Underwriter's Laboratory
UP	unsaturated polyester (resin)
UPVC	unplasticized poly(vinyl chloride)
UV	ultraviolet (light)
VA	vinyl acetate
VAE	vinyl acetate ethylene copolymer
VLDPE	very low density polyethylene (density < 0.92 g/cm^3)
VLLDPE	very low-density linear polyethylene
VOC	volatile organic compound
VUV	vacuum ultraviolet
XPE	cross-linked polyethylene
XPS	X-ray photoelectron microscopy (spectroscopy)
XRD	X-ray diffraction
WATS	weighted-average total strain
WAXS	wide-angle X-ray scattering (diffraction)
WLF	Williams–Landel–Ferry (equation)

A-Z Encyclopedia Outline

A

1. Abrasion Resistance

Also known as wear resistance, a characteristic describing the resistance of a material to wear, which takes place via the erosion of small particles from the surface as a result of the frictional forces exerted by a sliding member. A suitable measure of the rate of wear is provided by the ratio V/μ, where V is the volume abraded per unit sliding distance and μ is the coefficient of friction, which corresponds to the amount of abraded substance per unit energy dissipated in sliding. The abrasion resistance is measured in a number of different ways, according to the expected service conditions of a particular product. The most common method is to use abrasive paper mounted on a rotating drum, allowing contact with a specimen subjected to a constant load in order to bring about the formation of debris, which gives rise to loss of material by wear.

2. ABS

A blend or alloy consisting of a glassy styrene–acrylonitrile random copolymer matrix and dispersed microscopic rubbery butadiene–acrylonitrile random copolymer particles, which contain sub-micrometre glassy polymer inclusions. The three letters stand for acrylonitrile, butadiene and styrene. Acrylonitrile–butadiene–styrene (ABS) polymers are produced either by blending a styrene–acrylonitrile (SAN) copolymer (glassy major component) with a butadiene–acrylonitrile rubber (NBR) copolymer (rubbery component), or by a specially designed bulk polymerization method. The latter consists of dissolving the rubbery component in a mixture of the monomers for the formation of the glassy matrix, followed by free-radical polymerization, which produces the glassy SAN matrix, grafted onto the diene elastomer. At any early stage of the polymerization, there is an inversion of phases, leading to the precipitation of small glassy particles into the diene elastomer, which precipitate into larger particles, forming the characteristic morphological structure of these toughened polymer systems. The different particle structures obtained with the two methods of production are shown in the diagram.

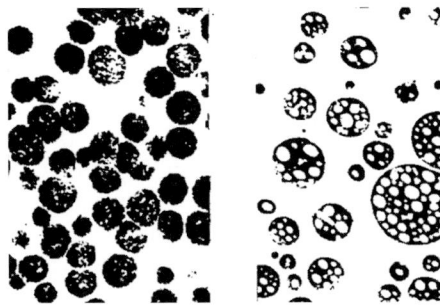

Structure of rubber toughening particles in ABS polymers. Left: System produced by blending of emulsions. Right: System formed by mass polymerization. Source: Schmitt (1979).

There are many grades of ABS available commercially, differing in rigidity characteristics, which are determined by the amount of rubber used, and also in the degree of plasticization of the glassy phase by the elastomer component. The glass transition temperature (T_g) values are in the region of 100–110 °C for the glassy phase and around -40 to -50 °C for the rubbery component. The ordinary polymer grades are opaque, owing to the particle size of the precipitated particles being greater than 1 µm and to the substantial difference in the refractive index values for the two components.

ABS is widely used in the automotive industry, as it combines its intrinsic high toughness with the ability to produce surfaces with a high gloss, and also because of the ease with which parts can be metallized by conventional electroplating methods. It is also available in the form of blends with polycarbonate, which exhibit better thermal oxidation stability and higher rigidity than conventional ABS grades. (See Styrene polymer.)

3. Absorbance

A coefficient denoting the fractional intensity of radiation absorbed by a body, that is,

$$\text{absorbance } \alpha = \frac{\text{flux of radiation absorbed}}{\text{total incident flux}}.$$

4. Accelerator

An additive used to speed up the rate of cross-linking reactions in the curing of thermosetting resins or elastomers. For polyester thermosetting resins, the accelerator is usually a cobalt naphthenate or cobalt octoate in solution. For the case of elastomers, the accelerator is a sulfur-containing compound, such as a thiourea, mercaptan and dithiocarbamate, or a non-sulfur-containing compound, such as a urea, guanidine and aldehyde diamine. The chemical structures of typical accelerators for sulfur curing of elastomers are shown.

2-mercaptobenzothiazole (MBT)

2,2′-dithiobisbenzothiazole (MBTS)

zinc diethyldithiocarbamate (ZEDC)

diphenylguanidine (DPG)

In many cases, mixtures of accelerators are used to obtain a synergistic effect. Typically, a thiazole type is used with smaller amounts of dithiocarbamate or an amine.

5. Acetal

A term used for poly(methylene oxide) (PMO), represented by the chemical formula $-(CH_2O)_n-$. Acetals are crystalline polymers with a melting point (T_m) in the region of 180–190 °C and a T_g around −40 °C. They are also available in the form of random copolymers containing small amounts of ethylene oxide units. Acetals are widely used in engineering applications for their high resistance to creep and wear under high loads. They also have a good resistance to solvents and low water absorption characteristics, which makes them suitable for applications requiring dimensional stability under moist environmental conditions. Blends with up to about 30% polyurethane elastomers have been reported to display a good balance of engineering properties with respect to stiffness, strength, toughness and solvent resistance.

6. Acid Number

A term used to denote the acid content of a polymer or resin, which is defined as the amount in milligrams (mg) of KOH (potassium hydroxide) required to neutralize 100 g of polymer or resin.

7. Acoustic Properties

Describe the response of a material to sound and particularly the absorption of sound. In polymers, the absorption of sound takes place through molecular relaxations. The parameter that denotes the capability of a polymer to dissipate mechanical energy through vibrations, known as the loss factor or damping factor (tan δ), is also used to describe the sound absorption characteristics of polymers. (See Viscoelasticity.) The methods used to measure acoustic properties are broadly divided into wave propagation, resonance and forced vibration methods.

Resonance and forced vibration methods are used to measure the Young's modulus and shear modulus as viscoelastic parameters comprising an elastic and a loss component. Wave propagation methods are used, on the other hand, to record the actual sound absorption characteristics of materials. Measurements are made by sending acoustic pulses with duration less than 1 ms through the specimen immersed in a liquid, and detecting them by another transducer on the opposite surface.

8. Acrylate Elastomer

(See Acrylic polymer.)

9. Acrylic

A generic term for monomers and polymers containing acrylate or methacrylate units. (See Acrylic polymer.)

10. Acrylic Polymer

Contains acrylic monomeric units in the molecular chains, represented by the general formula shown.

$$-(CH_2-\underset{\underset{COOR'}{|}}{\overset{\overset{R}{|}}{C}})-$$

Polymers where R is an H atom are called polyacrylates, and those where R is a methyl group are called polymethacrylates. Acrylics are generally amorphous polymers and are available as linear polymers or blends (thermoplastic), cross-linkable elastomers and cross-linkable adhesives and surface coatings. The properties of acrylic polymers are strongly dependent on the nature of the substituent R and R' groups, as illustrated in the table.

	Density (g/cm^3)	T_g (°C)
Poly(acrylic acid)	—	106
Poly(methyl acrylate)	1.22	8
Poly(ethyl acrylate)	1.12	−22
Poly(n-butyl acrylate)	<1.08	−54
Poly(t-butyl acrylate)	1.00	43
Poly(methacrylic acid)	—	130
Poly(methyl methacrylate)	1.17	105
Poly(ethyl methacrylate)	1.12	65
Poly(n-butyl methacrylate)	1.06	20
Poly(n-hexyl methacrylate)	1.01	−5
Polyacrylamide	1.30	165

Source: Data from Mascia (1989).

The data in the table clearly illustrate three important aspects of the structure–property relationship in polymers, respectively chain stiffness (energy required to rotate a group attached to the carbon atom in the backbone chain), internal plasticization (creating free volumes between polymer chains) and intermolecular forces. On this basis, one notes that polymethacrylates have a higher T_g than the corresponding acrylates. Increasing the length of the alkyl-group acrylate or methacrylate polymers brings about a large reduction in T_g through internal plasticization. The presence of a carboxylic acid or an amide group causes the opposite effect through the increase in intermolecular forces by the formation of inter-chain hydrogen bonds. The main characteristics of acrylic polymers available commercially are described below in alphabetical order of the monomer units within the polymer chains.

10.1 Acrylic Elastomer

Generally available as a random copolymer or terpolymer, typically poly(ethyl–butylene acrylate) with T_g within the range -30 to $-40\,°C$. These elastomers are well known for their resistance to oils, as well as for their oxidative stability at high temperatures and in UV light environment. They tend to absorb large quantities of water via hydrolysis of the ester groups. Vulcanization can be carried out by either peroxide curatives, for systems containing unsaturation in the main chains, or by metal hydroxides through salt formation via carboxylic acid groups.

$$\left[\text{CH}_2\text{CH}(\text{CN}) \right]_n$$

where the CN groups are organized in an atactic configuration, preventing the molecular chains from packing in a highly ordered crystalline lattice. The fibres are produced by solution spinning, because of the high viscosity and the rapid decomposition of the polymer at the temperatures that would be required for melt spinning. Thermal degradation takes place by cyclization involving the pendent CN groups along the polymer chains, which results in the formation of infusible ladder polymers, as shown by the reaction scheme, and confers on the fibres a certain degree of fire retardancy

Moreover, this particular feature makes PAN fibres suitable for the production of carbon fibres due to the dimensional stability that they acquire while they are heated up to the high temperatures required for graphitization. (See Carbon fibre.)

10.2 Acrylic Adhesive

(See Adhesive.)

10.3 Acrylic Fibre

Produced from polyacrylonitrile (PAN), which is represented by the chemical formula

10.4 Acrylic Flocculant and Hydrogel

Based on polyacrylamide, a water-soluble crystalline polymer represented by the formula shown.

$$\left[\text{CH}_2\text{CH}(\text{CONH}_2) \right]_n$$

10.5 Acrylic Impact Modifier

Based on methyl methacrylate–butadiene–styrene (MBS) terpolymer alloys, used widely as impact modifiers in rigid PVC formulations. They are produced mainly from emulsions of a styrene–butadiene rubber (SBR) elastomer and through graft polymerization of methyl methacrylate chemically bonded on the surface of the pre-formed SBR particles. The toughening action in poly(vinyl chloride) (PVC) compounds arises mostly from the miscibility of the acrylic outer layers of the dispersed particles with the PVC matrix. These are preferred to ABS impact modifier systems in applications requiring superior resistance to UV light.

10.6 Acrylic Plastic

The most important polymer in this category is poly(methyl methacrylate) (PMMA), a glassy polymer available either as a thermoplastic material or as a lightly cross-linked pre-formed product, such as sheets or rod castings. PMMA is widely for its high transparency and high resistance to UV light. It is represented by the formula shown and is usually available as a homopolymer grade. (See Acrylic polymer.)

$$\left[-CH_2-\underset{\underset{COOCH_3}{|}}{\overset{\overset{CH_3}{|}}{C}}- \right]_n$$

11. Activation Energy

A term in the Arrhenius equation widely used to denote the sensitivity of the rate of chemical or physical processes to changes in temperature. The activation energy (ΔH) is determined from the slope of the plot of the logarithm of the rate of reaction, or rate of physical change, against the reciprocal of the absolute temperature (T). This can be derived from the Arrhenius equation,

$$\text{rate } K = A \exp(-\Delta H/RT),$$

where A is a constant for the system R is the universal gas constant. (See Arrhenius equation.)

12. Activation Volume

A term in the Eyring equation used to describe the sensitivity of the yield strength of materials to changes in applied strain rate. It is an adaptation of the Arrhenius equation for which the activation energy term (ΔH) is replaced by the term ($\Delta H - v\sigma_Y$), where v is the activation volume and σ_Y is the yield strength. The product $v\sigma_Y$ represents the quantity by which the activation energy has been reduced by the application of the stress required to induce yield (plastic) deformations. (See Yield failure.) The Eyring equation is written as

$$d\varepsilon/dt = B \exp[(\Delta H - v\sigma_Y)/RT],$$

where $d\varepsilon/dt$ is the strain rate used in the test, corresponding also to the rate at which yielding deformations take place, and B is a material constant. (See Eyring equation.)

13. Activator

An additive used in conjunction with an accelerator for the vulcanization of elastomers. Activators consist of metal oxides or salts of lead, zinc or magnesium, often used in conjunction with stearic acid to enhance their solubility in the rubber mix.

14. Addition Polymerization

The conversion of monomer to polymer without the loss of other molecular species. This is contrary to the case of condensation polymerization, which takes place through the loss of small molecules, such as water. Typically, addition polymerization takes place by a free-radical mechanism or cationic polymerization.

15. Additive

A substance added to a polymer formulation in minor amounts to improve or modify specific characteristics related to manufacture and/or end use of products.

16. Additive Concentrate

A mixture of additive and a polymer powder (usually) where the amount of additive present is much larger than the quantity required in the final formulation. The concentrate is mixed with the neat polymer or resin to bring down the amount of additive to the required level. (See Master batch.)

17. Adherend

A term often used to denote the component in contact with the adhesive layer.

18. Adhesive

Substance used to stick (bond) two or more components of a product or structure. Adhesives can be divided into 'hot-melt adhesives', based on thermoplastic polymers, 'curable adhesives', based on thermosetting resin, and 'pressure-sensitive adhesives', based on elastomeric polymers. Apart from the bonding requirements, any difference in molecular structure or formulation details between polymer compositions used for 'bulk' components and those used for adhesives arises from differences in the way the two systems have to be processed or applied. The table lists some polymers used for adhesives.

Commercial name	Chemical nature
Thermoplastic hot melts	
Polyesters and polyamides	Low-crystallinity systems
Phenoxy	Poly(glycidyl hydroxyl ether)
EVA	Ethylene–vinyl acetate copolymer
EEA	Ethylene–ethyl acrylate copolymer
Thermosetting (cold cure and heat curable)	
Phenolics	Mostly modified with NBR
Epoxy	Mostly DGEBA resins
Acrylics	Several types available
Pressure-sensitive adhesives (PSAs)	
Most uncured elastomers	Contain tackifiers
Water-borne emulsions	
Poly(vinyl acetate)	Partially hydrolysed

The chemical compositions of polymers used for hot-melt adhesives are similar to those used for plastics, but they have lower molecular weights to meet the low-viscosity requirements for manufacturing purposes. Adhesive grades are usually copolymers containing small amounts of carboxylic acid or hydroxyl groups to enhance their affinity and bonding characteristics for the more common adherends. Oligomeric systems used for curable adhesives usually contain complex mixtures of auxiliary ingredients, and often have a heterogeneous morphology, as a means of obtaining a good balance of properties and also to satisfy specific requirements for a multitude of applications. Acrylic adhesives are usually based on low-volatility monomers containing peroxide initiators with a very long half-life. Some very innovative systems have been developed that take advantage of prevailing conditions to accelerate or induce curing in the adhesive. An interesting example is the anionic polymerization of cyanoacrylate adhesives, which takes advantage of the slightly basic nature of many inorganic substrates and the presence of water adsorbed on the surface to promote cure reactions within the adhesives.

19. Adhesive Test

Usually refers to a mechanical test for measuring the strength of bonded specimens, which is normally defined as the force required to separate the adherends of a joint, divided by the overall bonded area. Under ideal conditions, that is, those in which the actual intrinsic strength of the adhesive is to be measured, failure has to take place within the bulk of the adhesive (cohesive failure) and not at the interface between adherend and adhesive (interface failure). This is an essential requirement of structural adhesive, and it is for this reason that the surface of the adherend is often treated to increase the interfacial bond strength to the level required to ensure that failure takes place within the adhesive. The preparation of the surface of the adherend usually entails the generation of chemical groups that can react with the adhesive to produce chemical bonds across the interface. If the adhesive is mechanically strong (i.e. it requires high stresses for fracture), it is unlikely that cohesive failures can be achieved only through physical bonds at the interface, even when these are strong types such as hydrogen bonds, bearing in mind that these may be weakened through the absorption of water. Nevertheless, the strength values obtained depend on geometric factors, such as the dimensions of the bonded area and the thickness of the adhesive layer. Consequently, they cannot be used as fundamental parameters for use in theoretically based design procedures and have to be considered primarily as data for 'quality control and specifications'.

19.1 Tensile Test

Typical bonded specimens, known as butt joints, are shown. These are pulled at constant strain rate up to fracture. The bond strength is calculated by dividing the load to fracture over the bonded area.

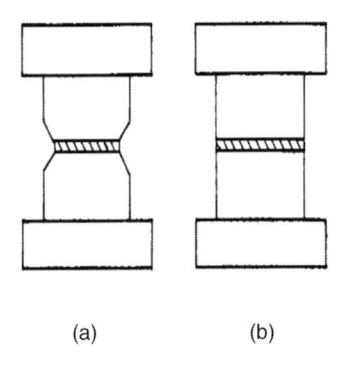

Typical axial loaded 'butt-joint' specimens: (a) wood-to-wood bond; (b) metal-to-metal bond.

19.2 Shear Test

Also known as 'lap shear' to describe the type of specimen used. Typical specimens types are shown.

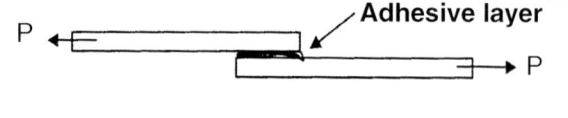

(a) Single lap shear

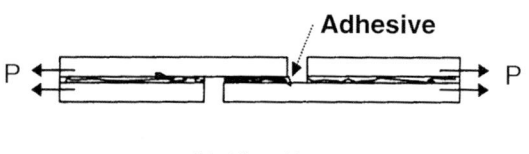

(b) Aligned lap shear

These are the more widely used tests, mainly because both the specimens and the experimental procedure are simple. The strength of the bonded specimen, known as the 'lap shear strength', is expressed as the load to fracture divided by the bonded area. In these tests the adhesive layer is subjected to a combination of tensile and shear stresses at the interface and within the adhesive layer, which vary in the opposite manner along the bonded length of the lapped area, as shown. The shear stress in the adhesive layer decreases from a maximum at the edge of the lapped area to zero in the middle when the bond line is sufficiently long. Completely the opposite to this is the change in axial tensile stress, which is maximum in the centre and smallest (zero) at the outer edges.

cantilever specimens. These are specimens with a very long bond line subjected to a 'crack opening' fracture mode by pulling the bonded cantilevers away from each other and recording the load during fracture propagation along the 'bonded line'. The value of the critical strain energy release rate G_c is calculated from a fundamentally derived equation, using the appropriate values for dC/da (i.e. the rate of increase of specimen compliance C with crack length a). (See Compliance and Fracture mechanics.)

Two types of specimens are usually used:

- the uniform double cantilever beam (UDCB) specimen, which is known also as the thin strip test;
- the tapered double cantilever beam (TDCB) specimen, which is known also as the wedged specimen.

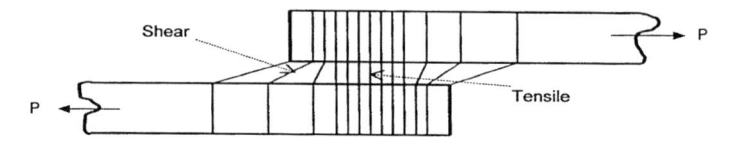

Variation of strain in the adhesive layer and the two adherends in a single lap shear test.

19.3 Fracture Toughness Test: General

The fracture toughness of bonded specimens is normally measured using double cantilever specimens. The description and geometry of these two specimen types are given below.

19.4 Fracture Toughness Test: UDCB Specimen

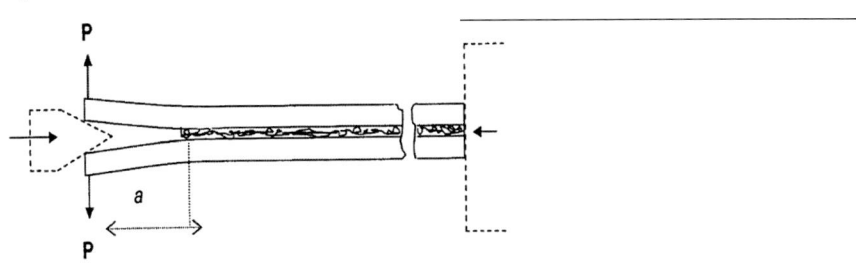

Double cantilever beam (thin strip) test. The load can be applied either directly at the edge of the strips or through a wedge at the open end of the bonded strips.

In UDCB testing, it is important that the two adherends undergo only elastic deformation during the test to ensure that all the energy imposed is used only to induce fracture of the adhesive layer. Note that, for a double cantilever beam, the compliance C is given by the expression

$$C = \frac{\Delta}{P} = \frac{64a^3}{EWB^3},$$

where Δ is the deflection, P is the load, a is the non-bonded distance (cf. crack length), E is the Young's modulus of the adherend, W is the width of the specimen and B is the thickness of the beam. Differentiating the above equation with respect to a gives an expression for dC/da that can be substituted in the generic equation for G_c, which can then be calculated from the recorded load to fracture (P_f), that is, the maximum load recorded in the test. From the above equation one obtains

$$G_c = \frac{96 P_f^2 a^2}{EW^2 B^3}.$$

Note that, when the adherend is very flexible (compliant), the test is known as the 'T-peel test'. If, on the other hand, one of the adherends is very rigid (i.e. very thick and/or the material has a very high modulus) and the other is very compliant (i.e. very thin and/or the material has a very low modulus), the test is known as the 'L- peel test'. Both tests can be used to measure the fracture toughness in terms of the G_c value, which is given by the formula

$$G_c = (P_f/B)(1+\varepsilon),$$

where B is the width of the flexible adherend strip and ε is the tensile strain ($\varepsilon = P/BWE$). When ε is very small, the value of G_c is obtained directly from the ratio P_f/B. The unit N/m corresponding to this ratio is equal to J/m², which is the appropriate unit for G_c.

19.5 Fracture Toughness Test: TDCB Specimen

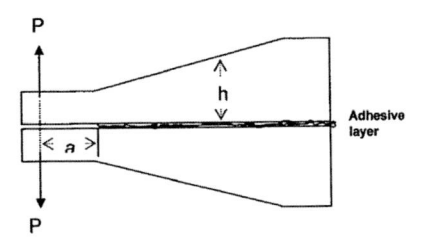

Tapered double cantilever specimen for fracture toughness testing of adhesives.

For TDCB specimens the value of dC/da is given by the expression

$$dC/da = 6m/EB,$$

where B is the thickness, E is the Young's modulus of the beam and m is a function of both the height h of the beam and crack length a, that is, $mh^3 - (1+v)h^2 - 4a^2 = 0$, where v is the Poisson ratio of the beam. The taper of the outer edge of the specimen is estimated by maintaining the value of m constant so that the height (depth) of the beam becomes larger as the non-bonded (crack) length increases. The solution of m gives a slightly concave contour for the outer edge, with the curvature becoming more pronounced as the non-bonded distance gets smaller. Therefore, by using large values of a, the contour of the specimen becomes approximately linear, that is, the gradient is constant. Typically, if the gradient angle is 11°, then

$$dC/da = 90/EBh_0,$$

where h_0 is the height (depth) of the beam at the crack tip. This simpler geometry makes it easier to produce specimens for testing. The advantage of the TDCB specimen is that the fracture load (P_f) remains constant during fracture propagation, making it possible to measure the rate of crack propagation along the bond line.

A slip–stick type of crack propagation occurs if fracture takes place in a mixed mode, involving cleavage through the bulk of the adhesive layer and debonding at the interface with the adherend.

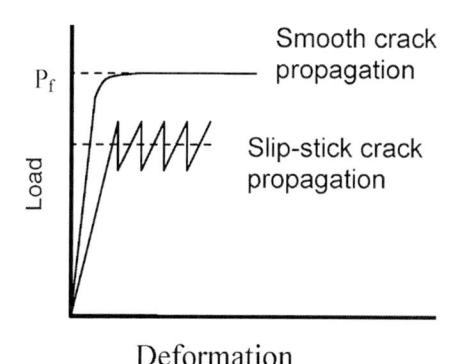

Typical force–deformation curves obtained in tests with TDCB specimens.

The TDCB test is particularly useful for fractures induced under fatigue (cyclic loading) conditions, so that the crack growth rate can be measured as a function of the frequency and magnitude of the applied load. Using UDCB specimens, on the other hand, the load to fracture decreases after crack initiation because the value of dC/da becomes smaller with increasing crack length. For many systems, the fracture toughness of structural adhesives has been measured also with respect to the long-term static and dynamic (fatigue) behaviour, through measurements of the increase in crack length with time and/or loading cycles.

20. Adhesive Wetting

Wettability is an essential characteristic of an adhesive to ensure that it completely covers the surface of the adherend. This condition is satisfied by ensuring that the adhesive and adherend have a similar surface energy. In quantitative terms, this requirement is to make sure that the so-called 'work of adhesion', W_A, is very small. This is related to the surface energies, γ, between the various phases by the equation

$$W_A = \gamma_{LV} + \gamma_{SV} + \gamma_{LS},$$

where the subscripts LV, SV and LS denote the three interfaces concerned, respectively liquid–vapour, solid–vapour and liquid–solid. The surface energy for the solid–liquid interface is lowest when the respective surface energies of the solid (substrate) and the liquid (adhesive) are very similar. Surface energies depend on the chemical constitution. Low polarity will give low values for the surface energy. Polytetrafluoroethylene (PTFE) has the lowest surface energy because of the absence of net dipoles in the structure. Polyethylene (PE) also has a very low surface energy, owing to the absence of net dipoles, but it is not as low as for PTFE. Polymers with very strong dipoles, such as those containing COOH or OH side groups or NH groups in the main chains (e.g. polyamides and polyurethanes), have very high surface energies. (See Surface energy.) Surface oxidation of PE or PTFE can introduce polar groups and increase accordingly the surface energy. These treatments would be used, therefore, to improve the bonding of polar adhesives, such as those based on epoxy resins. Chemical reactions across the interface bring about an increase in the interfacial bonding between adhesive and adherend, which represents the highest level of bonding that can be achieved.

21. Adiabatic Heating

A term used to describe the self-heating of a polymer taking place during intensive mixing and extrusion, resulting from the shearing action of the rotors of the mixer or the screw of an extruder.

22. Air-Slip Forming

A technique used in vacuum forming (also thermoforming) by which air is injected through fine orifices in the cavity of the mould in order to prevent sticking and rapid cooling of the polymer sheet during drawing. (See Thermoforming.)

23. Aliphatic

A term used in chemistry to describe the linear connection of carbon atoms to each other. This implies the absence of benzene rings, a compound containing which would be referred to as 'aromatic'. (See Aliphatic polymer.)

24. Aliphatic Polymer

A polymer containing aliphatic carbon atoms along the backbone of the molecular chains, for example, polypropylene.

25. Alkyd

A type of polyester resin. The term is derived from a combination of the words alcohol and acid to indicate that it is a product resulting from the reaction of an alcohol (multifunctional) and a dicarboxylic acid. The latter is usually a mixture of naturally occurring fatty acids and aromatic types. Alkyds are normally divided into oxidizing and non-oxidizing types. The oxidizing types are produced from unsaturated fatty acids, whose double bonds can react with oxygen from the atmosphere to produce free-radical species that can cause cross-linking reactions. The non-oxidizing types are cross-linked by reactions of the free OH groups in the chains with a urea formaldehyde or melamine formaldehyde resin, or with multifunctional isocyanates.

26. Allophonate

A type of chemical group formed from reaction between an isocyanate and a urethane group.

27. Alloy

(See Polymer blend.)

28. Alpha Transition Temperature

The alpha (α) transition temperature T_α is thermodynamically classified as a secondary transition, denoting the temperature at which the first partial derivative of a primary function, such as the volume or the enthalpy ($\partial V/\partial T$ or $\partial H/\partial T$), shows a discontinuity. T_α is also known as the glass transition temperature (T_g) insofar as it represents a reference temperature for the change in the deformational behaviour of a polymer from the glassy state to the rubbery state, which entails a reduction in modulus by a factor of 10^3–10^4 with increasing temperature.

29. Alternating Current (AC)

A current resulting from the application of a cyclic (sinusoidal) voltage. The cyclic variation of voltage and current takes place at specific frequencies. (See Dielectric properties.)

30. Aluminium Trihydrate $Al_2O_3 \cdot 3H_2O$

A functional filler used to impart flame-retardant and antitracking characteristics to a polymer. Particle size is in the range 2–20 μm and surface area in the region of 0.1–6 m^2/g. The amount of water corresponds to 34.5 wt%, which volatilizes very rapidly above 220 °C and reaches about 80% completion at around 300 °C. (See Flame retardant, Antitracking and Filler.)

31. Amine-Terminated Butadiene–Acrylonitrile (ATBN)

An oligomer, also known as liquid rubber, amine-terminated butadiene–acrylonitrile (ATBN) is used for the toughening of epoxy resins. Commercial systems are available with molecular weight between 2000 and 5000 and with acrylonitrile content around 25–35%. (See Epoxy resin, subsection 'Reactive toughening modifiers'.)

32. Amino Resin

A resin produced from the reaction of a multifunctional amine and formaldehyde. The most commercially important amino resins are urea formaldehyde (UF) and melamine formaldehyde (MF) resins. Both resins are water clear, unlike phenol formaldehyde (PF) systems. The network structure of the cured UF resins and the reactions involved are shown schematically.

$$\begin{array}{c} O H O N \\ \| | \| | \\   CH_2 O \\ -NH-C-N-CH_2-N-C-NH-CH_2-N-C-N \\ | \| | \\ CH_2 O CH_2 \\ NH-C-NH-CH_2-N-C-N-CH_2-N- \\ \| | | \\ O CH_2 CH_2 \\ | | \\ N -N- \end{array}$$

Dense network structure of cured urea formaldehyde (UF) resin.

These are formed from condensation reactions between $-CH_2OH$ of the basic resin to produce oxymethylene bridges, as shown in the scheme.

$$\begin{array}{c} H H \\ | | \\ NCH_2OH HOCH_2NCH_2OH NCH_2OCH_2NCH_2OH \\ | | | | \\ O{=}C O{=}C \rightarrow O{=}C O{=}C + H_2O \\ | | | | \\ NCH_2OH NCH_2OH NCH_2OH NCH_2OH \\ | | | | \\ H H H H \end{array}$$

Condensation reactions in the curing of UF resins.

Similar reactions take place in the cross-linking of MF resins, as shown.

$$\text{HOCH}_2\diagdown\underset{|}{\text{N}}\diagup\text{CH}_2\text{OH}$$

[melamine structure with HOCH₂ and CH₂OH substituents on nitrogen atoms around triazine ring]

$$-(\text{H}_2\text{O}, \text{CH}_2\text{O}) \rightarrow$$

A large number of these oxymethylene bridges undergo a loss of formaldehyde to produce a denser network.

completion, so that there will be a substantial number of CH_2OH groups present in addition to $-NCH_2-O-CH_2N-$ and $-NCH_2N-$ groups.

The use of cellulosic fillers for moulding powder or paper for laminates not only provides an efficient reinforcing and toughening function but also ensures that the water produced does not result in the formation of voids. The strong affinity of water with the structural units of amine resins, via hydrogen bonds, allows a substantial amount of water to remain dissolved in the network.

[Network diagram of highly cross-linked MF resin]

Network of a highly cross-linked melamine formaldehyde (MF) resin. Source: Ehrenstein (2001).

The moulding compounds of both UF and MF resins contain cellulose fillers and additives, such as pigments, curing catalysts, external lubricants and sometimes also small amounts of plasticizer. For white formulations, the cellulose filler is usually a bleached variety in order to eliminate the possibility of producing undesirable tints from impurities. Sometimes MF resins are butylated in order to increase their shelf-life when used as aqueous solutions or microsuspensions. In this case the condensation reactions will be preceded by the loss of butanol. For both UF and MF resins, however, the curing reactions rarely go to

33. Amnesia

A term (jargon) used in the technology of heat-shrinkable polymers to denote the extent by which the product fails to reach the dimensions it had before being stretched. (See Permanent set.)

34. Amorphous Polymer

The term is used to denote the lack of order at the supramolecular level. This implies that the molecular chains (thermoplastics) or the macromolecular networks (thermosets or cross-linked rubber) have a random configuration, as illustrated in the schematic diagram.

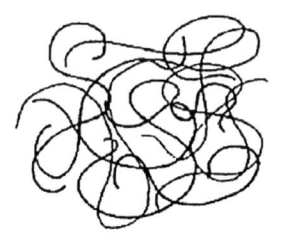

Linear polymers (left) and cross-linked polymers (right).

35. Anatase

A crystalline form of titanium dioxide (titania) used as the basic component of white pigments. (See Titanium dioxide.)

36. Anionic Polymerization

A type of polymerization that takes place via the growth of chains containing an anion in equilibrium with a small-sized counter-ion (cation) originating from the initiator, usually a metal alkyl (e.g. lithium butyl) or an alkyl metalamide (e.g. potassamide). The steps involved in the polymerization are as follows.

Step 1. Initiation: Addition of the anion from the initiator onto the monomer,

$$LiBu + H_2C=CHX \rightarrow Bu - H_2C - CHX^- Li^+$$

Step 2. Propagation: Rapid addition of a large number of monomer molecules,

$$Bu - H_2C - CHX^- Li^+ + n H_2C=CHX \rightarrow Bu - H_2C - CHX - (H_2C - CHX)_n^- Li^+$$

Step 3. Termination: Polymer chains reach full size when the monomer is exhausted or through secondary reactions with other species present, such as solvent.

37. Anionic Surfactant

(See Surfactant.)

38. Anisotropic

A product or specimen exhibiting anisotropy. (See Anisotropy.)

39. Anisotropy

The characteristic of a specimen or product in which the value of a given property is different in magnitude when measured in two perpendicular directions. In the case of cylindrical geometries such as rods, fibres or tubes, the two perpendicular directions of anisotropic behaviour are the longitudinal and circumferential directions. Anisotropy in polymer products results from the preferential alignment of polymer chains and crystallites in a specific direction. In most cases anisotropy is deliberately introduced to enhance the properties in one specific direction. In the case of fibres and tapes, the orientation is deliberately induced in the axial, or longitudinal, direction as a means of increasing the mechanical strength in the direction in which the product would be subjected to mechanical forces in service. (See Orientation.) In the case of composites, anisotropy arises from the different degree of fibre alignment in two perpendicular directions.

40. Annealing

A term denoting the thermal treatment (usually at constant temperature followed by slow cooling) of a specimen or product with a view to releasing the internal stresses set up during moulding or to developing the highest possible degree of crystallinity. This thermal treatment serves also to stabilize the dimensions and properties of a product prior to it being used in service. Annealing is carried out at temperatures very close to the glass transition temperature (T_g value) of the polymer if amorphous or just below its melting point (T_m) if the polymer is crystalline. Note, however, that annealing may induce embrittlement of products made from glassy polymers through physical ageing and those made from crystalline polymers through the thickening of the lamellae.

41. Anodizing

(See Metallization.)

42. Antiageing

(See Antioxidant, Stabilizer and UV stabilizer.)

43. Antibacterial Agent

(See Antifouling additive and Antimicrobial agent.)

44. Antiblocking Agent

An additive that produces roughness on the surfaces of a soft polymer product, preventing these from sticking to another through interfacial attraction forces. This phenomenon is particularly problematic in the case of flexible films used for plastic bags, as it makes it difficult to open them. The most widely used antiblocking additives consist of fine particles 20–50 nm diameter, usually inorganic fillers such as silica, talc, kaolin, calcium carbonate and zeolites.

45. Antidegradant

A generic name used primarily in relation to rubber formulations to denote an additive that improves the resistance to ageing. (See Degradation and Stabilizer.)

46. Antifoaming Agent

An additive used to prevent the formation of foams during the mixing of liquid systems. They are also known as defoamers or foam suppressants. Early systems were primarily vegetable or mineral oils. Nowadays antifoaming agents are complex mixtures in the form of hydrophobic solids containing active ingredients, consisting of a variety of compounds derived from water-soluble polymers. They normally contain a liquid-phase component such as mineral or vegetable oils, poly(ethylene glycol), silicone oils (polydimethylsiloxanes) and fluorosilicones (polytrifluoropropylmethylsiloxanes). These are particularly effective in non-aqueous system particles because of their low surface energy and immiscibility. The other component is a solid consisting of hydrophobic silica or hydrocarbon waxes. Antifoaming agents may also contain ancillary ingredients consisting of surfactants, coupling agents, stabilizers and carriers. The latter have the function of holding the ingredients together. In general, the defoamer has a surface energy lower than that of the foaming medium and should be immiscible and readily dispersible. The main uses of defoamers in polymer systems are in coating, latex and emulsion formulations.

47. Antifouling Additive

Known also as biocides and bactericides, these contain chemicals that are toxic to microorganisms. These include zinc and barium salts, as well as mixtures of zinc oxide or barium metaborate with thiazoles or imidazole compounds. They are widely used for marine coatings. More recently, extensive use has been made of the incorporation of sub-micrometre particles of silver, particularly for furniture and appliances for clinical application. (See Antimicrobial agent.)

48. Antimicrobial Agent (Biocidal Agent)

An additive with the ability to inhibit the growth of a broad range of microbes, such as bacteria, moulds, fungi, viruses and yeasts. Antimicrobial activity in polymers can be attained either by the incorporation of additives or through modifications of the chemical structure. Additives that can be used as antimicrobial agents include antibiotics, heavy-metal ions (silver or copper), cationic surfactants (quaternary ammonium salts with long hydrocarbon chains), phenols and oxidizing agents. An additive that is frequently used as an antimicrobial agent in polymers is metallic silver, in the form of fine (nano-sized) particles, even though it does not release metal ions as easily as copper. The use of copper is avoided, however, as it can have some devastating effects on the heat stability of the polymer. In order to increase the rate of release of ions, silver salts are sometimes embedded in hydrophilic carrier particles, such as silver-substituted zeolites. Macromolecular antimicrobial agents used in polymers include poly(2-*t*-butylaminoethyl methacrylate) (PTBAEMA) and N-halamines grafted onto polymer chains through acid or anhydride functionalization of the latter. N-halamines are compounds in which one chlorine atom is attached to the nitrogen atom. In both cases the antimicrobial activity derives from the formation of a quaternary ammonium salt, while the polymeric nature prevents them from being easily extracted by water.

49. Antimony Oxide

Corresponds to Sb_2O_3 and is used as a white pigment as a result of its high scattering power resulting from its high density (5.7 g/cm^3) and as a functional filler in conjunction with chlorinated or brominated compounds to impart fire-retardant characteristics to polymers. Optimal tinting and flame-retardant properties are achieved with particle size in the range of 0.2 to 1 μm. These systems are generally considered to produce toxic products under burning conditions. (See Flame retardant.)

50. Antioxidant

An additive within the general class of anti-ageing additives or stabilizers. An antioxidant reduces the rate of degradation of polymers resulting from the action of oxygen in the atmosphere on defective sites of a molecular chain. This creates free radicals, which initiate and propagate a series of reactions resulting in chain scission, cross-linking and formation of oxygen-containing chromophore groups, such as carbonyls. These reactions cause discolorations and embrittlement of the product. An antioxidant reacts with the free radicals to produce inert compounds, thereby preventing the propagation of degradation reactions. The antioxidant activity is regenerated through secondary reactions so that they can act as efficient stabilizers even at very low concentrations (0.1–0.5%). Typical antioxidants used in polymers are *ortho*- and *para*-substituted tertiary butyl phenols or aromatic amines. Amine antioxidants are less widely used, as they impart a dark discoloration to the products and they are rather toxic, which makes them less attractive for general uses. Typical examples of phenolic stabilizers are shown.

$$\left[C \left| CH_2-O-\overset{O}{\underset{\parallel}{C}}-CH_2-CH_2-\underset{tBu}{\overset{tBu}{\bigcirc}}-OH \right. \right]_4$$

$$\underset{tBu}{\overset{tBu}{HO-\bigcirc}}-CH_2-\overset{O}{\underset{\uparrow}{P}}(OC_{18}H_{37})_2$$

Example of phenolic antioxidants.

Note that the second example of phenolic antioxidant contains a phosphite group, which provides a synergistic effect with the phenolic unit by acting as a peroxide decomposer. The more common types of antioxidants do not have a built-in peroxide decomposer and, therefore, they require the presence of another additive to exert this function. These are typically phosphite and mercaptan compounds, which accelerate the decomposition of hydroperoxides and deactivates them through the formation of stable products.

Antioxidants intervene in the degradation reactions caused by the formation of free radicals in the polymer chains (P) through a series of reactions with oxygen, starting at sites with the weakest CH bond, such as tertiary or allylic carbons. The 'initiation' reaction.

$$PH + O_2 \rightarrow POOH \rightarrow PO^\bullet + {}^\bullet OH$$

is followed by rapid propagation reactions

$$P^\bullet + O_2 \rightarrow POO^\bullet$$
$$POO^\bullet + PH \rightarrow POOH + P^\bullet$$
$$PH + H^\bullet \rightarrow H_2 + P^\bullet$$
$$P^\bullet + P'H \rightarrow PH + P'^\bullet$$
$$PH + {}^\bullet OOH \rightarrow P^\bullet + H_2O_2$$
$$OH^\bullet + PH \rightarrow P^\bullet + H_2O$$

Propagation reactions are the most damaging reactions insofar as they affect a large number of polymer molecules. The antioxidant intervenes predominantly at this stage by reacting with the free radicals in the polymer chains, forming non-reactive products, a phenomenon known as quenching. Denoting by AH the antioxidant molecule, either amine or phenol type, the quenching reaction can be written as

$$P^\bullet + AH \rightarrow PH + A^*$$

where A^* represents a stable (non-reactive) radical. The mechanism for the loss of reactivity of the A^* radicals for the case of a phenolic antioxidants is as shown.

Note that the stability of the radicals is due to internal delocalization, making it inaccessible for interacting with polymer chains.

The reaction scheme for the action of secondary stabilizers as peroxide decomposers is:

$$\sim\!\!-\!\!\overset{|}{\underset{|}{C}}\!HOOH + P(OR)_3 \longrightarrow \sim\!\!-\!\!\overset{|}{\underset{|}{C}}\!HOH + O\!=\!P(OR)_3$$

$$\sim\!\!-\!\!\overset{|}{\underset{|}{C}}\!HOO\cdot + RSH \longrightarrow \sim\!\!-\!\!\overset{|}{\underset{|}{C}}\!HOOH + RS\cdot$$

$$\sim\!\!-\!\!\overset{|}{\underset{|}{C}}\!HOO\cdot + RS\cdot \longrightarrow \sim\!\!-\!\!\overset{|}{\underset{|}{C}}\!H\text{-}OOSR \quad \text{(stable products)}$$

Examples of mercaptan and phosphite compounds used as secondary stabilizers are dilaurylthiodipropanoate (DLTDP) and distearylthiodipropanoate (DSTDP).

$$(C_{12}H_{25}O-\overset{O}{\underset{\|}{C}}-CH_2-CH_2)_2S \quad (C_{18}H_{37}O-\overset{O}{\underset{\|}{C}}-CH_2-CH_2)_2S$$
$$\text{DLTDP} \qquad\qquad \text{DSTDP}$$

The structures of two phosphite secondary stabilizers are shown. These are known commercially as Weston 618 (left) and Ultranox 626 (right).

51. Antiozonant

An additive used primarily in diene elastomers to improve the resistance to attack from ozone in the environment, particularly the ozone generated in electrical motors and other devices through corona discharges. An antiozonant is often a wax capable of migrating to the surface to form an oxidation-resistant layer.

52. Antiplasticization

Although the term denotes a phenomenon with the opposite effect to plasticization, this must not be interpreted in terms of an increase in the glass transition temperature. Antiplasticization is a phenomenon that brings about an increase in modulus at low temperatures, as well as embrittlement at ambient temperatures. In PVC, this is brought about by the incorporation of small amounts of plasticizers (typically 5–10 parts per hundred) capable of exerting strong physical interaction, such as H– bonds with the Cl atoms in the chains. These interactions cause a depression of short-range relaxations, normally associated with rotational movements of side groups in a polymer chain. The effects of the addition of small amounts of plasticizer in PVC compounds on the variation of modulus with temperature, shown in the diagram, provides evidence that antiplasticization is essentially a phenomenon occurring at low temperatures and that normal plasticization takes places at higher temperatures.

The embrittlement effect due to antiplasticization is illustrated in the second diagram in terms of the reduction in 'critical crazing strain' as a function of the CH_3OH/H_2O ratio, representing the relative affinity of the environmental agent for the polar groups in the polymer.

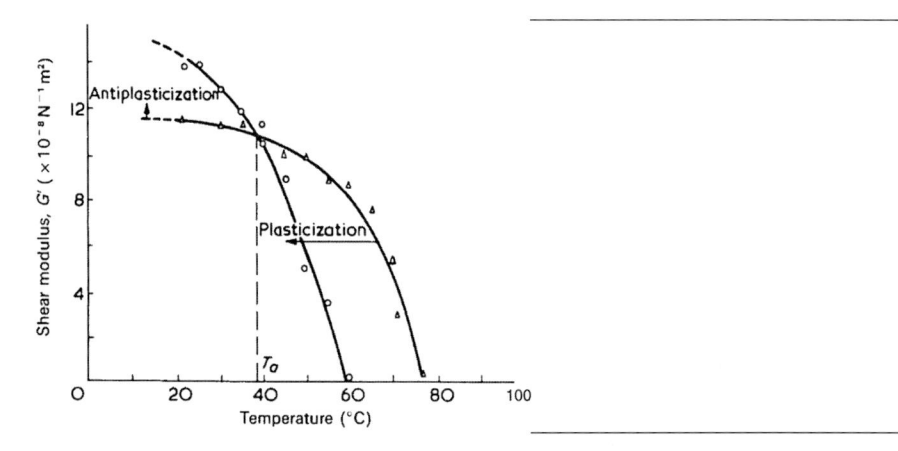

Plot of dynamic shear modulus against temperature for a PVC sample (curve on the right, triangles) and a sample containing 9 wt% tricresyl phosphate (curve on the left, circles). *Source: Mascia (1978).*

53. Antistatic Agent

An additive used in polymer formulations to reduce the build-up of static charges on the surface of products or structures. Antistatic agents are additives capable of migrating to the surface of a product, forming a conductive path through ionization resulting from the absorption of moisture in the

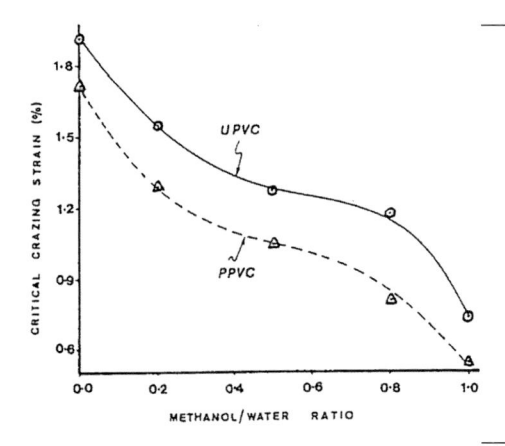

Illustration of effects of antiplasticization of a PVC compound. Upper curve: unplasticized PVC. Lower curve: PVC plasticized with 9 wt% tricresyl phosphate. *Source: Mascia et al. (1989).*

atmosphere. These are usually surfactants consisting of long-chain aliphatic amines, amides or quaternary ammonium salts, alkyl aryl sulfonates and alkyl hydrogenphosphates. Among other compounds used as antistatic agents are water-soluble ionically conductive polymeric compounds, such as hexadecyl ethers of poly(ethylene glycol). (See Surface resistivity.)

54. Antitracking

Preventing the formation of surface tracks on the surface of a dielectric, or insulator, subjected to a high voltage. The surface tracks consist of carbonaceous paths that are formed as a result of the chemical degradation of the polymer, thereby producing very conductive channels capable of leaking a current to ground. Tracking of insulators is often prevented by depositing suitable oils or greases on the surface, such as silicone types, as they are thermally stable and do not form carbonaceous residues.

55. Antitracking Additive

An additive that imparts antitracking characteristics to a polymer. These are substances that can release large quantities of water when the outer surface layers of the insulator reach temperatures in the region of 300–400 °C, as a result of the heat produced by the arcs generated by the applied high voltage. Arcs contain very reactive oxygen ions, which give rise to extremely rapid degradation reactions. The most widely used antitracking additive is aluminium trihydrate, $Al_2O_3 \cdot 3H_2O$, which loses all three H_2O molecules at around 350 °C, corresponding to about 35 wt% weight loss. (See Flame retardant.)

56. Apparent Shear Rate

The shear rate calculated on the basis that the polymer melt behaves as a Newtonian liquid. The value of the apparent shear rate, $\dot{\gamma}_a$, at the wall can be calculated from the flow rate Q and the dimensions of the channels, as follows.

- Circular channel: $\dot{\gamma}_a = 4Q/\pi R^3$, where R is the radius of the channel.
- Rectangular channel: $\dot{\gamma}_a = 6Q/WH^2$, where W is the width of the channel and H is the depth, valid for shallow channels, that is, when $W \gg H$.

A shape factor, S, has to be introduced for other situations. Correction has to be made by multiplying the value calculated from the above equation by the appropriate shape factor, $S = 1 - 0.65(H/W)$. (See True shear rate and Non-Newtonian behaviour.)

57. Apparent Viscosity

The value of the viscosity (η_a) calculated on the basis that the polymer melt behaves as a Newtonian liquid, that is $\eta_a = \tau/\dot{\gamma}_a$, where τ is the shear stress and $\dot{\gamma}_a$ is the value of the apparent shear rate.

58. Aprotic Polar Solvent

A solvent without the capability of undergoing hydrogen-bonding interactions through the involvement of protons.

59. Araldite

A tradename for a variety of epoxy resins.

60. Aramid

A term used to denote aromatic polyamide fibres, an example of which is shown.

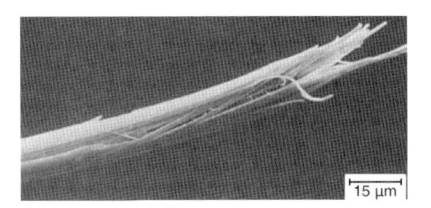

Example of an aramid fibre displaying fibrillation characteristics. Source: Ehrenstein (2001).

The chemical structure of a typical polyamide used for the production of aramid fibre is shown.

These are often identified by the tradename Kevlar. The molecular chains of aromatic polyamides are very rigid and are oriented in the axial direction of the fibres, forming strong attractions with other neighbouring molecules via hydrogen bonds. This provides a high strength and modulus, with a high level of ductility, relative to carbon fibres (which are extremely brittle) and even glass fibres. Aramids have a very high melting point (around 500 °C) but are susceptible to UV-induced degradation. (See Composite.)

61. Aromatic Anhydride

Frequently used compounds for the production of polyesters and polyimides and sometimes used as curing agents for epoxy resins.

62. Arrhenius Equation

The name of an equation derived by the chemist Arrhenius to describe the variation of the rate of a chemical reaction (K) as a function of the absolute temperature (T). This is written as

$$K = A \exp(-\Delta E/RT),$$

where A is a characteristic constant, ΔE is the activation energy for the reaction and R is the universal gas constant. This equation has been found to apply equally well to physical processes that involve movements of some structural constituents, for instance in the case of electronic or ionic conduction and gas diffusion.

63. Asbestos

A magnesium silicate fibre occurring naturally in four different forms. The 'chrysotile' variety has been used for reinforcement of thermosetting resins, particularly phenolic types, in engineering applications such as brake pads. It consists of fine fibrils packed into bundles for its high reinforcing efficiency due to the high modulus (see the micrograph).

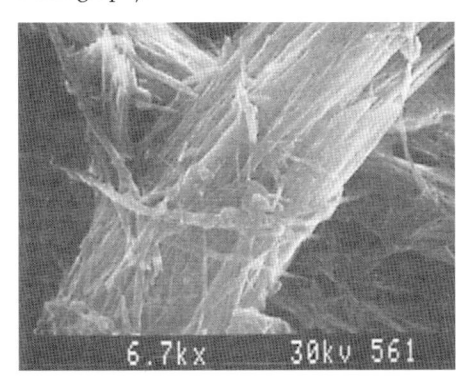

SEM micrograph of chrysotile asbestos fibres. Source: Wypych (1993).

Owing to toxicity issues, asbestos has been largely replaced by other high-performance reinforcing fibres. (See Composite.)

64. Aspect Ratio

The ratio of the length to diameter of fibres used in composites. Sometimes used also for platelet particles as the ratio of the nominal width to the thickness.

65. Atactic Polymer

A term used to denote the lack of a specific order in the chemical structure of a polymer, such as polypropylene. A polymer with an atactic chemical structure. (See Isotactic polymer and Tacticity.)

66. Atomic Force Microscopy (AFM)

A microscopic technique that produces images of the surface topology via the tip of a scanning probe that measures the atomic forces (attractive or repulsive) against the surface under examination. Atomic force microscopy (AFM) is particularly useful for examining surface features of very small dimensions, for example 10–1000 nm. The tip is placed on a cantilever that will deflect as a result of the atomic surface forces, as shown in the diagram.

proximity with the surface without touching it. In most cases a feedback mechanism is employed to adjust the tip-to-sample distance to maintain a constant force between the tip and the surface of the sample.

67. Attenuated Total Reflectance Spectroscopy (ATR Spectroscopy)

An infrared spectroscopy technique used to identify specific chemical groups present on the surface of a sample, supported on a prism with a very high refractive index, which reflects the incident infrared radiation transmitted through the thin film in contact with the prism. (See Infrared spectroscopy.)

68. Attenuation

(See Damping.)

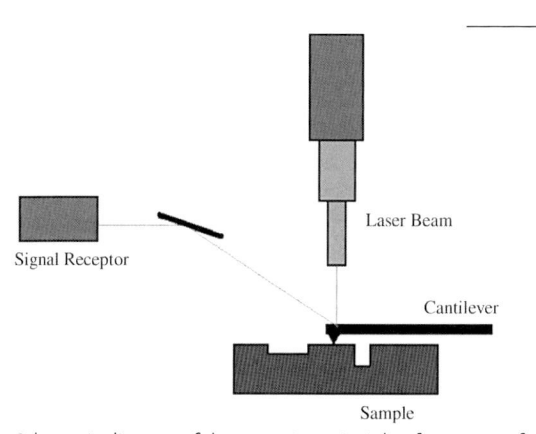

Schematic diagram of the operating principle of an atomic force microscope. Source: Lavorgna (2009).

The movements are detected by the reflection of a suitable focused laser beam. In tapping-mode AFM the cantilever is oscillated at a certain frequency with an amplitude that allows the tip to come into close

69. Avrami Equation

Originally developed to model the crystallization rate of metals during solidification, it has been found to apply equally well to the crystallization of polymers. The Avrami equation is usually written as

$$\ln(1-\varphi_t) = Zt^n,$$

where φ_t is the volume fraction of polymer crystals (also known as the degree of crystallinity) at time t, while Z and n are characteristic constants for the polymer.

70. a/W Ratio

The ratio of the length of the crack (or notch), a, to the width of the specimen, W, used in the evaluation of the toughness of materials using fracture mechanics principles. (See Fracture mechanics.)

71. Azeotropic Copolymerization

Conditions in which the composition of the molecular chains of a copolymer (i.e. the ratio and position of the two monomer units) remains the same throughout the polymerization process.

B

1. Back-Flow

The flow that can take place in the opposite direction to the inlet flow. Back-flow can occur sometimes through the gate of a mould cavity if the pressure is removed before the gate is frozen. It also takes place frequently in the form of leakage flow over the flights of the screw of an extruder. (See Extrusion theory.)

2. Back-Pressure

The pressure exerted against the screw of an injection moulding machine during its rotation for the plasticization and transportation of the melt to the front of the barrel.

3. Backscatter

Denotes the scattering of light from the surface of a flat object, such as a sheet or film, impinged by incident rays, causing a deterioration of clarity of objects seen through. The term is also used in energy-dispersive scanning (EDS) analysis by stereo scanning electron microscopy (SEM).

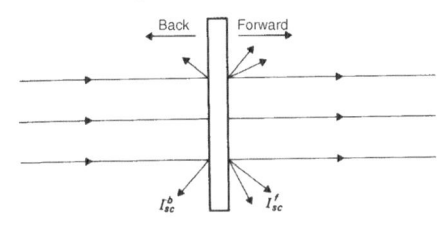

Backward and forward light scattering from the surface of a polymer film.

4. Bagley Correction

Represents the pressure drop experienced by a polymer melt at the entry of the capillary of a die used in rheological measurements with a capillary rheometer. In order to calculate the value of the shear stress at the wall of the capillary, τ_{wall}, the entry pressure drop (ΔP_{entry}) has to be subtracted from the total pressure drop, ΔP_{total}, measured by the transducer fitted in the heating chamber, that is,

$$\tau_{wall} = (\Delta P_{total} - \Delta P_{entry}) R/2L,$$

where L and R are the geometric parameters for the capillary die (L is the length and R is the radius).

5. Bagley Plot

A plot of the measured pressure (ΔP_{total}) against the ratio L/R in rheological measurements carried out for polymer melts with a capillary rheometer using dies of different ratios of length (L) to radius (R) and at different apparent shear rates (s^{-1}) for each plot (see diagram). The extrapolated P value for $L/R = 0$ corresponds to the entry pressure drop, which has to be subtracted from the measured pressure drop (ΔP_{total}) in calculating the shear stress at the wall, that is,

$$\Delta P_{capillary} = \Delta P_{total} - \Delta P_{entry}.$$

(See Bagley correction.)

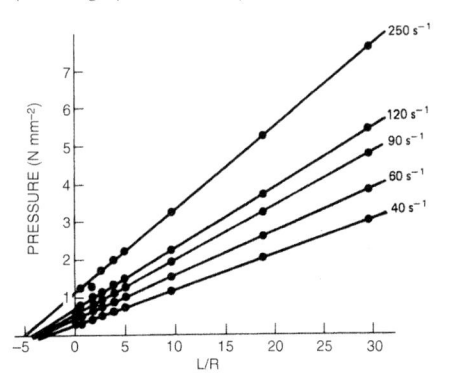

Example of Bagley plot for different shear rates.

6. Bakelite

Originally a tradename for phenol formaldehyde moulded products, after the inventor Leo Beakeland, it became widely accepted as a generic term. (See Phenolics.)

7. Ball-Drop Test

(See Falling-weight impact test.)

8. Bank

A term used to describe the beading of polymer melt along the entry side of the first nip of the rolls of a calender. This prevents the entrapment of air at the interface between melt and rolls. (See Calender and Calendering.)

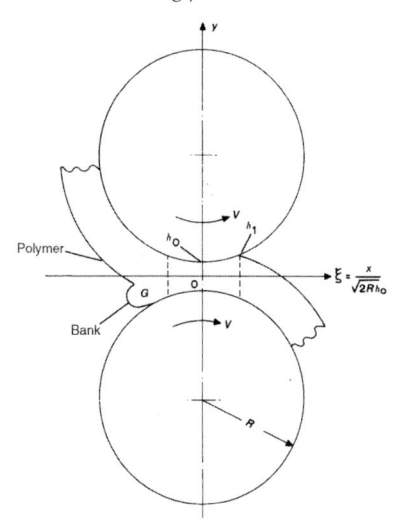

Formation of the rolling bank by the rotation of the bottom roll at the nip of the calender. Source: Mascia (1989).

9. Barrel

The cylinder enclosing the screw of an injection moulding machine or that of an extruder.

10. Barrier Flight Screw

Type of screw used in extrusion processes, designed to keep the solid bed separated from the melt pool within the transition zone. The flow path followed by the solid bed and the melt pool within the unwrapped spiralling channel of an ordinary screw is shown.

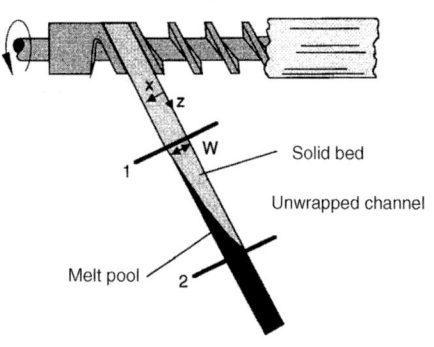

Solid bed and melt pool in an unwrapped screw channel. Source: Osswald (1998).

Barrier screws prevent the break-up of the solid bed and the creation of local perturbations of pressure and temperature in the transition zone, which eliminates the possibility of a pulsating flow occurring at the die, known as 'surging'. There are two flight designs to achieve these requirements: 'constant solid-bed channel width', known as the Maillefer screw, and 'constant channel depth', simply known as the barrier screw. The separation of the solid bed from the melt pool and the cross-sections of the channels are illustrated.

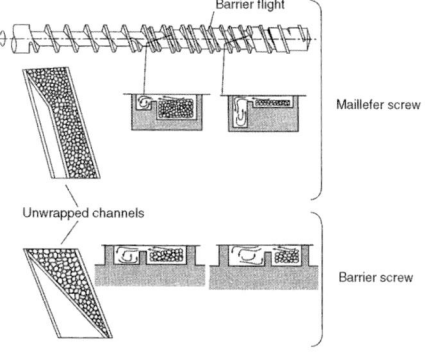

Schematic diagram of screws with different barrier flights. Source: Osswald (1998).

11. Barrier Polymer

A polymer with the ability to restrict the permeation of gases, vapours or liquids. The barrier characteristics of polymers are expressed in terms of their permeability coefficient, which is defined as the molar flux of penetrant through a film per unit thickness. (See Barrier properties.)

The barrier characteristics of polymers are related to their molecular and morphological structure. This is illustrated in the table, which compares the effects of functional group and chain packing on the oxygen permeability P_{O_2} value (wherein mil = 1/1000th of an inch, and atm = atmosphere).

Effects of functional group

Nature of X in $-(CH_2-CHX)_n-$	P_{O_2} (cm³/mil 100 inch² day atm)
–OH	0.01
–CN	0.04
–Cl	8.0
–F	15
–COOCH$_3$	17
–CH$_3$	150
–C$_6$H$_5$	420
–H (LDPE)	480

Effects of chain packing

Structure	P_{O_2} (cm³/mil 100 inch² day atm)
$-(CH_2-CH_2)_n-$ (HDPE)	110

$-(CH_2-CH)_n-$
          |
          CH$_3$
	150

$-(CH_2-CH)_n-$
          |
          CH$_2$
          |
          CH
         /   \\
      CH$_3$ CH$_3$
 (P$_4$MePe-1)
	4000

Source: Data from Mascia (1989).

In the first series is shown the effect of decreasing the polarity of the polymer chains, and in the second is shown the effect of decreasing the molecular chain packing within the crystals forming the lamellae. It is known that the polymers with the highest barrier characteristics towards oxygen, carbon dioxide and water are ethylene–vinyl alcohol (EVOH) copolymers, poly(vinylidene dichloride) (PVDC) and liquid-crystal polymers (LCPs). All polyolefins, on the other hand, have poor barrier properties towards the same diffusants. Blending small amounts of polyamides with polyolefins, through appropriate compatibilization methods, has been used as a method for improving the barrier properties of the latter polymers for packaging. The inclusion of exfoliated nanoclays in a polymer is another effective way of reducing the permeability of gases through polymer films or coatings, in view of their intrinsic low permeability, which provides an effective barrier through a dilution effect, as well as through the creation of long tortuous paths for the diffusion of penetrants, due to their high aspect ratio (area/thickness ratio). A chemical approach has also been used to reduce the diffusion of oxygen through polymer films resulting from the oxygen scavenging characteristics of certain additives, such as cobalt carboxylate salts.

12. Barrier Properties

Properties of packaging films or containers denoting the resistance to permeation of gases or liquids. The barrier properties of polymers are described by the solubility (S), the permeability coefficient (P) and the diffusion coefficient (D). These parameters are related to each other by the expression $P = DS$. The SI units for permeability are mol m/m² s Pa, while a widely used unit is the barrer, which corresponds to 7.60×10^{-9} cm³ cm/cm² s atm. The most widely used penetrants for measurements of the barrier characteristics of packaging films and containers are oxygen, water and carbon dioxide.

13. Barus Effect

The swelling of extruded melt at the die exit. (See Die swell ratio.)

14. Baryte

A filler based on barium sulfate, $BaSO_4$, used mostly to achieve sound attenuation characteristics, arising primarily from its high density (4.5 g/cm^3). The particle size is in the region of 2–15 m. For toxicity reasons, the amount of water-soluble barium salts have to be kept at very low levels.

15. Beer–Lambert Law

Describes the absorption of radiation of materials with the equation

$$\log(I_0/I) = \varepsilon c l$$

where I_0 is the intensity of the incident beam of light, I is the intensity of the transmitted beam, ε is the extinction coefficient (a property of the material), c is the concentration of the absorbing species and l is the path length. The Beer–Lambert law is the basis of radiation spectroscopic techniques, which makes it possible to calculate the concentration of particular chemical groups present in a sample.

16. Bending Moment

A fundamental concept used for the bending theory of beams. The bending moment (M) is defined as the product of the applied load (P) and the distance from the point of support.

17. Bending Test

Also known as flexural test, is usually carried out in the form of three-point bending, where a beam is supported at the two ends and is loaded in the centre. (See Charpy impact strength and Flexural properties.)

18. Bentonite

A type of clay used as filler, which contains exfoliatable platelets. (See Nanocomposite and Filler.)

19. BET Isotherm (Brunauer–Emmett–Teller Isotherm)

A theory and a method for measuring the total surface area of powders based on the absorption of gas/vapour molecules, through the relationship between volume of gas adsorbed and gas pressure at constant temperature,

$$\frac{p}{V(p_0-p)} = \frac{1}{V_m c} + \frac{(c-1)p}{V_m c p_0},$$

where V is the volume of absorbed vapour, V_m is the monolayer capacity of the solid surface, p is the partial pressure of the vapour, p_0 is the saturation vapour pressure and $c = \exp(\Delta H_A - \Delta H_L)/RT$, in which ΔH_A is the heat of adsorption, ΔH_L is the heat of vapour condensation, R is the universal gas constant and T is the absolute temperature.

20. Beta Transition Temperature

The beta (β) transition is a minor transition occurring in glassy polymers at temperatures substantially below the alpha (α) transition, which is known as the glass transition (T_g). The β transition is associated with short-range relaxations, arising primarily from rotations of side groups attached to the main polymer chains. The change in modulus of a polymer at the β transition temperature, however, is quite small: much less than one order of magnitude as compared to a change of three or four orders of magnitude taking place above the α (glass) transition temperature. These are illustrated in the dynamic mechanical spectra.

23. Bingham Body

A liquid or melt requiring a minimum shear stress level at the walls of a channel before flow starts, known as the yield shear stress.

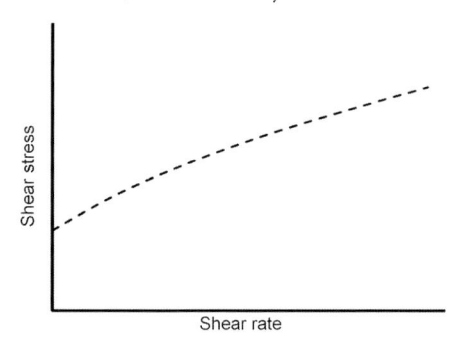

Schematic representation of a Bingham body flow behaviour for polymer melts.

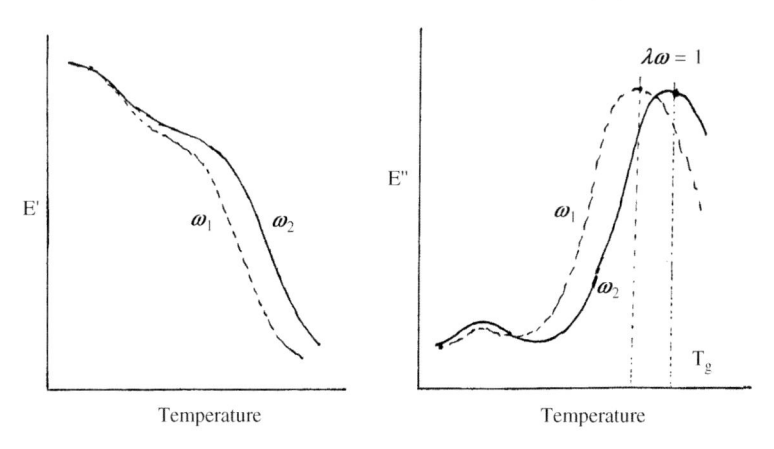

Dynamic mechanical spectra at two frequencies of a glassy polymer showing the α and β transitions. Source: Mascia (1989).

21. Biaxially Oriented Film

(See Orientation.)

22. Binder

A term widely used for the binding together of fibres in mats or fabrics by means of a low-melting-point solid resin or film-forming polymer deposited from a water emulsion.

24. Bioactive Filler

A filler that imparts biocompatible characteristics to a polymer. Typical bioactive fillers are hydroxyapatite and starch. (See Biopolymer.)

25. Biodegradable Polymer

A polymer or a polymer composition that can be attacked by microorganisms in the environment, causing a breakdown of the molecular structure.

26. Biopolymer

A biodegradable or biocompatible polymer, which can be classified into natural and synthetic materials. The majority are naturally occurring polymers, such as 'polysaccharides' (e.g. starch, cellulose, lignin and chitin), 'polypeptides' and 'proteins' (e.g. gelatin, casein, wheat gluten, silk and wool), and 'lipids' (e.g. plant oils and animal fats). The most widely researched synthetic polymers are 'modified cellulose' (e.g. cellulose acetate, cellulose propionate and cellulose acetate butyrate), 'polyesters' (e.g. polyhydroxybutyrate, polyhydroxyalkanoate, poly(lactic acid) and polycaprolactone), as well as mixed aliphatic–aromatic polyesters. Other systems include poly(vinyl alcohol) and 'modified polyolefins' (e.g. ethylene–carbon monoxide copolymers containing both photo- and biodegradation catalysts).

27. Birefringence

The difference in the refractive index in two perpendicular directions, that is, $\Delta n = n_1 - n_2$, where n is the refractive index, and 1 and 2 refer to the directions. This arises from the anisotropic structure of polymer products resulting from the presence of aligned molecular chains. Measurement of the birefringence is widely used to assess quantitatively the degree of molecular orientation in an anisotropic polymer product, such as fibres and tapes, and also in injection-moulded products to determine the level of residual stresses through the thickness, as shown in the diagram. (See Orientation.)

Birefringence is measured directly on specimens using polarized electromagnetic (EM) radiation by measuring the optical path difference (or relative retardation, D) with the aid of a characteristic colour chart, known as the Michel-Lévy chart, or an optical compensator (i.e. a device with precalibrated and adjustable birefringence characteristics). This is a straightforward method insofar as the relative retardation is directly proportional to the birefringence, that is, $D = b\Delta n$. A spectrophotometric technique is used to measure the relative retardation in those cases where the birefringence is very high.

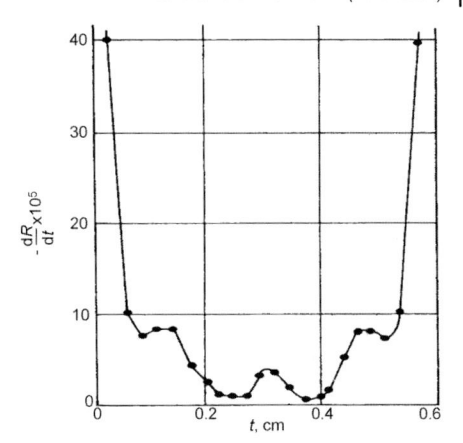

Example of variation of birefringence through the thickness of an injection-moulded specimen. Source: Murphy (1969).

28. Bismaleimide Resin (BMI Resin)

A type of resin used as the matrix for composites to obtain a high glass transition temperature and a high resistance to degradation of mechanical properties in high-temperature environments. A variety of different types are available commercially, varying from simple resins, such as bis(maleimidophenyl)methane, to more complex types, such as adducts (see diagrams).

Bis(maleimidophenyl)methane.

Adduct of maleimide to trimethylphenylindane.

The curing and grafting reactions take place via a free-radical mechanism. Often these systems are mixed with vinyl ester type resins to reduce costs and optimize performance. (See Vinyl ester.)

29. Bisphenol

The bisphenols are a class of raw materials widely used primarily for the production of epoxy resins and some aromatic polyesters, including polycarbonate, sometimes used as an antioxidant. There are three types of bisphenols, namely bisphenol A, bisphenol F and bisphenol S, consisting of two phenol units bridged at the *para* position by, respectively, a propylidene group, a hexafluoropropylidene group and a sulfone group.

30. Biuret

Chemical group formed from the reaction between a urethane group and a urea group in the production of polyurethanes.

31. Bleeding

A term used for the loss of additive from a polymer when immersed in an aqueous environment. The term 'leaching' is also used to describe the same phenomenon.

32. Blend

(See Polymer blend.)

33. Blistering

A phenomenon observed in surface coatings, especially when used for corrosion protection, and in polyester resin–fibre composites (glass-reinforced plastic, GRP), particularly at the interface between the outer gel coat layer and the fibre composite. In most cases blistering is associated with the difference in osmotic pressure between the water contained in defect areas, such as voids or microcracks, and that of the water medium in the surrounding areas. The osmotic pressure differential arises from disparities in concentration of ions, which produces a thermodynamic drive for equilibration of ion concentration through the transportation of water vapour and ions. For the case of surface coatings on metals, blistering can occur also through corrosion, which can take place cathodically, when the surrounding medium is acidic, or anodically, for high-pH environments. In these cases the transport of water into the blister areas is driven by electro-osmosis, that is, electropotential gradients rather than osmotic pressure differentials.

34. Block Copolymer

A copolymer in which the two constituent monomer units are joined in blocks. These are divided into consecutive AB types, end-

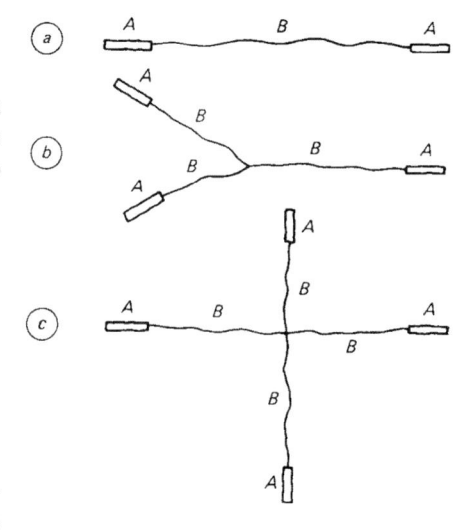

Schematic of ABA block copolymers, where A represents rigid polystyrene blocks (low molecular weight) and B denotes soft polybutadiene blocks (high molecular weight): (a) linear blocks, (b) branched blocks, and (c) star blocks.

block ABA types and alternating ABAB types. They may consist of linear blocks of two different chains or star blocks comprising more than two linear blocks attached to a common branched point, as illustrated for styrene–butadiene–styrene systems.

An important feature of block copolymers is their morphological structure arising from the lack of miscibility of the various blocks in the structure. According to the composition and chain lengths of the blocks, the two domains assume different spatial organization, as indicated.

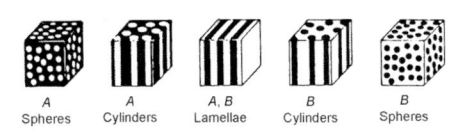

A Spheres | A Cylinders | A, B Lamellae | B Cylinders | B Spheres

Increasing A Content
Decreasing B Content

Changes in spatial organization of immiscible domains of block copolymers according to A/B ratio.

At the same time, the composition of the block copolymers can impose different behaviour with respect to physical properties. There are two aspects that have received considerable attention in industry. An important area of interest of these systems is the elastomeric behaviour resulting from the large difference in T_g of the two blocks, one producing a rigid dispersed phase (glassy or crystalline), and the other consisting of soft (rubbery) domains. The presence of rigid domains acts as an internally built reinforcement for the rubbery phase, so that the resulting polymer acquires the characteristics typical of a conventional rubber without cross-links in the polymer chains and therefore exerts a thermoplastic behaviour. For this reason, they are referred to as 'thermoplastic elastomers'. Typical thermoplastic elastomers based on block copolymers found commercially are shown in the table. (See Thermoplastic elastomer.)

Name	Rigid phase blocks	Soft phase blocks
SBS	Polystyrene (glassy)	Polybutadiene
SIS	Polystyrene (glassy)	Polyisoprene
SEBS	Polystyrene (glassy)	Poly(ethylene–butylene)
Polyester	Poly(butylene terephthalate) (crystalline)	Poly(tetramethylene oxide)
Polyurethane	Polyurethane from methylene diisocyanate (glassy)	Aliphatic polyester or poly(tetraethylene oxide)
Polyamide	Polyamide (crystalline)	Poly(tetramethylene oxide)

35. Blocking

A phenomenon describing the sticking of flexible films to each other, as in plastic bags. This is caused by attractive surface forces, of van der Waals type, operating when the surface of the films is very smooth, which allows the touching surfaces to have intimate contact. (See Antiblocking agent.)

36. Blooming

A term used to describe the migration of additives, usually solid types, from the bulk to the surface of a product.

37. Blow Moulding

A process used for the production of hollow articles, such as bottles and containers. These can be produced either by extrusion blow moulding or by injection blow moulding. The operation takes place in two stages, either in-line or separately: production of a parison and inflation into a cold mould. (See Extrusion.) Examples are shown of two widely used extrusion blow moulding processes. The first is a conventional multiple-cavity process for relatively small bottles. The second is a process, used for larger containers, involving the accumulation of melt at the front of the extruder so that the screw can rapidly extrude the parison through a rapid forward movement, thereby preventing excessive cooling and sagging of the parison before the subsequent blowing operation.

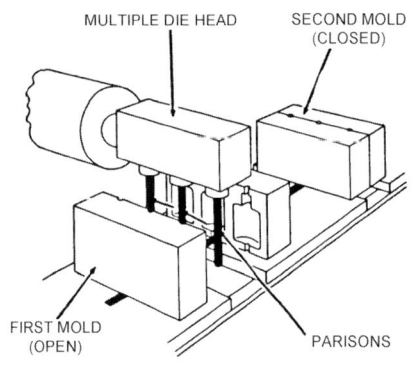

Example of extrusion blow moulding operation with multicavity mould. Source: Muccio (1994).

38. Blowing Agent

An additive used for the production of foams (cellular products). Blowing agents exert their function by rapidly producing large quantities of gas at specific sites within the bulk of a resin or molten polymer, where nucleation of the cells takes place. They are divided into physical and chemical types, depending on whether the blowing action results from a rapid evaporation of the additive or through chemical decomposition of the additive, producing large quantities of gases (usually nitrogen) for the nucleation and subsequent growth of cells. Typical physical blowing agents are low-boiling-point hydrocarbons, such as pentane (boiling point, b.p. $= 30$–$40\,°C$) and heptane (b.p. $= 65$–$70\,°C$). The use of fluorinated hydrocarbons, known as freons (b.p. $= 25$–$50\,°C$), is now generally discouraged because of environmental issues. The direct induced infusion of gases in the polymer, such as N_2 or CO_2, is a method by which physical blowing can be induced through a sudden release of the pressure. The more common types of chemical blow-

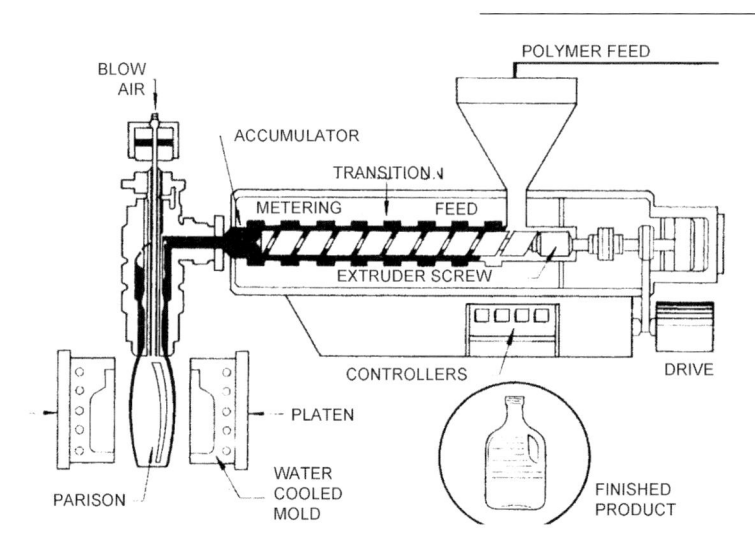

Example of extrusion blow moulding with accumulator head. Source: Lee (1998).

ing agents used for the foaming of elastomers and thermoplastics, together with their chemical structure and range of decomposition temperature, are shown in the table.

Name	Structure	Decomposition range (°C) (maximum rate)
Azodicarbonamide	$NH_2-CO-N=N-CO-NH_2$	160–200
Azobisbutyronitrile	$\begin{array}{c}CH_3 \quad CH_3 \\ NC-C-N=N-C-CN \\ CH_3 \quad CH_3\end{array}$	90–115
Benzenesulfonylhydrazine	C6H5–SO2–NH–NH2	95–100
p-Toluenesulfonyl semicarbazide	CH_3–C6H4–SO_2–NH–NH–$CONH_2$	210–270

39. Blown Film

Also known as tubular film, is produced by extrusion using circular dies. The entire set-up for the production of blown films is illustrated in the diagram.

The essence of the tubular film process is the blowing of a bubble by the inlet of air through the mandrel of the die and maintaining a constant pressure inside the bubble by folding the film via two nip rolls. The expansion of the bubble stops at the so-called 'frost line', which is formed when the temperature of the melt reaches the glass transition, for the case of glassy polymers, or the onset crystallization temperature, for crystalline polymers. A degree of biaxial orientation is introduced in the film due to a certain amount of stretching that takes place within the rubbery state of the polymer. The balance between longitudinal and circumferential (transverse) orientation can be controlled through adjustments in blow-up ratio, stretch ratio imposed by the take-off rolls and cooling rate.

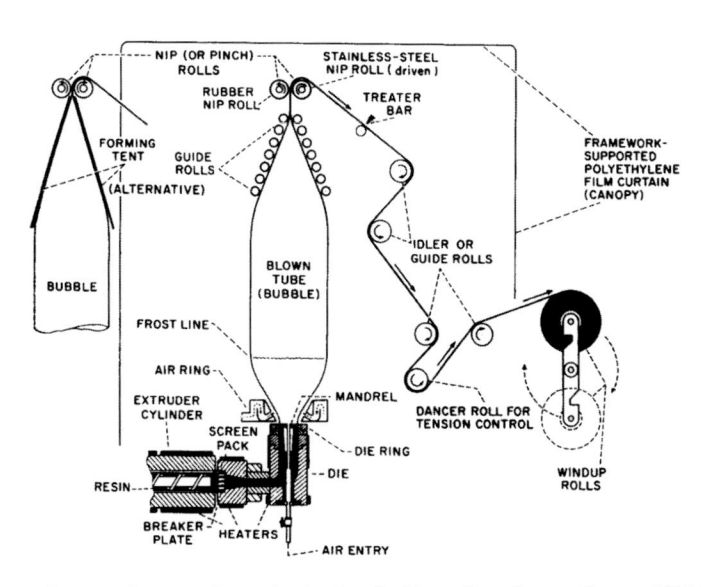

Schematic diagram of a production line for blown films. Source: Rosato (1998).

The highest level of orientation is obtained by ensuring that most of the stretching of the bubble takes place just below the frost line in order to minimize the extent of molecular relaxations.

40. Blow-Up Ratio (BUR)

This is the ratio of the diameter of the bubble to the diameter of the hollow cylindrical die, in the tubular film extrusions of polymer films. The BUR is usually in the region of 2–4. In this process a tubular film exiting from the die is blown out while still in the melt state by the injection of air through the centre of the die before being cooled and collapsed by the nip rolls acting as sealant for the bubble.

41. Boltzmann Superposition Principle

A procedure for determining the accumulated strain in a specimen, or in an area of a product, subjected to variable stresses in time. This is based on the principle that an increase or reduction in the magnitude of a stress at a certain time during creep can be accounted for in the design of the article, on the basis that resulting change in strain is independent of the previous history. The principle is illustrated in the diagram, which shows the incremental increase in strain resulting from a stepwise increase in applied stress.

This implies also that the reduction in the magnitude of an applied tensile stress will result in a partial recovery and can be considered to be equivalent to the addition of a compressive stress of the same magnitude as the removed tensile stress. (See Creep.)

42. Boundary Condition

The extreme values that a variable can reach in constrained situations. Particularly used in calculations involving the integration of a function in heat transfer and transport phenomena.

43. Breakdown Voltage

Represents the value of the voltage applied on a polymer insulator, in the form of specimen or electrical insulation, that causes the local breakdown of the chemical structure, resulting in the formation of conductive paths through the bulk for the current to leak to earth. (See Tracking.)

44. Breaker Plate

A disc containing a large number of orifices, placed between the barrel of an extruder and the die. The function of the breaker plate is to act as a support for 'screen packs', that is, wire gauzes that filter the melt before it reaches the die. (See Extruder.)

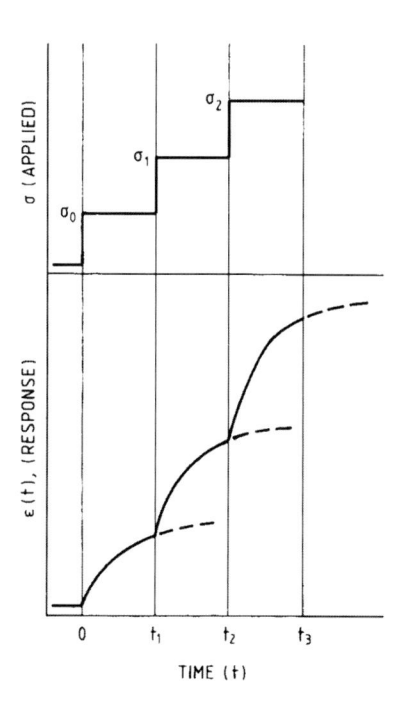

Incremental accumulation of the strain resulting from a stepwise increase in stress level.

45. Brittle Fracture

A type of failure that results in the formation of two smooth fracture surfaces, which takes place at low strains. It occurs in components that contain geometrical discontinuities or defects such as cracks, which have the effect of increasing the stress at the tip of the crack and preventing the onset of large-scale deformations (ductile failures). A pictorial description of the deformations occurring in a tensile test specimen of both brittle and ductile polymer is shown. The graphical representation of brittle and ductile failures is shown in the graph.

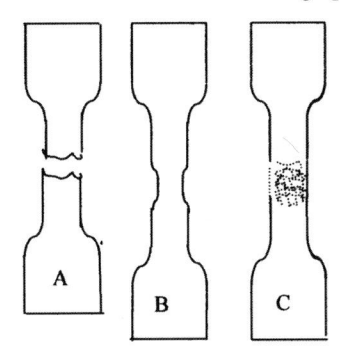

Changes in tensile specimens after failure: A, brittle fracture; B, ductile failure with necking and cold drawing (typical for rigid polymers); C, ductile failure with microvoiding (applicable mostly to failures of polymer blends).

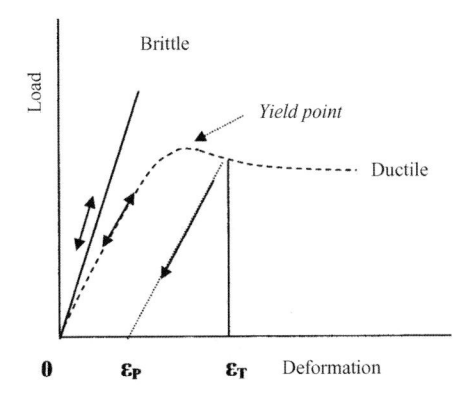

Schematic representation of brittle and ductile deformational behaviour of materials up to failure.

46. Brittle Point

The failure mode of polymers changes from brittle to ductile with increasing temperature, and vice versa. The temperature at which this takes place is known as the brittle point or the tough–brittle transition temperature.

47. Brittle Strength

The value of the stress recorded in tensile tests in which the material fails in a brittle manner is sometimes known as the brittle strength. This is not a fundamental property because the value recorded depends on the dimensions of internal flaws, usually in the form of cracks. (See Fracture mechanics and Griffith equation.)

48. Brittle–Tough Transition

A term used to denote the temperature (T_B) at which there is a change in deformational behaviour from brittle to ductile. This is usually determined from impact tests carried out at different temperatures and is identified by a rapid change in the recorded fracture energy, using the mid-point to identify the T_B value. This does not correspond to the glass transition temperature (T_g), and depends on the test method used, particularly on the notch tip radius in the case of Izod or Charpy tests. If fracture mechanics principles are used to measure the G_c value as the fracture toughness parameter, then the T_B value becomes an invariant parameter that depends only on the nature of the material, and hence is a fundamental property of the material, like the T_g.

49. Brominated Compound

These are widely used in polymer formulations as fire-retardant additives in conjunc-

50. Brookfield Viscometer

An instrument used to measure the nominal viscosity of pastes, suspensions and resins. Consists of a spindle rotating in the liquid medium contained in a beaker and recording the torque developed to maintain a constant rotation speed. The viscosity of the liquid is determined from calibration procedures using reference liquids of known viscosity. The viscosity data obtained are not accurate for fluids exhibiting non-Newtonian behaviour due to the variations in shear rate within the fluid.

51. B-Stage

A term used to describe an intermediate stage in the cure of thermosetting polymer products (usually phenol formaldehyde types), which is characterized by a very high viscosity but still capable of flowing under high pressures to allow shaping operations to be carried out before the final (C-stage) cure.

52. Bulk Modulus

A parameter denoting the resistance of a material to volumetric deformations, resulting from applied pressure or hydrostatic tension. (See Modulus.)

53. Bulk Moulding Compound (BMC)

A moulding composition containing about equal amounts of styrene, an unsaturated polyester resin, short glass fibres (approximately 5 mm long) and an inorganic filler, usually calcium carbonate. This is sometimes referred to as dough moulding compound (DMC).

54. Bulk Polymerization

Polymerization of liquid monomers carried out in the absence of solvents or water as a dispersing medium. This is also known as 'mass polymerization'.

55. Butyl Rubber

Butyl rubber, also known as isobutylene–isoprene rubber (IIR), consists of about 97–99.5% isobutylene and 0.5–3.0% *trans*-1,4-isoprene units, which can be represented with the chemical formula shown.

$$\left[-CH_2-\underset{\underset{CH_3}{|}}{\overset{\overset{CH_3}{|}}{C}}-CH_2CH=\underset{\underset{CH_3}{|}}{C}-CH_2- \right]_n$$

Commercial grades have an average molecular weight (MW) in the region of 300 000–500 000 and a broad molecular-weight distribution (MWD). The lack of unsaturation and the total absence of *tert* C–H bonds in the molecular chains give rise to high thermal oxidation stability, while the lack of dipoles provides very good dielectric properties. A significant characteristic of butyl rubber is its ability to crystallize under stress, like natural rubber, which provides a high resistance to tearing. Another important feature of IIR is the low oxygen permeability. These characteristics make IIR a suitable material for cables and inner liners for tubeless tyres.

C

1. Calcium Carbonate (CaCO₃)

A low-cost filler for many different polymer compositions, such as plastics, elastomers and paints. The main sources are calcite and limestone refined into particulate fillers through various routes. CaCO$_3$ from natural sources is often referred to as ground calcium carbonate (GCC) to differentiate it from the purer version, known as precipitated calcium carbonate, which is obtained by a synthetic route. In all cases, the filler used for polymer formulations is invariably coated with stearic acid to prevent agglomeration and to improve the dispersion. CaCO$_3$ has a density of 2.96 g/cm^3 and an average particle size around 2–3 µm, giving a surface area in the region of 2–10 m^2/g. The particles have an irregular geometry, with a wide distribution of shape and size. Some grades can have a particle size as low as 0.03–0.05 µm and a surface area of 30–90 m^2/g.

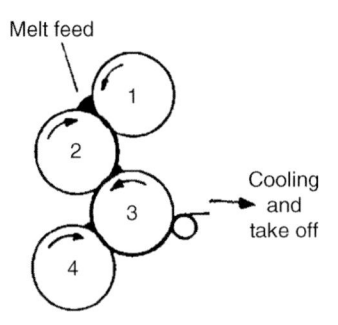

Inclined Z stacking of the rolls of a calender.

3. Calendering

The process used to produce films or sheets using a calender. A calendering production line includes several units for the preparation of the mixes and for the feeding of the melt into the nip of the first set of rolls, as well as other units for cooling and take-off systems (see diagram).

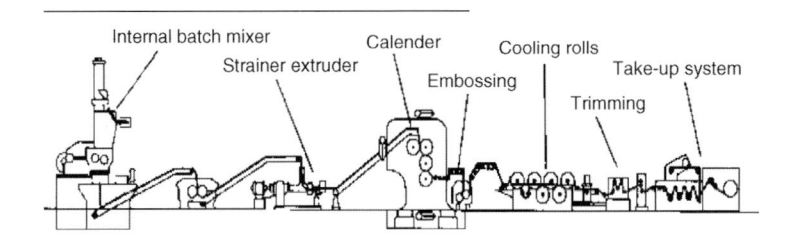

Typical calendering line for the production of films. The arrangement of rolls is an inverted L type. Source: Osswald (1998).

2. Calender

Equipment used to produce sheets or films by passing a melt though a series of highly polished heated rolls and then through cooling rolls before being wound or cut to the required length. The most common roll configurations of a calender are the inverted L type and the inclined Z type (shown).

The output of a calendering operation, that is, the flow rate of the material passing through the nip of the rolls, is related to the velocity of the rolls (U), the gap between the rolls (h) and viscosity of the melt (η) by the expression

$$Q = 2h\left[U - \frac{h^2}{3\eta}\left(\frac{\delta P}{\delta x}\right)\right]$$

Polymers in Industry from A–Z: A Concise Encyclopedia, First Edition. Leno Mascia.
© 2012 Wiley-VCH Verlag GmbH & Co. KGaA. Published 2012 by Wiley-VCH Verlag GmbH & Co. KGaA.

where Q is the flow rate per unit width of the rolls and $\delta P/\delta x$ is the pressure gradient in the nip. The velocity $U = 2\pi RN$, where R is the roll radius and N is the roll rotation speed in revolutions per unit time. In addition to calendering as an entire process, calendering rolls in vertical stacks of three are used as polishing and cooling devices in a sheet extrusion line.

4. Calorimetry

A characterization technique using heat as the medium to bring about changes in a small quantity of a polymer sample. (See Differential scanning calorimetry.)

5. Cambering (Roll Crowning)

A term used in calendering processes to describe the slight concavity of the outer surface of the rolls along the axial direction to counteract the differential thickening of the produced sheet at the exit due to swelling. (See Calender and Calendering.)

6. Camphor

A natural product, originally, used in cellulose nitrate compositions to reduce the melt viscosity. The chemical formula of camphor is:

7. Capacitance

Defined as the ratio of the charge stored (Q) to the voltage applied (V) on a specimen placed between the electrodes of a capacitor, that is, $C = Q/V$. The unit of capacitance is the farad (F). (See Permittivity and Dielectric constant.)

8. Capillary Rheometer

An instrument to characterize the rheological (flow) behaviour of polymer melts. It consists of a cylindrical chamber, in which the polymer is fed, heated and then forced by a driven piston to flow through an aligned capillary die.

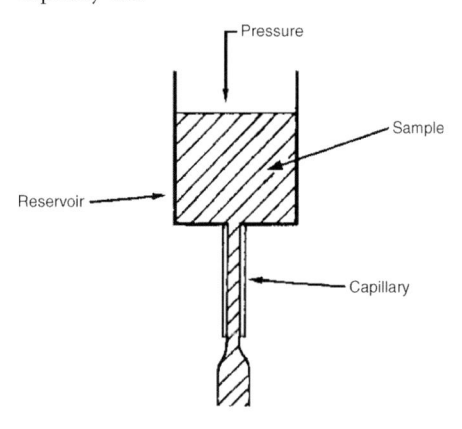

Schematic diagram of a capillary rheometer.

The pressure variation that occurs during flow of the melt from the heating chamber (reservoir) to the exit of the die is as shown. The pressure, however, can only be measured by means of a transducer fitted at the bottom of the reservoir, just before the melt enters the capillary die. The flow rate, Q, is estimated from the velocity of the piston, V, used to force the melt through the die, that is, $Q = VA$, where A is the area of the piston. (See Rheology and Momentum equation.)

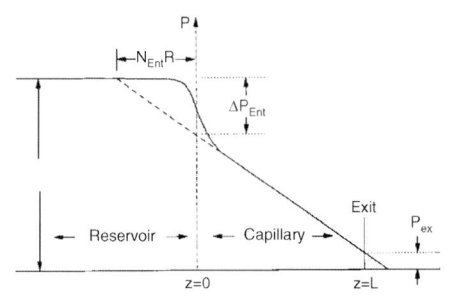

Pressure profile for the flow of the melt in a capillary rheometer.

The principle for the measurement of the melt viscosity is based on the possibility of relating the measured pressure to the shear stress at the wall of the capillary through the application of the momentum equation and correcting for the entry pressure drop (ΔP_{entry}) by the Bagley correction method. (See Bagley correction.) The relationship between shear stress at the wall (τ_{wall}) and the pressure drop (ΔP) along the die capillary is

$$\tau_{wall} = \Delta PR/2L,$$

where R is the radius and L is the length of the capillary. Note that the small pressure drop at the die exit (ΔP_{exit}), associated with die swelling, can be neglected, as it is quite small. It is noted that the entry pressure drop decreases if the entry angle from the reservoir to the capillary is reduced from 90° to smaller values, say 45°. At the same time, if the length of the capillary is very large with respect to the radius, say $L/R > 30$, the entry pressure drop becomes quite small relative to the total pressure drop, so that it can be neglected, making the Bagley correction unnecessary.

From the velocity of the piston, it is possible to calculate the shear rate at the wall ($\dot{\gamma}_{wall}$) through the relationship between velocity gradient (dV/dt) and flow rate. The velocity gradient is calculated from the flow rate through the capillary, which is the same as the flow rate induced by the piston in the reservoir. The relationship obtained on the assumption that the behaviour is Newtonian is

$$\dot{\gamma}_{wall} = 4Q/\pi R^3.$$

By describing the behaviour of the melt with a power-law equation, that is,

$$\tau = k\dot{\gamma}^n,$$

where k is known as the fluidity index and n as the power-law index, the shear rate at the wall becomes

$$\dot{\gamma}_{wall} = \frac{(3n+1)}{(4n)} \frac{4Q}{\pi R^3},$$

where the $4Q/\pi R^3$ term is known as the apparent shear rate, on account that it is the value calculated on the basis of Newtonian behaviour. This modification of the equation is also known as the 'Rabinowitsch correction'.

Further corrections can be made to increase the accuracy of the calculated shear rate through a slip analysis, which makes it possible to estimate the velocity of the melt at the wall (V_{slip}) and is used to calculate the true velocity gradient. (See Wall slip and External lubricant.) This is done by carrying out experiments with a series of dies with constant L/R ratio, possibly choosing a large L/R value (say $L/R \sim 40$) and plotting the values of $4Q/\pi R^3$ against $1/R$ at each pressure (i.e. for each wall shear stress value).

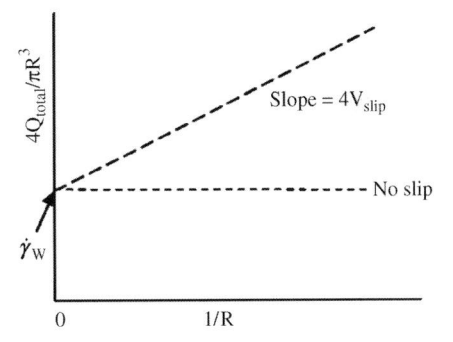

Plot of $4Q_{total}/\pi R^3$ against $1/R$ at pressure P.

The rationale for this plot (also known as the Mooney analysis) is that the total flow rate Q_{total} is the sum of the slip flow rate $Q_{slip} = V_{slip}A$ (where A is the cross-sectional area of the channel, πR^2) and the shear flow rate. The latter can be obtained by taking the average velocity $\bar{V}_{shear}$, which gives $Q_{shear} = \bar{V}_{shear}A$ (obtained by integrating the velocity gradient across the channel section). Therefore,

$$Q_{total} = V_{slip}\pi R^2 + \int f(\dot{\gamma})dr.$$

This equation can also be written as

$$\frac{4Q_{total}}{\pi R^3} = \frac{4(V_{slip} + \bar{V}_{shear})}{R} = \frac{4V_{slip}}{R} + \dot{\gamma}_{wall}.$$

In the absence of wall slip (i.e. $V_{slip} = 0$), the shear rate at the wall is equal to $4Q/\pi R^3$ irrespective of the radius of the capillary used. When, on the other hand, slip takes place (i.e. for conditions when $1/R = 0$), the shear rate at the wall is the intercept.

From measurements made at different pressures, a series of plots can be made for the viscosity against shear rates and slip velocity as a function of shear stress. It is generally found that the slip velocity increases with shear stress at the wall according to a power law, that is,

$$V_{slip} = b\tau_{wall}^m,$$

where b and m are constants for the polymer melt. Note that the incorporation of external lubricants will increase the values of V_{slip} and may also have an effect on the values of the two constants, b and m.

9. Capillary Viscometer

An instrument to measure the viscosity of polymer solutions as a means of determining the molecular weight.

Measurements are made by recording the time for the solution to flow from an upper reservoir through a capillary leading to the lower reservoir. The recorded time values are compared with those obtained for the pure solvent to flow through the same capillary.

The viscosity of the solution can be calculated from the relationship between flow rate (Q = volume/time) and viscosity (η) given by the Poiseuille equation,

$$Q = K\Delta p/\eta,$$

where K is the geometric constant for the capillary and Δp is the pressure drop along the capillary.

10. Caprolactone Polymer (Polycaprolactone, PCL)

A polymer obtained by the ring-opening polymerization of caprolactone. The chemical structure can be represented with the formula $-(OCH_2CH_2CH_2CH_2CH_2CO)_n-$. PCL has been developed primarily as a biocompatible polymer for medical application and as a biodegradable polymer. It has a low melting point ($T_m = 60-62\,°C$), with a degree of crystallinity around 50–60% and a T_g in the region of $-60\,°C$. At room temperature, the properties are similar to those of a low- to medium-density polyethylene, that is, a Young's modulus of approximately 0.5 GPa, tensile yield strength in the region of 15–18 MPa and elongation at break >500%. The water absorption is approximately 0.3%.

11. Carbon Black

Often referred to in the rubber industry simply as 'black'. It consists of agglomerates of nanosized carbon particles with surface area varying from around 25 up to 1450 m^2/g and size approximately within the range 15–75 nm. The density is in the region of 1.85 g/cm^3. Structurally similar to graphite, consisting of large sheets of polynuclear rings of carbon atoms, which are separated by electrons and are responsible for the high electrical conductivity. There are a variety of different types of carbon blacks, varying in chemical and morphological structure. They are accordingly classified as 'low structure' when the agglomerated structure is compact, and as 'high structure' if the structure is porous with a much higher surface area. Carbon black is usually obtained by the controlled combustion of hydrocarbon gases or acetylene. The more common types are referred to as 'furnace blacks', 'lamp blacks' or 'thermal blacks' depending on the manufacturing method. 'Channel blacks' are obtained from natural gas, now an obsolete process, and are known to have a surface with an acidic character, which has a repressing effect on sulfur vulcanization.

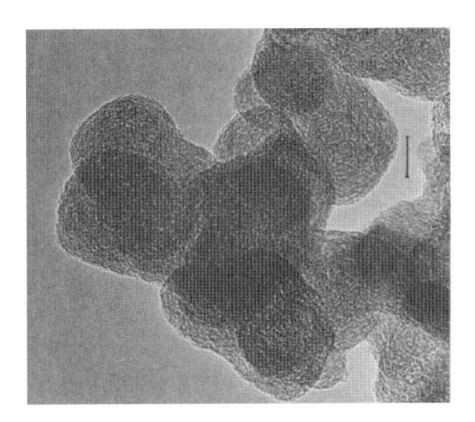

High-resolution phase-contrast TEM micrograph of a typical 'furnace black'. The scale bar represents 10 nm. Source: Kraus (1978).

Chemical reactions can take place between functional groups on the surface of carbon black, mostly quinonic type, and the free radicals present in the system during compounding and curing operations. For this reason, carbon black tends to retard the curing reactions but provides considerable protection against oxidative degradation. When carbon black is heated at 2700 °C or higher in an inert atmosphere, all surface functional groups are removed and the structure is converted into a crystalline graphitic type. The loss of functional groups on the surface reduces the ability to form chemical bonds at the interface, thereby reducing the reinforcing efficiency with respect to low strain properties, such as modulus and abrasion resistance. This does not have a significant effect, however, on the ultimate strength. About 90% is used by the rubber industry as a reinforcing filler. The rest is used in the plastics industry and in printing inks as a pigment or to produce semiconducting products.

12. Carbon Fibre

Widely used as reinforcing fibres for the production of composites. (See Composite and Reinforcement.) Such fibres have a graphene structure resulting from the controlled pyrolysis of suitable organic fibres, such as polyacrylonitrile or cotton, carried out in an inert atmosphere. (See Graphene.) The final diameter of the fibres is in the region of 6–10 µm. Some low-cost carbon fibres are produced from pitch, a tar-like material. Their mechanical properties are largely related to the level of crystallinity. Fibres with a higher modulus and higher strength are often referred to as graphite carbon fibres. The morphological structure of carbon fibres is shown in the diagram and micrograph.

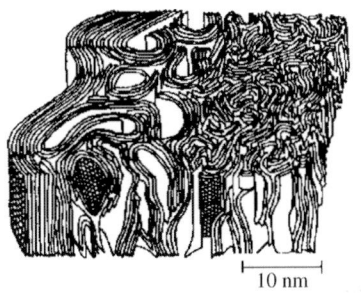

Pictorial description of the internal structure of carbon fibres.

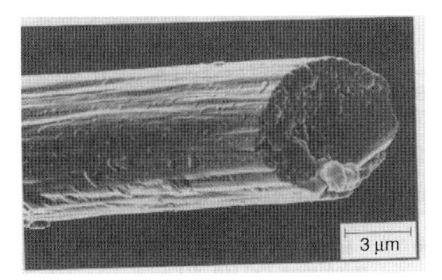

Micrograph of a typical carbon fibre. Source: Ehrenstein (2001).

Since the surface of carbon fibres is virtually free of any reactive (functional) groups, special chemical treatments have to used to introduce some OH, COOH and NH_2 groups onto the surface, as a means of producing chemical bonds with the matrix. To facilitate handling of the fibres and to improve their wettability with matrix resins, the surface of carbon fibres is usually coated with an extremely thin layer of a solid epoxy resin. This reacts with the surface functional groups to produce chemical coupling with the resin matrix, normally another epoxy resin or a different resin that is compatible with epoxy resins. The epoxy coating acts also as a binder for the 'tows' so that they remain compact and this makes them easier to handle during the production of prepregs.

13. Carbon Nanotube (CNT)

CNTs are extremely fine tubes with a graphene structure, produced by chemical vapour deposition method from unsaturated hydrocarbons, such as ethylene or acetylene. They are available as single-walled carbon nanotubes (SWNTs), consisting of a single hexagonal layer of graphitic carbon with a diameter in the region of 20–50 nm, or as multiple-walled carbon nanotubes (MWNTs), with 50–100 nm diameter.

The orientation of the graphene layers is in the axial direction and is symmetrical with the central axis in both CNT systems. CNTs are usually used at very low concentrations (<3 wt%) owing to their very high reinforcing efficiency and their ability to produce electrically conductive paths through percolation of the high-aspect-ratio nanotubes. The organization of the graphene layers of an MWNT and a TEM micrograph of actual MWNT bundles are shown in the diagram and micrograph, respectively.

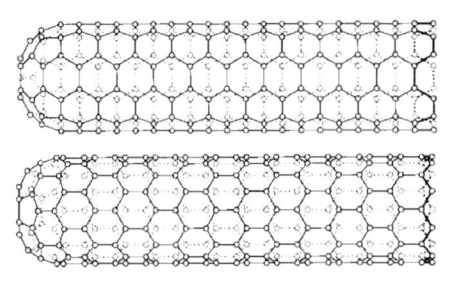

Graphene layers in an MWNT: (top) armchair stacking; (bottom) zig-zag organization. Source: Bell et al. (2006).

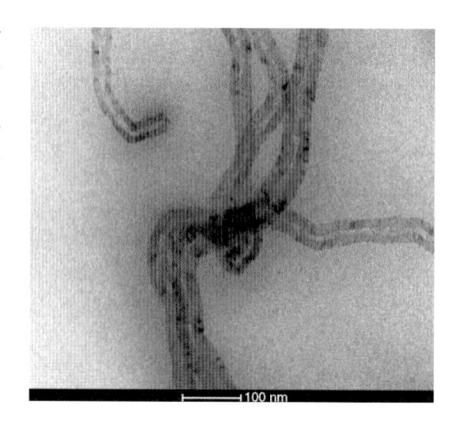

Double-walled carbon nanotubes (diameter 30 nm) from Shenzen (China). Courtesy of D. Acierno, University of Naples, Italy.

14. Carbonyl Index

A parameter used to describe the extent of oxidation of a polyolefin resulting from degradation reactions induced by UV light and/or by heat. It is simply defined by the equation $CI = 100A/t$, where A is the infrared absorbance at wavenumbers in the region of 1715–1750 cm^{-1} and t is the thickness of the film used in the experiment.

15. Carothers

Wallace Carothers was the inventor of nylon polymers and is well known for his fundamental work on polymers produced by condensation reactions.

16. Carreau Model

A mathematical model to describe the variation of the viscosity (η) of a polymer melt with shear rate, as shown in the graph.

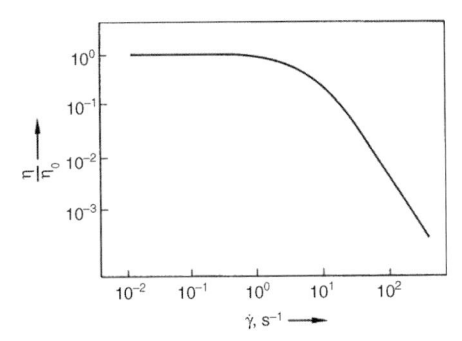

Variation of relative viscosity of polymer melts (η/η_0) with shear rate according to the Carreau model.

The Carreau equation is usually written as

$$\eta - \eta_0 = \frac{\eta_0 - \eta_\infty}{[1 + (\lambda \dot{\gamma})^2]^{(1-n)/2}},$$

where η_0 is the viscosity extrapolated to zero shear rate, hence known as the 'zero-shear viscosity', η_∞ is the limiting viscosity at very high shear rates, λ is a constant for the melt (known as the characteristic time because of its units), $\dot{\gamma}$ is the shear rate and n is an exponential index (constant) for the melt.

17. Cartesian Coordinates

Also known as rectangular coordinates. A mathematical representation of variables in space with linear coordinates at right angles to each other. They are also used to represent the direction of three-dimensional space.

18. Case II Diffusion

A type of diffusion of small molecules through a solid, which takes place via an advancing front layer that gradually increases in thickness until it traverses the entire section of the medium. The actual diffusion begins after a certain induction time is reached in order to allow an accumulation of diffusing species on the surface of the medium. In case II diffusion the mass absorbed (M_t) increases approximately linearly with time (t), whereas for Fickian diffusion M_t is proportional to $t^{1/2}$. (See Diffusion.)

19. Casein

A plastic material with the same name as the protein found in milk and cheese, from which it is produced. Originally obtained from straight polymerization by treatment with acid, it was later improved by treatment with formaldehyde. It has been used primarily for the production of buttons to replace horn. It is now largely superseded by other rigid plastics derived from petrochemicals.

20. Casting

A technique for the production of articles by pouring a liquid resin, a monomer or a paste into an open mould and allowing it to solidify through chemical reactions or diffusion. The term is also used in film extrusion to denote a technique by which the extrudate is rapidly quenched on chilled rolls. This latter technique is called 'chill roll casting'.

21. Cationic Polymerization

A type of polymerization carried out with the use of Lewis acid catalysts, such as BF_3, $AlCl_3$, $TiCl_4$ and $SnCl_4$. It can be used for the polymerization of both unsaturated or cyclic monomers or oligomers, which become active in association with a proton donor auxiliary. For example

$$BF_3 + H_2O \rightarrow H^+ + BF_4OH^-$$

or

$$BF_3 + R_2O \rightarrow R_2O - BF_3$$

Cationic polymerization takes place in three stages.

Step 1. Initiation: This takes place by the addition of H or R to the double bond, or to the oxygen of an oxirane ring in the monomer or resin. For example

$$RCH = CH_2 + BF_3/H_2O \rightarrow RCH_2-CH_2{}^+ BF_4OH^-$$

Step 2. Propagation: For example

$$RCH_2-CH_2{}^+ BF_4OH^- + nRCH = CH_2$$
$$\rightarrow (RCH_2-CH_2)_{n+1}{}^+ BF_4OH^-$$

Step 3. Termination: For example (transfer of ions to monomer)

$$(RCH_2-CH_2)_{n+1}{}^+ BF_4OH^- + RC$$
$$= CH_2 \rightarrow (RCH_2-CH_2)_n + RCH_2-CH_2{}^+ BF_4OH^-$$

However, this is only one of many possible termination reactions. Others include abstraction of a proton from other species present, such as the solvent.

22. Cationic Surfactant

(See Surfactant.)

23. Cavity

The component of a mould that receives the melt for the shaping of a moulded part.

24. Cavity Filling

A term that refers to the flow of a polymer melt into the cavities of an injection mould.

24.1 Cavity Filling Mechanism

The mechanism by which the melt enters and flows into a mould cavity is determined primarily by the type and position of the gate. Jet filling takes place when the gate is positioned at the opposite end of the cavity so that the cavity fills up in a zig-zag fashion. When the melt emerging from the gate immediately finds an obstruction or a sharp change of direction, the turbulence created produces a laminar flow filling mechanism. The two mechanisms are illustrated in the diagram.

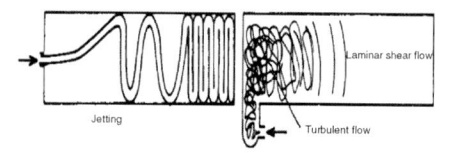

Cavity filling mechanisms in injection moulding. Source: Mascia (1989).

The cold walls of the cavity of the mould produce a frozen solid layer of 'oriented polymer' chains, owing to the setting-up of elongational stresses at the flowing front of the melt, as shown.

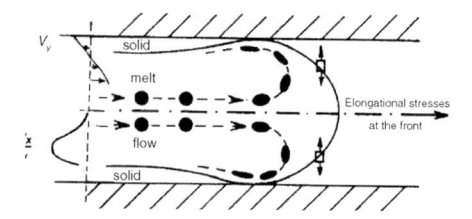

Flow mechanism of polymer melt in the cavities of an injection mould. Source: Tadmor (1974).

24.2 Pressure Requirements for Cavity Filling

Owing to the complexity of the flow path in the cavities of injection moulds, it is difficult to obtain estimates of the pressure requirements to fill the cavity of a mould without the aid of analytical techniques such as 'finite element analysis'. Other difficulties arise from the non-isothermal nature of the flow and the non-Newtonian behaviour of polymer melts, which make it difficult to specify single values for the viscosity.

25. Cavity Packing

The flow of polymer melt entering the cavity of an injection mould as a result of the volumetric shrinkage taking place while cooling. Shrinkage is particularly large for the case of crystalline polymers and can be as high as 20% at atmospheric pressure. A considerably larger pressure is required for cavity packing than for cavity filling to counteract the volumetric shrinkage. This can be estimated from the pressure–volume–temperature (PVT) diagram of particular polymers and makes it possible to design a pressure profile for the moulding cycle. (See PVT diagram.)

26. Cellophane

Film cast from a water-dispersed 'regenerated cellulose', obtained by treating cellulose with sodium hydroxide solution in order to swell the fibrils by allowing water to penetrate between polymer chains. The product is then reacted with carbon disulfide to form the sodium salt, known as 'cellulose xanthate'. The aqueous dope of regenerated cellulose is cast into films to produce cellophane.

27. Cellular Polymer

Another term for polymer foams. (See Foam.)

28. Celluloid

Originally a tradename for cellulose nitrate plastic products.

29. Cellulose

A polydisperse linear polysaccharide with the chemical structure shown.

Chemical structure of cellulose, $n = 2000$–$10\,000$.

Cellulose is the most abundant natural organic compound, which is found in cotton (95%), flax (80%), jute (60–70%) and wood (40–50%). It is insoluble in most organic solvents and decomposes before reaching the melt state. However, it can be processed from aqueous dispersions of regenerated cellulose to produce glossy films (known as cellophane) and fibres (rayon). Cellulose has played an important role in the development of plastics materials, first through the development of cellulose nitrate and subsequently for the production of other esters, such as cellulose acetate (CA), cellulose propionate (CP), cellulose acetate butyrate (CAB) and cellulose acetate propionate (CAP). These are often mixed with plasticizers, such as triethyl citrate or dioctyl adipate, to enhance their processability through a reduction in melt viscosity. Attractive features of cellulosic polymers include their biodegradability and their controlled-release characteristics for membranes and coatings.

30. Cellulose Acetate (CA)

The first thermoplastic material that could be moulded and extruded in the melt state. CA is a glassy amorphous polymer, usually containing small amounts of plasticizer to improve its processing characteristics and to increase its ductility. Produced by the acetylation of the hydroxyl groups present in cellulose, aimed at interrupting the formation of hydrogen bonds, responsible for the strong intermolecular forces that prevent the reaching of the melt state before the onset of degradation reactions. (See Cellulose.)

31. Cellulose Acetate Butyrate (CAB)

A glassy polymer produced by the partial substitution of acetyl groups with butyric groups in the esterification of cellulose. This substitution brings about a more efficient internal plasticization than can be achieved by acetylation alone, which makes it unnecessary to use an external plasticizer to improve the processing characteristics of the polymer. CAB also has a lower level of water absorption than cellulose acetate.

32. Cellulose Nitrate (CN)

A polymer obtained by the nitration of cellulose. It represents the first plastic material made available commercially under tradenames of Parkesine and Celluloid. CN is used mostly in the form of mixtures with camphor.

33. Cellulose Propionate (CP)

A glassy polymer obtained by esterification of cellulose with propionic acid. The properties of CP are intermediate between those exhibited by cellulose acetate and cellulose acetate butyrate.

34. Cellulosic

A generic term for all cellulose-derived plastics.

35. Ceramer

A term derived from the combination of two truncated words, namely, ceramic and polymer. It has been used to describe organic–inorganic hybrid materials intended for use in coatings.

36. Chain Configuration (*cis* and *trans*)

Denotes the spatial arrangement of two different substituent groups on C=C bonds, such as the CH_2 groups in polybutadiene. The *cis* configuration refers to the position of equivalent groups on both sides of the C=C bond, while *trans* refers to configurations where equivalent groups are positioned on opposite sides, as shown.

$$\left[\begin{array}{c} CH_2 \\ \diagdown \\ H \end{array} C=C \begin{array}{c} CH_2 \\ \diagup \\ H \end{array}\right]_n$$

cis-1,4-

$$\left[\begin{array}{c} CH_2 \\ \diagdown \\ H \end{array} C=C \begin{array}{c} H \\ \diagup \\ CH_2 \end{array}\right]_n$$

trans-1,4-

37. Chain Extension

A term used to describe reactions that lead to an increase in the length of a polymer chain.

38. Chain Flexibility

It describes the ease with which a polymer chain can rotate about the constituent C–C bonds, which is related to the internal energy required for the rotation of segments of polymer chains at temperatures above the glass transition.

39. Chain Folding

Mechanism by which lamellae are formed during the crystallization of polymers. (See Crystallinity.)

40. Chain Scission

The breaking of a polymer chain, associated with degradation reactions.

41. Chain Stiffness

The opposite of chain flexibility.

42. Chain Stopper

An additive or impurity that reacts with a growing polymer chain during polymerization, thereby controlling chain length.

43. Chain Transfer

Mechanism by which a free radical attached to a growing polymer chain, during polymerization, abstracts hydrogen from a different molecule, usually a chain transfer additive, producing a different free radical.

44. Chalking

A term (jargon) used to describe the formation of a brittle layer on the surface of a polymer product as a result of weathering-induced degradation reactions.

45. Channel Black

A grade of carbon black widely used as a UV absorber additive.

46. Char

A carbonaceous residue formed on the surface of a burning polymer, either cellulosic in nature or containing aromatic groups along the backbone chains, as in polysulfones (thermoplastic) or phenol formaldehyde (thermoset).

47. Charge Density

The quantity of charges present on a unit surface area of the electrodes of a capacitor containing a dielectric.

48. Charge Exchange Capacity

A generic term that quantifies the capability of nanoclays to exchange structural cations with others of similar charge derived from the surrounding medium, expressed as meq/100 g clay.

49. Charpy Impact Strength (and Izod Impact Strength)

Denote the amount of mechanical energy per unit cross-sectional area beneath the notch that is consumed when a specimen is fractured under impact conditions (see diagram). The use of the term 'strength' in this case is anomalous insofar as strength is normally expressed in terms of the stress, and not energy, required to fracture a specimen. The accurate term should be 'impact toughness' for consistency of nomenclature.

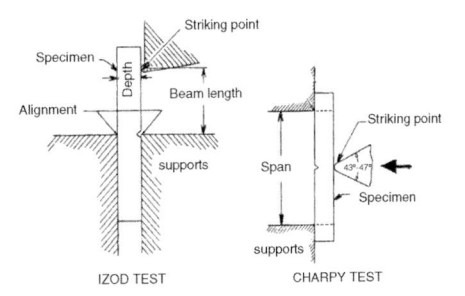

Geometry of specimen and loading mode in pendulum impact tests.

In these impact tests, the load is delivered at high speed to the specific loading point by a mass placed at the extremity of a pendulum. The apparatus records the energy used in fracturing the specimen by registering the loss of potential energy (ΔU) from the reduction in the height reached by the striking mass after fracturing the specimen. The loss of energy of the pendulum is, therefore, given by the expression

$$\Delta U = mg(h_o - h_f),$$

where m is the mass of the pendulum, h_o and h_f are the respective heights before and after fracture of the specimen. (See Fracture test and Pendulum impact test.)

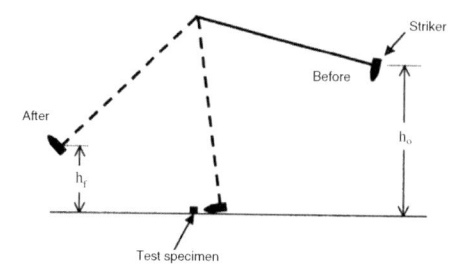

Principle of pendulum impact tests.

50. Chelating Agent

Additive used to produce complexes or co-ordination products with metal ions in order to prevent their catalytic action on the degradation caused by thermal or UV-induced oxidative reactions.

51. Chemical Blowing Agent

Additive used for the formation of cells in the production of foams. A gas, usually nitrogen, is formed at a very fast rate over a relative narrow range of temperatures. Typical chemical blowing agents and the corresponding blowing temperature range are: azo-bis-dibutyronitrile (90–115 °C), azodicarbonamide (160–200 °C) and p-toluenesulfonylsemicarbazide (210–270 °C).

52. Chemical Resistance

Denotes the capability of a polymer to resist either swelling by solvents or attack by chemicals.

53. Chemical Shift

A term used primarily in nuclear magnetic resonance (NMR) analysis to denote the shifting of the magnetic resonance frequency to values lower than the applied external magnetic field as a result of the interaction with the electrons of the molecules of the substance examined. (See Nuclear magnetic resonance.)

54. Chill-Roll Casting

A processing method for the production of slit films (see diagram). This process makes it possible to produce films at a very high production rate by operating with the die heated to high temperatures and with a large drawdown ratio.

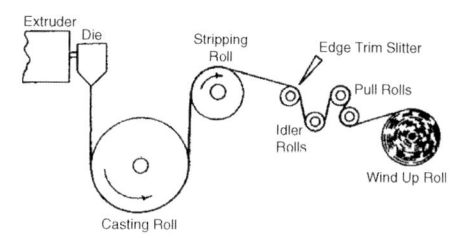

Schematic illustration of the chill-roll casting process for the production of films. Source: Teegarden (2004).

55. Chitin

An abundant natural product, corresponding to poly(N-acetyl-D-glucosamine), found in the exoskeleton of crustaceans and insects, as well as in the cell walls of fungi and microorganisms. The structure of chitin can be drawn as

or with a more precise special configuration as

Chemical structure of chitin.

The main commercial sources of chitin are fish shells such as crabs, shrimps and lobsters.

56. Chitosan

A polymer obtained by the partial deacetylation of chitin, having the chemical structure shown, where the residual acetyl groups vary between 5% and 70%.

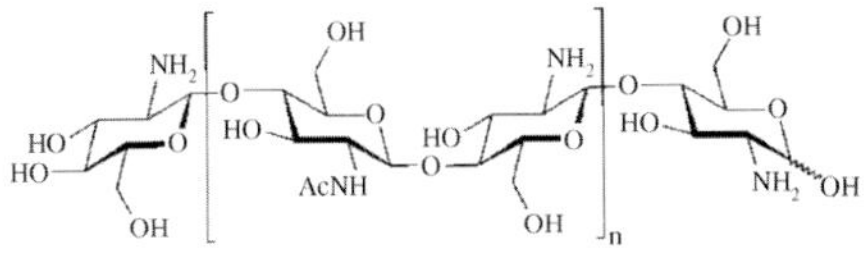

Chemical structure of chitosan.

The molecular weight of chitosan varies according to the degree of deacetylation but is generally found to vary from about 1×10^5 to 5×10^5. Chitosan is a biopolymer soluble in dilute acids such as acetic acid or formic acid, and can be plasticized with polyols, such as glycerol, sorbitol and poly (ethylene glycol), as well as with fatty acids, such as stearic and palmitic acids. Chitosan degrades readily at high temperatures and, therefore, cannot be processed in the melt state. Normally chitosan is used for coatings and films cast from dilute acid solutions.

57. Chlorinated Fire Retardant

Examples are chlorinated paraffins, containing 30–70% chlorine, chlorinated alkyl phosphates and chlorinated cycloaliphatics such as dodecachlorodimethanodibenzo cyclooctane.

58. Chlorinated Polyethylene

A grade of polyethylene obtained by the chlorination of the polymer chain with chlorine gas to improve the chemical resistance towards polar solvents.

59. Chlorinated PVC

A grade of poly(vinyl chloride) (PVC) obtained by the introduction of more chlorine atoms in the polymer chain as a means of improving the chemical resistance.

60. Chlorosulfonated Polyethylene

A grade of polyethylene (PE) containing a small number of SO_2Cl side groups along the polymer chains. These provide the polymer with the ability to become cross-linked by reaction with magnesium oxide and lead oxide in conjunction with an accelerator.

61. Chopped Strand Mat (CSM)

Mats consisting of chopped glass fibre rovings, usually about 5–10 cm long, held together with a binder that will readily dissolve in the resin in the manufacture of composites. A solid unsaturated polyester resin is often used as the binder for liquid polyester resins containing styrene as the cross-linking agent.

62. CIE Chromaticity Diagram

The abbreviation for the Commission International de l'Eclairage, used to describe the plot of the values x, y or z, as fractions of the sum of the three primary colours, R (red), G (green) and B (blue), used in colour matching, that is

$x = R/(R + G + B)$,

$y = G/(R + G + B)$,

$z = B/(R + G + B)$,

where $x + y + z = 1$ (see diagram). This makes it possible to specify the desired colour for a product in terms of the x, y and z values. (See Colour matching.)

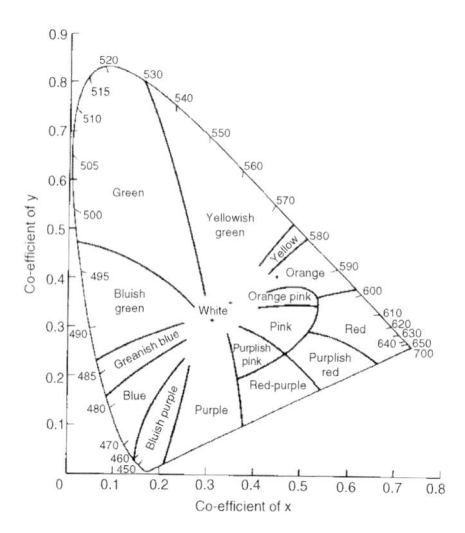

CIE chromaticity diagram.

63. Clamping Force

The force that has to be exerted onto the mould of an injection moulding machine in order to prevent the melt escaping through the parting line of the mould to form a 'flash'. The simplest approach to calculate the clamping force is to multiply the pressure on the melt by the projected area of the mould. To take into account the increase in viscosity during cooling, a more elaborate procedure is required through the use of various dimensionless parameters, such as

$$\beta = \alpha(T_i - T_m),$$

where α is the thermal diffusivity ($k/\rho C_p$), T is the temperature, and subscripts i and m stand for injection and mould, respectively. The other important parameter is the dimensionless time, defined as

$$\tau = t_{fill} k / h^2 \rho C_p,$$

where t_{fill} is the mould filling time, h is the thickness, k is the thermal conductivity and C_p is the specific heat. The relationship between clamping force (F) and the parameter τ is available for different values of β.

64. Clamping System

The mechanical system used in injection moulding machines to lock the mould during the injection of the melt into the cavity. This may be either hydraulic or mechanical (toggle), or a combination of the two. (See Injection moulding.)

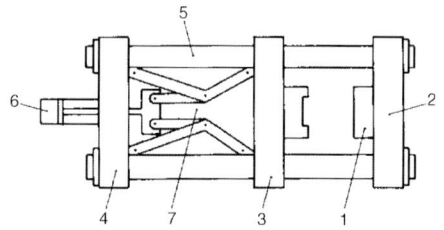

Typical toggle clamping unit fitted to injection moulding machines: 1, sprue half of mould; 2, fixed platen; 3, moving platen; 4, adjustable platen; 5, tie-bar; 6, hydraulic cylinder; 7, toggle mechanism.

65. Clarity

The ability of a product or a specimen to allow the details in an object to be resolved in the image observed. For this reason it is sometimes referred to as 'see-through clarity'. The loss of clarity is related to light scattered while passing through the object.

66. Clash and Berg Test

An empirical test usually used to determine the temperature at which a rubbery polymer, such as plasticized PVC or elastomers, acquire a predetermined level of flexibility when heated from low temperatures. The temperature at which this takes place is known as the 'cold-flex temperature', which is closely related to the glass transition temperature.

67. Clausius–Mossotti Equation

Relates the permittivity of a dielectric, ε, to its density, ρ, by the expression

$(\varepsilon-1)/(\varepsilon+2) = K_\rho,$

where K is a constant for the material.

68. Clay

A generic name for a variety of silicate minerals used as fillers. The density is about 2.3 g/cm^3 with particle size usually within the range 0.2–10 μm and surface area around 10–20 m^2/g. The more widely used varieties are 'bentonite', 'kaolin' and 'China clay'. According to the origin and treatments, clays can have different morphological structure – many are characterized by a platelet geometry susceptible to delamination.

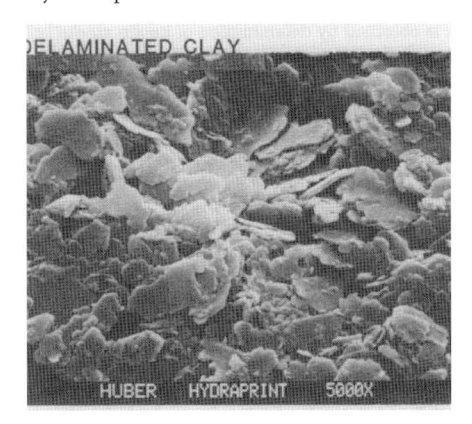

Micrograph of a clay from J. M. Huber Corporation. Source: Wypych (1993).

Montmorillonite is the principal constituent of bentonite, widely used for the production of nanocomposites in view of the possibility of exfoliating the platelets into laminae down to a few nanometres in thickness. (See Exfoliated nanocomposite and Nanoclay.)

69. Cluster

A term used to describe agglomerations of small structural entities into domains of nanoscopic dimensions. Examples are the clusters in ionomers consisting of anions hanging from polymer chains and metal cations added from an external source.

70. Co-Agent

A generic term for an additive used as an auxiliary, as in the case of curatives for elastomers.

71. Coagulum

A large agglomerate found in a latex due to coalescence of primary suspended particles.

72. Coaxial Cylinder Rheometer

A rheometer consisting of an inner cylinder, known also as the 'bob', which rotates within a static outer cylinder, thereby shearing the fluid in the gap between the two cylinders. The shear rate corresponds to the velocity gradient through the gap. It is also known as the Couette rheometer, named after the inventor.

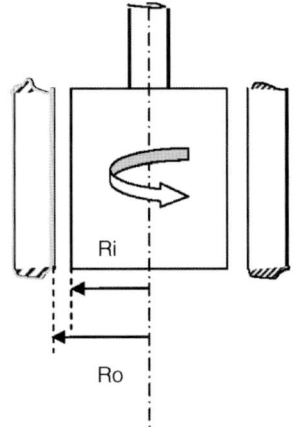

Principle of coaxial cylinder rheometer.

The shear rate is given by the velocity gradient across the gap $(R_o - R_i)$, that is,

$$\dot{\gamma} = r\frac{dV}{dr},$$

which gives

$$\dot{\gamma} = R_o \frac{\Omega}{R_o - R_i},$$

where Ω is the angular velocity. The shear stress τ is calculated from the torque developed by the rotating cylinder to maintain the imposed velocity, using the equation

$$\tau = \frac{M}{2\pi R_o^2},$$

where M is the torque per unit radius distance. The viscosity η is calculated as the ratio $\tau/\dot{\gamma}$.

73. Cobalt Naphthenate and Cobalt Octoate

Used in a dilute solution as accelerators for the cold curing of unsaturated polyester resins.

74. Co-Continuous Domain (Co-Continuous Phase)

A term used to describe the spatial configuration of the constituent components of polymer blends or organic–inorganic hybrids. The diagram shows an idealized representation of the silica domains in an organic–inorganic hybrid.

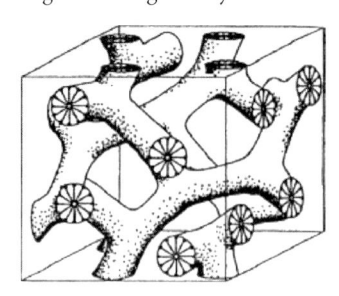

Bicontinuous phases in creamers or organic–inorganic hybrids.

75. Co-Extrusion

Extruding two or more polymers through a common die to produce multilayered sheets, films and tubular products. The principle is shown in the diagrams for a sheeting die and a tubular die. Note that in sheet dies the flow rate of each polymer can be controlled by individual choke-bar channels. Similarly, the uniformity of the thickness of each film can be controlled by individual spiralling channels in tubular film dies. (See Extrusion.)

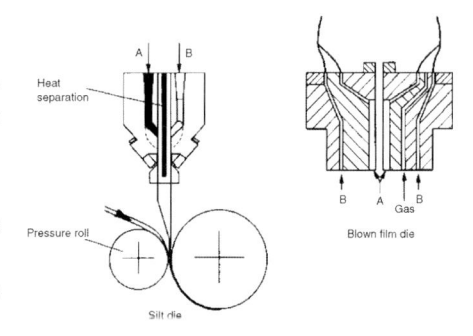

Die systems that bring together the two melt streams outside the die. Source: Baird and Collias (1998).

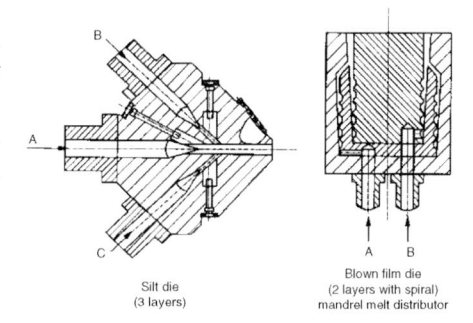

Die systems that bring together the two melt streams within the die just before reaching the die lips. Source: Baird and Collias (1998).

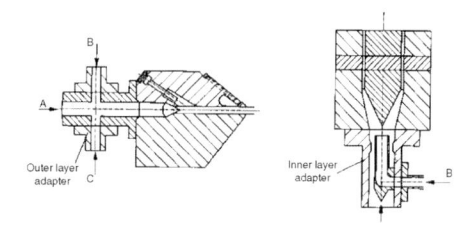

Die systems that bring together the two melt streams at the die adapter and then spread out and flow through the die in a laminar fashion. Source: Baird and Collias (1998).

76. Cohesive Energy Density (CED)

A concept used to determine the solubility of a polymer in a solvent or the mutual miscibility of two polymers. Defined as the difference between the molar heat of vaporization and the molar work of expansion, that is

$$\mathrm{CED} = (\Delta H^{\mathrm{vap}} - RT)/V_{\mathrm{m}},$$

where ΔH^{vap} is the latent heat of vaporization, R is the universal gas constant, T is the ambient (absolute) temperature (K) and V_{m} is the molar volume of the liquid.

77. Co-Injection Moulding

(See Injection moulding.)

78. Cold Crystallization

A term used in thermal analysis to denote the crystallization that occurs during the heating scan.

79. Cold Curing

Curing of a thermosetting resin or an elastomer carried out at ambient temperature.

80. Cold Drawing

A term used to describe the stretching stage after the formation of a neck (yield point) in tensile tests of ductile thermoplastics. The events that follow the yield point are illustrated.

Cold drawing introduces monoaxial orientation of the polymer chains and crystallite entities (where applicable), which is stabilized by the increase in intermolecular forces, resulting from the reduction in intermolecular distances.

81. Cold-Flex Temperature

A term used in the technology of plasticized PVC to denote the temperature at which the

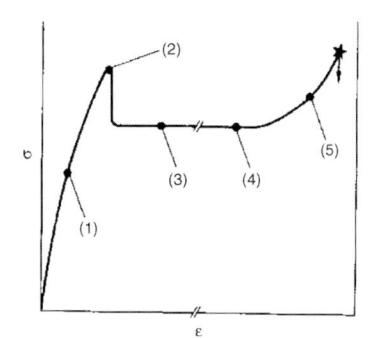

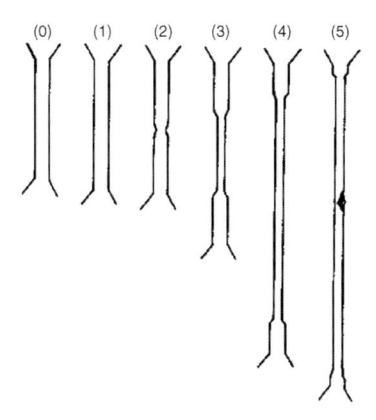

Sequence of observations in tensile tests on ductile polymer. (Top) Stress–strain curves (indicator 2 corresponds to the yield point). (Bottom) Tensile specimens showing necking deformations and cold drawing (indicators 3, 4 and 5).

material reaches a certain level of flexibility when heated up from its glassy state. The term derives from a standard test, known as the Clash and Berg test, using a torsion pendulum apparatus.

82. Cold Flow

Large deformations taking place at ambient temperature. This is sometimes referred to as 'creep'.

83. Cold Forming

An operation, such as extrusion or compression moulding, carried out at temperatures below the melting point of the polymer. For glassy polymers, cold forming is carried out at temperatures only just above the glass transition temperature.

84. Cold Plasma

Gas converted to plasma by the application of a radiofrequency (RF) field at ambient temperature and at relatively low pressures (1–100 Pa). This causes the ionization of gas molecules by collision with the high-energy free electrons generated by the RF field through a cascade process leading to the formation of plasma. At atmospheric pressure and in the absence of an RF field, a gas would reach temperatures in the region of 3000–5000 K to acquire the level of energy associated with plasma. The type of plasma generated depends on the nature of the gas used and can be classified into 'noble gas' plasma and 'active' plasma. The excitation reactions occurring in a noble gas, such as argon, are well known and can be represented as follows:

i) $e^- + Ar = Ar^+ + 2e^-$
ii) $e^- + Ar = Ar^* + e^-$
iii) $Ar^* = Ar + h\nu$
iv) $Ar^+ + e^- = Ar + h\nu$

Reaction (i) represents the ionization process by collision. A cascade will result if these reactions occur at a faster rate than those in reaction (iv), which represents the recombination accompanied by electromagnetic radiation emission in the vacuum ultraviolet (VUV) region. Reaction (ii) represents an interaction during which a collision produces not ionization but excitation of the Ar atom, that is, Ar acquires energy but this is not sufficient to cause ionization. The excited atoms (Ar*) thus produced will decay to the ground state, as shown in reaction (iii) with emission of radiation in the visible and UV regions of the spectrum.

Note that the net total electrostatic charge on the ionized gas is actually zero, due to the electrostatic balance.

In 'active' plasma there are many possible reactions that can take place. It has been estimated, for instance, that in oxygen plasma there are more than 30 possible reactions. Within the pressure range 1–100 Pa the electrons have energy of the order of 1–10 eV and are present at densities of 10^{15}–10^{18} per cubic metre. In this pressure range the mean free path of the electrons (i.e. distance travelled without undergoing collisions) is sufficiently large for them to gain energy from the applied electric field at a greater rate than that at which they lose energy by colliding with atoms and molecules in the gas. The electron energies may be expressed in terms of a 'Boltzmann equivalent temperature' (T_e) defined by the relation

$$\frac{1}{2}M_e V^2 = \frac{3}{2}kT_e,$$

where M_e is the mass of the electron, V is the root mean square (RMS) velocity of the electrons and k is the Boltzmann constant. Note that neutral atoms and molecules are not accelerated by the electric field and, therefore, their energy is determined primarily by the ambient temperature, with a small contribution from the collisions with electrons. Since the RMS velocity of neutral atoms and molecules is much lower than that of the electrons, the gas temperature (T) will also be much lower. Typically, the ratio T_e/T may be in the region of 10–100, making RF plasma a suitable expedient for the treatment of lower-melting-point materials, such as polymers. This means that highly reactive ions and free radicals can be produced at quite low temperature, both in the plasma and on the surface of the materials to be treated. The reactive interactions of plasma with polymers have been exploited commercially for various types of surface treatments, such as cleaning, etching, cross-linking and grafting reactions.

The reactive species produced on the surface of polymers are often sufficiently long-lived to allow further reactions to be induced after the plasma treatment, thereby permitting the functionalization of surfaces with chemical species that would decompose in a plasma atmosphere, such as unsaturated monomers. There are, however, substantial limitations in the treatment of the inner surface of articles such as tubing or open cells of foams, insofar as the mean free path of the ionized gas species could be longer than the diameter of internal cavities so that they will become 'deactivated' in the entrance regions through collisions before reaching the inner surface of the cells or tubing. A sketch of the vacuum chamber and set-up to receive gas and for generating the RF field for producing cold plasma is shown.

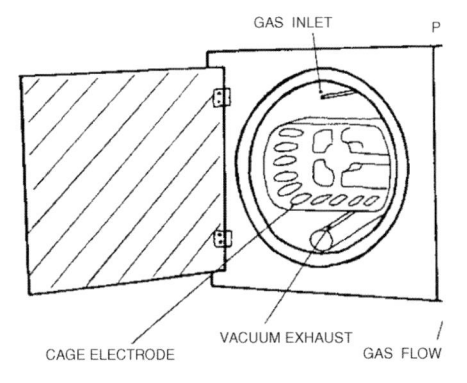

A plasma chamber for the treatment of polymer surfaces.

85. Cold Set (Compression Set)

A term widely use to denote the lack of recovery of a rubber or elastomeric product after compression.

86. Colligative Properties

Properties of solution depending only on the number of solute particles and not on their chemical nature. The main colligative properties are lowering of vapour pressure, depression of freezing point and lowering of osmotic pressure. These properties can be used to determine the molecular weight (molar mass) of a polymer. The relationship between osmotic pressure (Π) and number-average molecular weight (M) is

$$\lim_{c \to 0} \frac{\Pi}{c} = \frac{RT}{M},$$

where c is concentration, R is the gas constant and T is absolute temperature (K).

87. Colloidal

From Latin *colla*, meaning glue. Denotes the structureless state of very viscous fluids or soft solids, consisting of sub-micrometre particles or globules suspended in a fluid or dispersed in another solid. Protective colloids are agents that prevent the agglomeration of dispersed particles or coalescence of droplets. (See Suspension polymerization.)

88. Colorant

An additive used to impart colour. Dyes are used to produce the required coloration while retaining transparency. Pigments are used to produce opaque coloured formulations, created by internal light scattering. (See Dye and Pigment.)

89. Colour Matching

Adjusting the dye or pigment composition to the exact colour of a reference article or sample. Colour develops as a result of the absorption of visible light of specific wavelength. The colour seen corresponds to the wavelength of the light that has not been absorbed. For instance, red corresponds to wavelengths in the region of 750–610 nm, green in the region 550–510 nm, and yellow in the region 590–575 nm.

89.1 Dye

For dyeing purposes, the relative amounts of dyes to be used can be estimated from spectrographic measurements of the absorbance (also known as the optical density) of the colour to be matched,

$$d = \log(I_0/I_t),$$

where I_0 is the intensity of the incident light and I_t is the intensity of the transmitted light. For a three-colour mixture (the maximum number normally required) the optical density at any given wavelength is related to the concentration of each individual dye by the additivity rule, that is,

$$d = \alpha_A C_A + \alpha_B C_B + \alpha_C C_C,$$

where α_A, α_B and α_C are the absorbance values of dyes A, B and C, defined by Beer's law, $\log(I_0/I_t) = \alpha C x$, C_A, C_B and C_C are the molar concentrations of the dyes to be used, and x is the sample thickness.

89.2 Pigment

For pigmenting, it is necessary to take into account the scattering coefficient (δ) in addition to the absorbance (α) to calculate the reflectance (R) at the required wavelength, that is,

$$\frac{2R}{1-R^2} = \frac{\delta_A C_A + \delta_B C_B + \delta_C C_C}{\alpha_A C_A + \alpha_B C_B + \alpha_C C_C}.$$

Note, however, that in the case of crystalline polymers some scattering takes place also from the surface of the crystal lamellae, which may create considerable errors in the theoretical predictions of the above equation. (See CIE chromaticity diagram.)

90. Commingled Fabric

A fabric in which the constituent yarns or tows contain a mixture of reinforcing fibres and polymer fibres. The latter will melt to form the matrix of the composite when the temperature is increased above the melting point of the polymer and forced to flow around reinforcing fibres by the application of an external pressure.

91. Comonomer

The monomer present in minor amounts in a mixture of two monomers used for the production of copolymers.

92. Compact Tension Specimen (CT Specimen)

Denotes a type of specimen used to determine the fracture toughness of a material. (See Fracture mechanics.)

93. Compatibility

A term often used to denote the partial miscibility of polymers in blends.

94. Compatibilize

Making two polymers compatible, usually implying that either the chemical structure has been modified or an additive has been used to bring about a fine dispersion when mixed together. (See Polymer blend.)

95. Compatibilizing Agent

An additive or a third component of a polymer blend used specifically to enhance the compatibility of the two major components. Denoting the two major polymers in a blend by A and B, the structure of the compatibilizing agent is often a block copolymer of the type A–B or A–B–A, so that blocks A will provide the thermodynamic drive for mixing with polymer A, and similarly blocks B will be miscible with polymer B. (See Block copolymer.)

96. Complex Compliance (and Complex Modulus)

The term 'complex' for the compliance and modulus (known also as dynamic modulus and dynamic compliance) derives from the viscoelastic nature of polymers. It implies that both parameters consist of two components, a 'real' part (elastic or storage component) and an 'imaginary' part (viscous or loss component), that is,

$$E^* = E' + iE''$$

and

$$D^* = D' - iD'',$$

where E^* and D^* are the 'complex' modulus and complex compliance, E' and D' are the corresponding elastic components, while E'' and D'' are the loss components, and $i = \sqrt{-1}$, that is, the complex number.

The denomination of complex modulus and complex compliance derives from the viscoelasticity theory for cyclic loading conditions. That is to say, when a sinusoidal symmetric strain is applied to a specimen, there will be a corresponding sinusoidal stress response over the same cycle with the same period and frequency. If the material has an elastic behaviour, then the stress/strain ratio (i.e. the modulus) and the strain/stress ratio (i.e. the compliance) are constant. This means that the stress and strains are in phase, so when the stress is maximum, the strain is also maximum; and when one of these is zero, the other will also be zero. This situation can be elaborated through the appropriate functions. That is, if the applied strain is

$$\varepsilon = \varepsilon_0 \sin(\omega t),$$

then the resulting stress has to be

$$\sigma = \sigma_0 \sin(\omega t)$$

for the ratio of the two to remain constant. Here t is the time and ω is the angular velocity, insofar as the corresponding force and deformation are considered vectors that rotate in circular motions through positive and negative quadrants at angular velocity ω. The respective maximum values for stress and strain during each cycle are σ_0 and ε_0 as shown.

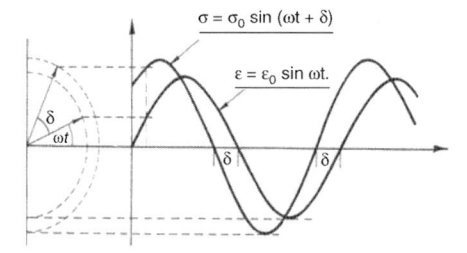

Response of a viscoelastic material to dynamic loads.

Since, for a viscoelastic material, the stress/strain or strain/stress ratios cannot be constant over the period in which the load is applied, the only way this situation can be sustained is for the stress and strain curves to be displaced with respect to each other by a phase angle δ, as shown in the diagram. This means that the stress associated with an applied sinusoidal strain at angular velocity ω will be described by the equation

$$\sigma = \sigma_0 \sin(\omega t + \delta),$$

which indicates that the strain vector lags behind the stress vector by an angle δ. From mathematical considerations, the stress function $\sigma_0 \sin(\omega t + \delta)$ can be decomposed with respect to the strain function $\varepsilon_0 \sin(\omega t)$ into an in-phase component (elastic component) and an out-of-phase component (loss component). The mathematical analysis produces expressions for the modulus and compliance in terms of complex parameters, respectively $E^* = E' + iE''$ and $D^* = D' - iD''$, where the magnitude of E'' and D'' provides a measure of the deviation of the behaviour of a polymer from elastic behaviour, which is normally associated with the behaviour of metals and ceramics. Alternatively, this deviation can be expressed in terms of the ratio of the loss component to the elastic component, that is,

$$\tan \delta = E''/E' = D''/D',$$

where $\tan \delta$ is also known as the 'loss factor'.

97. Complex Permittivity and Specific Impedance

A concept applied to the capacitance characteristics of polymers using similar principles as those used for the complex modulus and compliance in relation to the viscoelastic behaviour of polymers. In alternating current circuits, the electrical stress is out of phase with respect to the charge density of the capacitor component. The properties of a dielectric that define its ability to store charge are the permittivity (ε) and specific impedance (Z_p), which are defined as follows. In general, permittivity ε is

$$\varepsilon = \frac{\text{charge density}}{\text{electrical stress}}$$

and relative permittivity ε_r is

$$\varepsilon_r = \frac{\text{permittivity of dielectric}}{\text{permittivity of vacuum (or air)}},$$

which is also known as the dielectric constant. The complex permittivity is given by

$$\varepsilon^* = \varepsilon' - i\varepsilon'',$$

the complex specific impedance is expressed as

$$Z_p^* = Z'_p + iZ''_p$$

and the loss angle as

$$\tan \delta = \varepsilon''/\varepsilon' = Z''_p/Z'_p.$$

Note that

$$Z_p^* = \frac{1}{\omega \varepsilon_r C_1},$$

where ω is the angular frequency ($2\pi f$) and C_1 is a constant defined as $C_1 = C_0(d/A)$. In this case, C_0 is the capacitance of vacuum (or air), d is the distance between the electrodes (i.e. the thickness of the dielectric) and A is the area of contact between the dielectric and the electrode.

98. Compliance

A term denoting the deformation characteristics of a structure or a specimen, corresponding to the inverse of the stiffness. For the case of a beam, for instance, the stiffness (S) can defined in terms of the magnitude of the load (P) required to obtain a specified level of deflection (Δ). Hence, the ratio $S = P/\Delta$ is a parameter that defines the rigidity of a structure in terms of its resistance to deformations. Since the compliance (C) is the reciprocal of the stiffness, then $C = \Delta/P$ and, therefore, C represents a parameter that defines the flexibility of a structure. The term 'compliance', however, is also used to designate the inverse of the Young's modulus. (See Creep compliance and Complex compliance.)

99. Composite

A material consisting of a dispersion of high-modulus fibres or fillers within a linear polymer or macromolecular network matrix. The fibres may be short and randomly dispersed or continuous and laid in a highly ordered manner

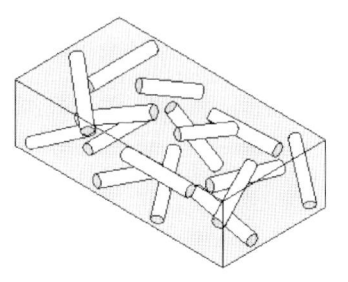

Random short-fibre composite.

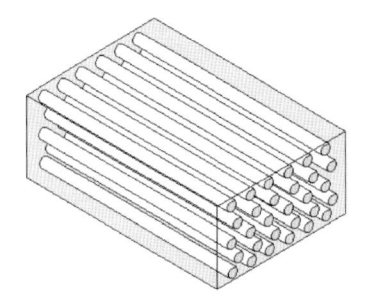

Unidirectional continuous-fibre composite.

Type of fibre	Density (kg/m³ × 10⁻³)	Tensile strength (N/m² × 10⁻⁶)	Young's modulus (N/m² × 10⁻⁹)
Glass, E type[a]	2.50	2400	72
Glass, S type[a]	2.44	2800	86
Carbon[b]	1.38	1700	190
Graphite[b]	1.38	1700	250
Aramids[a]	1.44	3600	127

[a] Fibre diameter = 16–20 µm.
[b] Fibre diameter = 6–10 µm.

In practice, the reinforcing component of composites can also be in the form of webs or fabrics, where yarns or tows of fibres are disposed (woven) at 90° (as warp and weft) in equal or different amounts, as shown.

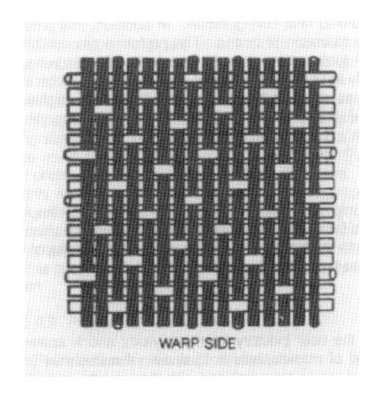

Fabric-type continuous-fibre reinforcement.

100. Composite Fibre

The reinforcing component of polymer matrix composites is in the form of high-modulus and/or high-strength fibres. The most widely used fibres and their main properties are shown in the table. (See Fibre reinforcement theory.)

Particulate fillers are much less efficient than fibres in enhancing the mechanical properties of the polymer matrix but are easily processed, particularly when used in combination with thermoplastic matrices.

101. Composite Manufacture

Short-fibre composites are processed by compression moulding or injection moulding. In the latter case particular care has to be taken in the design of the equipment and moulds in order to minimize fibre breakage during processing. This usually involves the omission of non-return valves in the barrel of injection moulding machines, particularly for thermosetting systems, and the use of large-cavity gates in the mould. For long-fibre composites most manufacturing methods are carried out at low pressures. Typical processes are shown schematically.

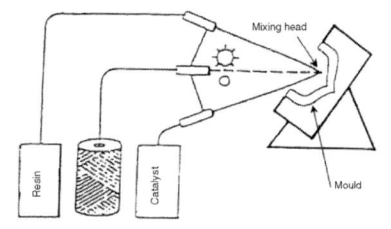

Wet layup technique with spray gun.

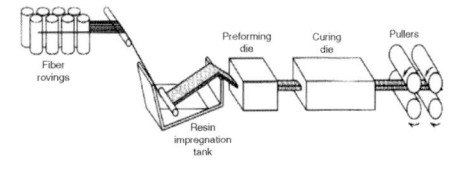

Pultrusion process with fibre impregnation.

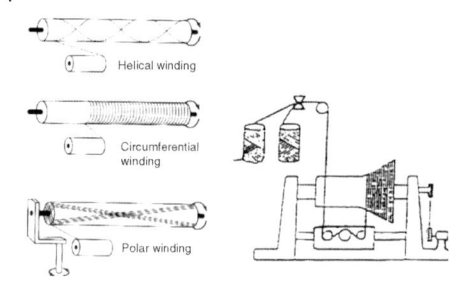

Filament winding with fibre impregnation.

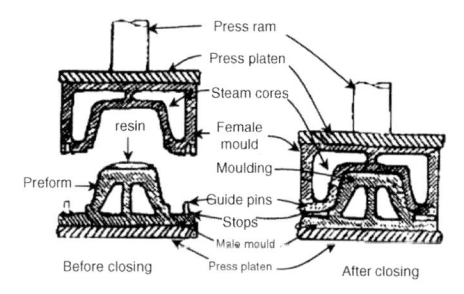

Matched-die moulding with resin spreading.

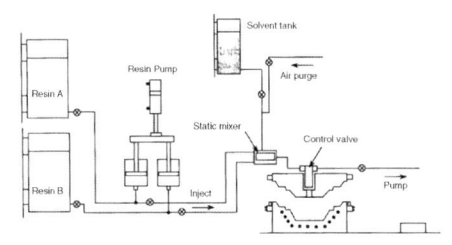

Resin transfer moulding process.

102. Composite Matrix

The component that embeds the reinforcing fibres or fillers in a composite. The most widely used polymer matrices in composites are shown in the table.

Thermoplastic polymers

Commercial name – Chemical nature	T_m (°C)	Tensile strength (MPa)
Coupled PP – Polypropylene grafted with maleic anhydride	165	30
Nylon – Polyamides 6 and 66	220–260	60
PPS – Poly(phenylene sulfide)	280	50
PEEK – Poly(ether ether ketone)	330	70

Thermosetting resins

Commercial name – Chemical nature	T_g (°C)	Tensile strength (MPa)
UPE – Unsaturated polyesters	80–100	50–60
Epoxy – Glycidyl ether resins	60–200	50
PF – Phenol formaldehyde resin	100	30

103. Composite Property Prediction

(See Fibre reinforcement theory.)

104. Composite Test

Usually refers to the measurement of mechanical properties. The tests used vary according to the type of composites. Long-fibre (or continuous-fibre) composites require special considerations with respect to the size of the specimens relative to the length and orientation of the fibres. In addition to these requirements, there are also tests specifically designed for long-fibre laminated systems, as outlined below.

104.1 Interlaminar Shear Strength (ILSS)

Tests widely used for quality control purposes and as screening tests for the evaluation of matrices, fibre reinforcement and fibre–matrix interactions. The ILSS test is carried out in three-point bending using a very short span (S) relative to the thickness (W). This allows the maximum shear stress to be set up between the fibre plies so that failure occurs within the matrix and/or at the fibre–matrix interface. For practical reasons, usually the recommended S/W ratio is 5 and the ILSS value is calculated from the first peak of the load (P_{max}) recorded in the test, which corresponds to the interlaminar shear failure, using the formula

ILSS = $0.75 P_{max}/BW$,

where W is the thickness and B the width of the specimen. Usually specimens are 1 cm wide and 3–4 mm thick. A schematic illustration of the ILSS test is shown.

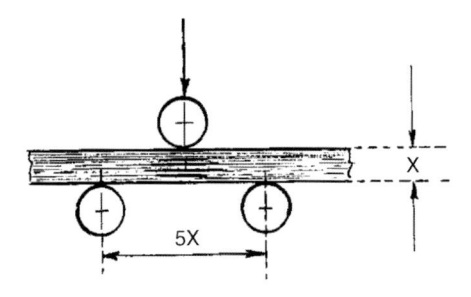

Interlaminar shear strength test by three-point bending.

The values obtained for the ILSS of continuous-fibre composites are in the region of 15–30 MPa for polyester resin laminates with 50% glass fibre content. ILSS values, however, may depend on the thickness of the specimen and fabric construction, as these will affect the complexity of the stresses that are set up within the laminae, owing to unaccounted interference from normal stresses in the calculation of the failure shear stress using the above equation.

104.2 Compression Strength Measurement

Tests often carried out on high-performance composites. A practical difficulty experienced in performing compression tests on relatively thin specimens, such as laminates, is maintaining the axial alignment during the test. Special arrangements are often made to achieve this through the use of guided grips, which prevent the occurrence of buckling instabilities in the test specimen during the test prior to fracture. Failures in compression can occur at stress levels that are much lower than those experienced in tensile and flexural strength. Typical compression strength values for woven fabric laminates with 50% glass are in the region of 150–250 MPa. The reason for this is that failure can take place within the matrix and at the interface through various types of internal events, such as microbuckling resulting from fibre–matrix debonding and microcracking of the matrix between the aligned fibres. In this respect it must be borne in mind that the 'slenderness ratio' of the fibres is extremely high and, therefore, buckling of the fibres can occur at very low compression strains.

104.3 Fracture Mechanics Applied to Failure in a Composite

The very important role of interlaminar failures in high-performance composites, and the inherent deficiencies of the ILSS tests (mentioned earlier), has stimulated considerable interest in the use of fracture mechanics to obtain more appropriate data. In particular, measurements have been made to obtain values for the critical strain energy release rate in both crack-opening mode (G_{Ic}) by the double cantilever beam (DCB) method and the in-plane shear failure mode (G_{IIc}) using the end-notched flexure (ENF) specimens, as illustrated.

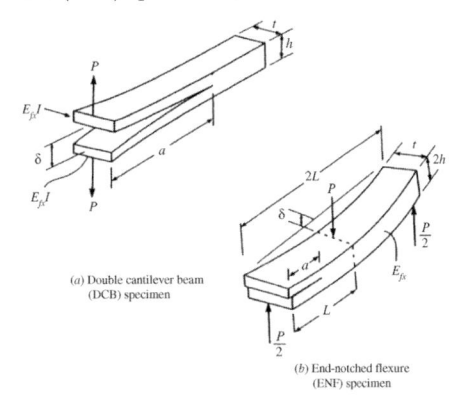

(a) Double cantilever beam (DCB) specimen

(b) End-notched flexure (ENF) specimen

Geometric features of specimens and loading mode for measuring the interlaminar shear toughness of composites.

The double cantilever beam test is the same as that described for adhesives. The same formula can be used to calculate the G_{Ic} value from the recorded peak load to fracture (P_f), substituting the Young's modulus of the adherend with the flexural modulus of the laminate (E_{fx}). (See Adhesive test.) In measurements of G_{IIc}, the compliance of ENF specimens is given by the formula

$$C = \frac{2L^3 + 3a^3}{8E_{fx}th^3},$$

from which is obtained

$$G_{IIc} = \frac{9P_f^2 a^2 C}{2t(2L^3 + 3a^3)},$$

where all the terms are as defined in the diagram. Note that the flexural compliance C can be measured directly on the ENF specimen. The two tests can also be used for measuring crack growth rate under both dynamic load (fatigue) and static load (creep) conditions. The values for G_{IIc} are usually larger than for G_{Ic}, ranging from about 80 to 2000 J/m^2. The larger values are obtainable particularly with the use of thermoplastic matrices.

104.4 Damage Tolerance Test

These tests measure the damage caused by accidental loading situations, such as impacts by falling objects or projectiles, in terms of deterioration of some specific property within the affected area. Compression tests are often performed on specimens containing different extents of damage induced by subjecting coupons of laminate to different levels of impact energy by means of falling projectiles. (See Falling-weight impact test.) The damage tolerance of the laminate can be derived from the strength retention after impact by constructing plots of compression strength versus applied impact energy.

105. Composting

According to the definition provided by the US Environmental Protection Agency, the term 'composting' denotes the controlled biological decomposition of organic materials in the presence of air to form humus-like substances. (See Biodegradable polymer.)

106. Compounding

A term used to describe the mixing of a polymer with additives or other auxiliaries, such as plasticizers and fillers, mainly in the melt state of the polymer. Compounding methods can be divided into batch mixing techniques and continuous extrusion, usually twin-screw extrusion. The typical equipment used for industrial-scale compounding is shown.

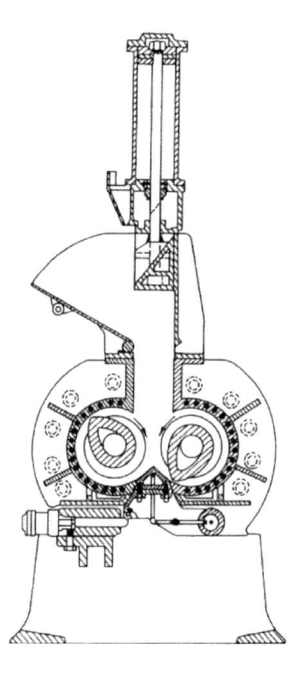

Banbury-type batch mixer. Source: Todd (2010).

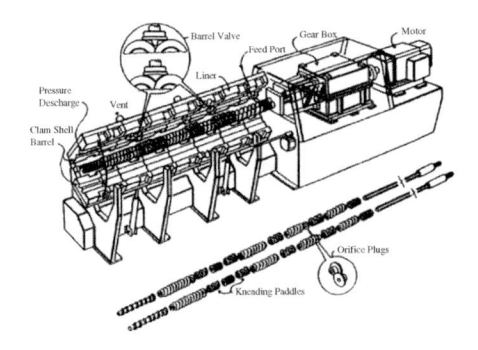

Twin-screw mixing extruder. Source: Todd (2010).

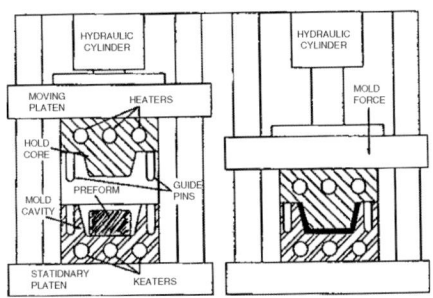

Example of compression moulding of thermosetting polymers. Left: feeding of a compacted preform of polymer mix. Right: cavity filled with melt undergoing curing.

107. Compression

The application of forces that bring about a reduction in the dimensions in the direction in which they act. In mechanics, compressive stresses and tensile stresses are distinguishable only for their sign, that is, $\sigma =$ tension and $-\sigma =$ compression.

108. Compression Moulding

A moulding process carried out in a vertical press in which the raw material, in the form of a prepreg, moulding compound, powder or pellets, is placed directly into the cavity and subsequently shaped by the application of pressure. Normally only thermosetting plastics and vulcanizable elastomers are processed by compression moulding, using a hot mould to allow curing reactions to take place in the shortest possible time. Long-fibre thermoplastic composites are also processed in this manner by placing a pre-heated feedstock in a 'cold' mould and applying the required pressure before the onset of solidification through crystallization or vitrification in the case of glassy polymer matrices.

109. Compression Ratio

The reduction in volume per unit length of the channel of an extrusion screw within the transition (or compression) zone. Since the width of the channel along the axial length of the screw is constant, then the compression ratio can be calculated as the ratio of the height of the channel in the feed zone to the height of the channel in the metering zone. (See Extruder and Extrusion.)

110. Compression Set

A term used to describe the lack of recovery characteristics of a rubber or elastomer under compressive loading conditions. Typically the measurement consists in determining the extent to which a cylindrically shaped specimen remains permanently deformed after being subjected to compression stresses for a predetermined period.

111. Compression Test

Test carried out in compression. These are not widely used for testing polymers, except

112. Compression Zone

The second section of the screw of an extruder, in which the root diameter gradually increases to compress the melting polymer and to remove the entrapped air. (See Extrusion.)

113. Concentric Cylinder Rheometer

(See Coaxial cylinder rheometer.)

114. Condensation Polymerization

Also known as polycondensation, refers to the reaction or process in which monomers or oligomers react to produce linear polymers, or polymer networks, through condensation reactions. Examples of condensation polymerization are those used for the production of polyesters, polyamides and polyethers, and phenol formaldehyde, urea formaldehyde and melamine formaldehyde networks. One important feature of polycondensation is that the polymer chains grow in steps so that all the reacting monomer species take part in the growth of the polymer chains right from the start. This is different from addition polymerization reactions, where a chain grows rapidly to the final size and a quantity of free monomer is always present until full conversion is reached.

115. Condensation Reaction

A chemical reaction that involves the joining of one or more molecules, with the elimination of a smaller molecule, usually water. An example of a condensation reaction is the formation of an amide from the reaction of a carboxylic acid (RCOOH) with an amine ($R'NH_2$), that is,

$$RCOOH + R'NH_2 \rightarrow RCONHR' + H_2O$$

Another example is the formation of esters from the reaction of a carboxylic acid (RCOOH) and an alcohol ($R'OH$), that is,

$$RCOOH + R'OH \rightarrow RCONHR' + H_2O$$

116. Conductance

Denotes the ratio of the current to the voltage applied on a dielectric material. By converting the current I into current density I/A, where A is the cross-sectional area, and the voltage V into stress V/L, where L is the path length, it is possible to calculate the 'conductivity' as the ratio of the current density to the applied stress. Conductivity is a property of the material and is equal to the reciprocal of the resistivity.

117. Conduction

Represents the flow of heat through the bulk of a solid medium along the path set up by the temperature gradient. It also denotes the transportation of electrons, or ions, to produce an electric current through the application of a voltage gradient across the medium.

118. Conductive Polymer

A polymer capable of conducting an electrical current. Such polymers are divided into 'extrinsically' conductive polymers, when the conductivity arises from the addition of conductive fillers, and 'intrinsically' conductive polymers, when the conductivity develops from within the polymer structure through the dislocation of electrons along polymer chains.

118.1 Extrinsically Conductive Polymer

The fillers most widely used to develop electron conductivity in commercial polymers are grades of carbon black with a high surface area. The amounts required have to exceed the critical concentration to create percolation paths for the movement of electrons. The particles do not have to make physical contact but have to be sufficiently close to allow electrons to hop across gaps about 2 nm wide. This phenomenon is often referred to as 'tunnelling'. Using conductive grades of carbon blacks, the amounts required to bring down the volume resistivity of typical polymers, around 10^{15}–10^{17} Ω m (ohm metre), to a lower plateau in the region of 100–300 Ω m, that is, a conductivity of about 10^{-2} S/m (siemens per metre), varies from values as low as 1.5 to more than 10 wt%, depending on the systems used. (See Law of mixtures.)

118.2 Intrinsically Conductive Polymer

These are all based on conjugated polymers. The most widely studied in this category is polyacetylene in its 'doped' form. This consists of polymer chains containing free radicals formed by controlled oxidation reactions that can recombine to form charge carriers along the chains.

Example of charge carriers on polyacetylene chains.

In effect, the controlled oxidation adds extra electrons, creating 'holes', which represent a position where an electron is missing. When such a hole is filled by an electron jumping in from a neighbouring position, a new hole is created and so on, allowing the charge to migrate a long distance. The 'doped' form of polyacetylene has a conductivity of 10^5 S/m, which is much higher than that achieved with the addition of carbon black to an 'ordinary' polymer and compares with values of around 10^8–10^9 S/m for copper. However, there are major drawbacks for these systems that prevent them from being used widely, namely their extreme susceptibility to oxidative degradation when exposed to air and their poor processability. Other widely researched conductive polymers are listed below.

118.3 Polypyrrole

The structure of the raw polymer is shown, but to achieve a high level of conductivity it has to be doped with strong acids, such as toluenesulfonic acid.

118.4 Polyaniline

The structure corresponds to that of emeraldine, which can be doped with strong acid. Alternatively, it can be self-doped through the incorporation of sulfonic acid groups along the polymer chain, as shown.

119. Conductivity

Usually refers to either thermal conductivity or electrical conductivity. Thermal conductivity, K, is the coefficient that relates the heat flux (Q) to the temperature gradient ($\delta T/\delta x$), that is $Q = K(\delta T/\delta x)$. Electrical conductivity, ρ, is the coefficient that relates the current density ($\delta I/\delta A$) to the voltage gradient ($\delta V/\delta L$), that is,

$$\delta I/\delta A = \rho(\delta V/\delta L).$$

(See Conduction and Conductance.)

120. Cone-and-Plate Rheometer

Operates by imposing a drag flow onto a fluid placed between a rotating or oscillating shallow cone and a flat plate in order to provide a constant shear rate within the gap. This is achieved by ensuring that the angle formed by the cone with the plate is less than 10°. Under these conditions, the shear rate is given by $\dot{\gamma} = \Omega/\alpha$, where Ω is the angular velocity and α is the angle between the two surfaces (see diagram).

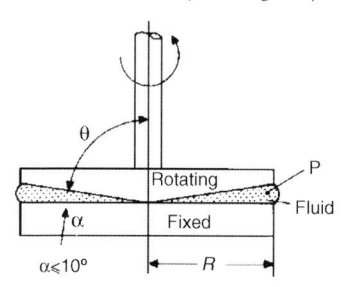

Schematic illustration of the principle of the cone-and-plate viscometer.

The cone-and-plate rheometer is used primarily to evaluate the rheological properties of polymer systems with a relatively low viscosity, such as resins, solutions, emulsions and pastes. The shear stress acting on the fluid can be calculated from the torque (T) developed by the rotating spindle using the relationship

$$\tau = 3T/2\pi R^3,$$

where R is the radius of the rotating cone. The viscosity, η, can be calculated as the ratio $\tau/\dot{\gamma}$ for measurements made at various shear rates and temperatures.

For operations in the rotational mode, it is possible to measure also the vertical thrust on the plate, which can be used to evaluate the melt elasticity characteristics in terms of the first normal stress difference (N_1) from the relationship

$$N_1 = 2F'/\pi R^2,$$

where F' is the 'net force' acting on the melt, that is,

$$F' = F - P_a \pi R^2,$$

where F is the measured force (thrust) and P_a is the atmospheric pressure exerted on the melt at the edge. From the obtained N_1 value it is possible to calculate the first normal stress coefficient (ψ_1), that is,

$$\psi_1 = N_1/\dot{\gamma}^2.$$

Vertical thrust measurements, however, are rarely made. (See Normal stress difference.)

For operations made in an oscillatory mode, the data are displayed in terms of 'real' shear modulus G' and 'imaginary' shear modulus G''. The latter corresponds to the 'real viscosity' (η') calculated as the ratio G''/ω, where ω is the angular velocity of the oscillatory motion. The viscosity data are presented as a 'complex viscosity' (η^*), which comprises a real component (η') and an imaginary component (η''), that is,

$$\eta^* = \eta' + i\eta'',$$

where $i = \sqrt{-1}$ (the imaginary number). The imaginary term, η'', is directly related to G', that is $\eta'' = G'/\omega$. (See Complex compliance.)

121. Cone Calorimeter

An apparatus that measures the heat evolved by a burning plaque, using the technique of the oxygen depletion calorimeter. This is to say, the heat released by a burning sample is directly proportional to the quantity of oxygen used in the process. The name derives from the shape of the truncated conical heater that is used to irradiate the test specimen.

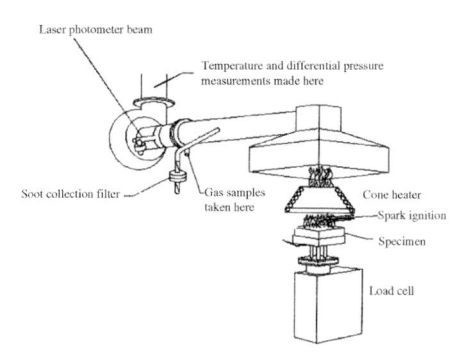

Cone calorimeter for measuring the burning resistance of polymers. Source: Ashton (2010).

122. Configuration

A term used the denote the spatial position of groups attached to a polymer chain, which gives rise to stereoregularity or tacticity designations, such as isotactic polypropylene. (See Tacticity.)

123. Conformation

A term denoting the configuration of a molecule derived from the rotations within a molecular chain. (See Chain conformation (*cis* and *trans*).)

124. Conjugated Double Bond

A term to describe the double bonds that appear in alternated sequences with single bonds, as in aromatic rings, or along linear aliphatic chains, as in polyacetylene. The electrons in the π orbitals of double bonds are said to be dislocated insofar as they can move from one π orbital to another within an aromatic ring or along the conjugation double bonds in a polymer chain. (See Conductive polymer.)

125. Conservation Law

A fundamental law of nature. Conservation laws are concerned with the conservation of momentum and energy, forming the basis for the setting of a constitutive equation for flow analysis and heat transfer in polymer processing.

126. Consistency Index

The coefficient relating shear stress to shear rate in the description of the power-law behaviour of the flow of pseudoplastic liquids, such as polymer melts,

$$\tau = K(d\gamma/dt)^n,$$

where τ is the shear stress, K is the consistency index, $d\gamma/dt$ is the shear rate and n is known as the power-law index.

127. Contact Angle

The angle formed by a liquid droplet between the tangent to the curvature and the plane on which it rests. It is related to the interfacial free energy and determines the ability of a liquid or a highly deformable body (D), such as a polymer melt, to wet the surface of a substrate (S) in air (A). The basic definition and the forces related to each surface energy term are shown.

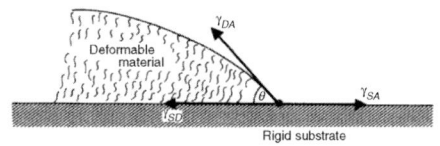

The contact angle (θ) is related to the interfacial energy (γ_{SD}) and the surface energy of the other two phases (γ_{DA} and γ_{SA}). The contact angle is obtained from the relationship

$$\gamma_{SA} = \gamma_{SD} + \gamma_{DA} \cos \theta.$$

128. Continuum Mechanics

A branch of engineering science dealing with definitions and relationships between stress, strain and strain rate, which form the basis for structural and flow analysis.

129. Converging Flow

Describes the flow of a fluid in channels whose cross-sectional area decreases along the flow path. The most common type of converging flow in polymer processing is that through conical sections of dies or through wedged sections of moulds or dies. The occurrence of converging flow is an important issue in polymer processing insofar as it produces an acceleration of the front of the flowing melt.

130. Coordination

Refers primarily to the formation of strong physical bonds between the chemical groups of an additive or a polymer chain with metal ions, often present as impurities. Coordination is also known also as complexation or chelation.

131. Copolymer

A polymer containing two different monomer units along the chains. The two units (say, A and B) can be arranged in three possible ways and the copolymers are named according to their disposition in the chains. A random or statistical copolymer is one in which the two units are arranged at random, or are statistically distributed along the chains. A block copolymer is one where the two units are arranged in blocks of specific lengths. An alternating copolymer is one where the two monomer units are in an alternating sequence along the chains.

- **Random copolymer:**

AAAAABBBAABBBBABBAAABAABBB

- **Block copolymer:**

AAAAAAAABBBBBBB

or

AAAAAAABBBBBBBBAAAAAAA

- **Alternating copolymer:**

ABABABABABABABABABABABAB

132. Core-and-Shell Structure

A type of structure of polymer particles consisting of two or more different types of polymers, arranged in such a manner that one polymer produces a core or nucleus and another polymer forms the outer skin or a shell of the particle.

133. Corona Discharge

Electrical discharges generated by the decomposition of gases, normally air, under the influence of a high voltage gradient. Their glowing appearance is caused by the formation of highly charged gaseous ions. These discharges can occur in air pockets of a high-voltage insulation and may cause the deterioration of the insulation through thermal oxidative reactions. Corona discharges are deliberately induced on the surface of polyolefin films used in packaging, such as polyethylene or polypropylene, as a means of producing oxygenated groups on the surface of the film in order to enhance their adhesion to printing inks. (See Cold plasma.)

134. Co-Rotating

A term used to denote the configuration and the relative direction of two parallel screws used in twin-screw extrusion. Co-rotation implies that the two screws rotate in the same direction. The concept is used primarily in the extrusion of rigid or unplasticized poly(vinyl chloride) (UPVC). (See Counter-rotating.)

135. Couette Flow

Type of flow, named after the homonymous scientist, that takes place between a stationary and a moving surface, also known as 'drag flow'.

136. Counter-Rotating

A term used to denote the configuration and the relative direction of two parallel screws used in twin-screw extrusion. Counter-rotation implies that the two screws rotate in the opposite directions, one clockwise and the other anticlockwise. This type of screw configuration is most widely used in extruders used for compounding.

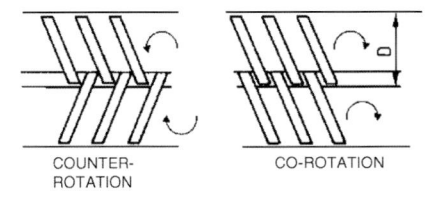

COUNTER-ROTATION   CO-ROTATION

Schematic diagram of intermeshing screws.

137. Coupled Polypropylene

A grade of polypropylene that contains anhydride grafts along the polymer chains to make them reactive towards silane coupling agents used in sizes for glass fibres or for the surface treatment of fillers.

138. Coupling Agent

Reactive additives used to produce a chemical bond between the surface of an inorganic substrate, such as that of glass fibres or fillers, with a polymeric matrix. Coupling agents are usually alkyl functional trialkoxysilane compounds, designed to allow the alkoxysilane groups to hydrolyse and undergo condensation reactions with the surface of the inorganic substrate, while the functional groups attached to the alkyl chain can react with the organic matrix, as indicated by the reaction scheme.

Hydrolysis and condensation reactions of alkoxysilanes with the surface of an inorganic substrate.

The hydrolysis reactions are catalysed primarily by acids, while the condensation reactions are much faster under basic conditions. The most commonly used silane coupling agents for typical polymer matrix composites are shown in the table.

Name	Formula	Matrix type
Vinyl triethoxysilane	$H_2C=CH(SiOEt)_3$	Polyester resins
γ-Methacryloxypropyl trimethoxysilane	$H_2C=C(CH_3)COO(CH_2)_3Si(OMe)_3$	Polyester resins
γ-Aminopropyl triethoxysilane	$H_2N(CH_2)_3Si(OEt)_3$	Epoxy resins, nylons, coupled PP
γ-Glycidoxypropyl trimethoxysilane	$H_2C-CHCH_2-O-(CH_2)_3Si(OMe)_3$ (epoxide)	Epoxy resins, nylons, coupled PP
γ-Mercaptopropyl trimethoxysilane	$HS-(CH_2)_3Si(OMe)_3$	Diene elastomers and EPDM for silica reinforcement

There are other types of coupling agents that are sometimes used in polymer matrix composites, particularly the particulate types. These are usually esters of titanium or zirconium, forming a bond with the surface of inorganic fillers through proton coordination interactions. They are used primarily for their ability to improve the dispersion of fillers, making it possible to incorporate larger quantities in a given formulation. This is particularly useful for non-halogenated fire-retardant formulations using magnesium hydroxide or aluminium trihydrate, which require large quantities of fillers to develop the necessary fire-retardant characteristics.

139. Crack Initiation

A phenomenon describing the initial formation of a crack or the extension of an existing crack, or notch, in a specimen subjected to a fracture toughness test.

140. Crack Length

The length of an edge notch of a specimen – single edge notch (SEN) or double edge notch (DEN) – used for the evaluation of the fracture toughness of materials.

141. Crack-Opening Displacement (COD)

Denotes the increase in the separation distance between the two surfaces of a notch in tests carried out on very ductile materials. The yield zone formed at the crack tip can bring about a quite large separation of the two surfaces of the notch, which can be readily measured. The resistance to fracture of very ductile materials is often expressed in terms of a critical crack-opening displacement (COD), representing the increase in separation distance at which the cracks begin to grow. (See Fracture mechanics.)

142. Crazing

The formation of crazes, consisting of flat ellipsoidal voids (axial length varies from about 50 nm to 1 mm) lying in the direction perpendicular to the applied stress (predominantly tensile). The upper and lower surfaces of the walls of a craze are connected by fibrils 10–40 nm in diameter, as illustrated in the diagram.

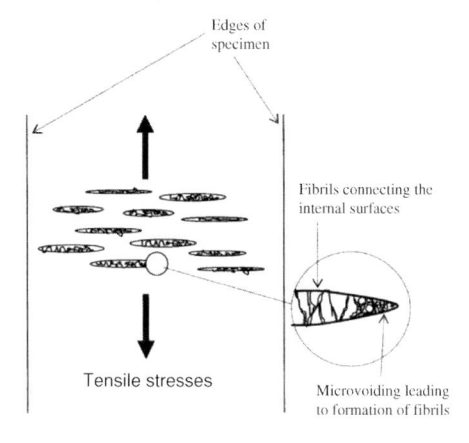

Formation and structure of crazes in glassy polymers.

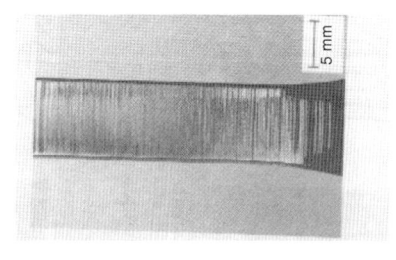

Crazes observed in a polystyrene specimen tested in tension.

Crazing occurs in glassy polymers at temperatures substantially lower than their glass transition temperature. Crazing takes place only when the polymer is subjected to predominantly dilatational stresses. In other words, crazing does not take place under compression or pure shear stresses. The phenomenon becomes more prominent in the presence of a liquid environment with miscibility parameter similar to that of the polymer. (See Critical crazing strain.)

143. Creep

A phenomenon denoting an increase in deformation with time when a structure or a specimen is subjected to a constant load or a constant stress. In the case of polymers, this type of behaviour is associated with viscoelastic deformations.

144. Creep Compliance

A parameter (C) that denotes the ability of a structure to deflect when subjected to external loads. It is defined, therefore, as the ratio of the applied load (P) to the resulting deflection (Δ), that is $C = P/\Delta$. The compliance of a structure is equal to the reciprocal of its stiffness (S). The term 'compliance' is also used as a property of the material (D), which corresponds to the reciprocal of the Young's modulus, that is, $D =$ strain/stress. The value of the compliance of a material is obtained from the gradient of the plot of strain against stress. For the case of viscoelastic materials, the compliance increases with the time of duration of the application of the stress. The standard linear model is often used to describe the variation of the compliance as a function of time, which is given by the expression

$$D(t) = D_0 + (D_\infty - D_0)[1 - e^{t/\lambda_c}],$$

where D_0 is the instantaneous compliance, that is, at time zero, D_∞ is the equilibrium compliance, that is, at time infinity, and λ_c corresponds to the retardation time, which is a material constant that determines the rate at which the compliance increases from D_0 to D_∞. (See Viscoelasticity.)

145. Creep Curve

Plots of the strain as a function of time, normally as linear–log plots, derived from experiments carried out on specimens subjected to a wide range of stresses at constant temperature. Normally the tests are carried out in tension or in a three-point bending mode. Typical curves obtained are shown in the graph for a grade of polypropylene tested at room temperature.

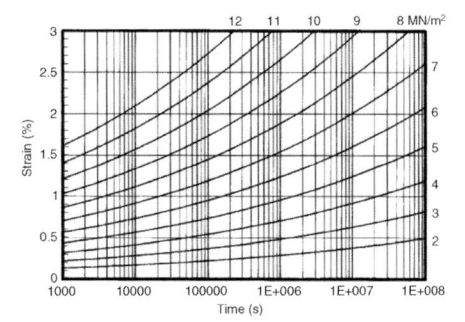

Creep curves for a typical grade of polypropylene.

146. Creep Modulus

Corresponds to the ratio of the applied stress to the resulting variable strain measured in creep experiments. The creep modulus of polymers is a decreasing function of the duration of the applied stress. The value of the creep modulus corresponds to the reciprocal of the creep compliance, which is regarded as the fundamental property that characterizes the behaviour of polymers under creep testing conditions. (See Creep compliance.)

147. Creep Period

Corresponds to the part of a creep experiment in which the stress remains constant before the start of the recovery period. (See Recovery.)

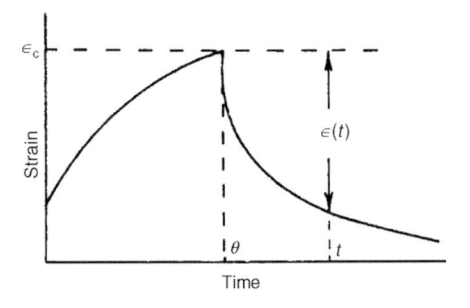

Illustration of variation of strain with time during the creep period and subsequent creep recovery.

148. Creep Rupture

Refers to the fracture of a specimen or an article resulting from the application of a constant stress as in creep experiments. The time required to induce failure by yielding or fracture of a specimen at different stress levels is sometimes measured to assess the time-dependent fracture/yield resistance of materials, as illustrated.

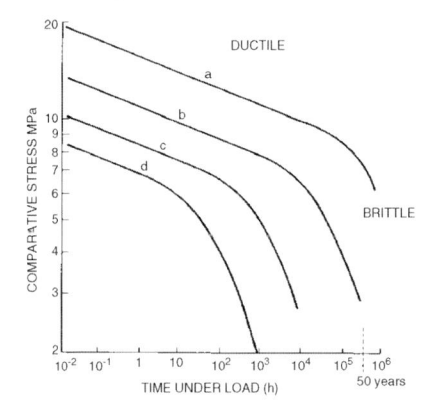

Failure of polyethylene pipes under pressure at different temperatures (a < b < c < d). Source: Birley et al. (1991).

149. Creep Test

Tests carried out at constant stress in which the strain is measured as function of the duration of the applied stress.

150. Crepe Rubber

The coagulated latex of natural rubber, usually obtained in the form of sheets

151. Criterion

A rationale used to determine the conditions under which a certain event may take place. For instance, the Tresca criterion and the Von Mises criterion are used to determine, respectively, the state of stresses and the magnitude of the distortional energy required to bring about failure of products by yielding. (See Yield strength.)

152. Critical Chain Length (Z_c)

The length of polymer chains above which entanglements are formed. This is an important parameter with respect to the relationship between viscosity and molecular weight. Below Z_c the zero-shear viscosity is directly proportional to chain length (Z), that is, $\eta_0 = K_1 Z$. Above Z_c the relationship is $\eta_0 = K_2 Z^{3.5}$, where K_1 and K_2 are proportionality constants, which depend on the chemical nature of the polymer and on the temperature at which the viscosity is measured. A log–log plot of the zero-shear viscosity against weight-average molecular weight is shown.

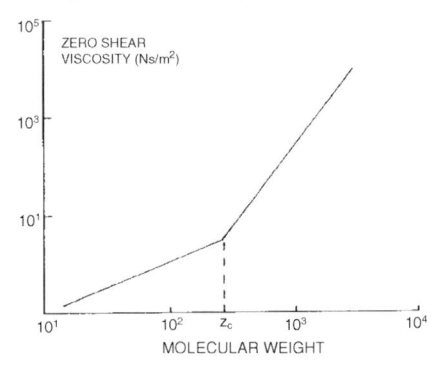

Graphical representation of the relationship between zero-shear viscosity and chain length of polymers.

153. Critical Crazing Strain

The value of the tensile strain in the direction of the maximum principal stress at which crazing takes place in a glassy polymer. (See Crazing.)

154. Critical Fibre Length (L_c)

Corresponds to the minimum length of the fibres in a unidirectional composite that is required for the tensile stress acting along the length of the fibres to reach the maximum possible value, which corresponds to the tensile strength of the fibres. For fibre lengths greater than L_c the stress distribution along a fibre (σ_f) rises from a minimum at the fibre ends and reaches a plateau along the middle section, where the conditions are isostrain, that is, the strain in the fibre is equal to the strain in the matrix. As the fibres get shorter, the portion of the fibres under isostrain conditions also becomes smaller, so that, when the length of the fibres decreases, and becomes equal to the L_c value, the maximum stress is reached only in the middle point (see diagram).

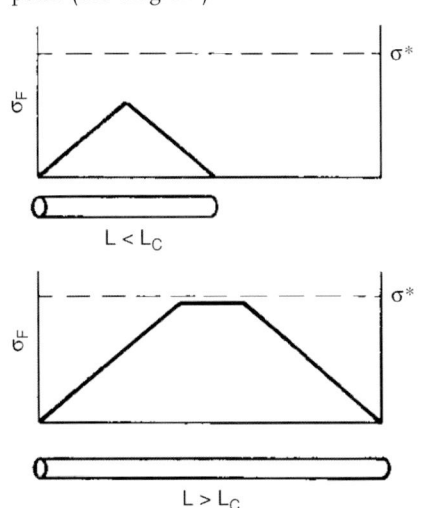

Variation of tensile stresses acting on the fibres of a unidirectional fibre composite.

Increasing the level of the external stress, the maximum stress in the middle section reaches the tensile strength value (σ^*), which causes the fibres to break and the fracture to propagate rapidly through the matrix. This is illustrated in the first micrograph, taken on the fracture surface of a thermoplastic composite tested in tension, which shows that fibres lying along the direction of the applied stress break first and are then pulled out of the matrix over a distance equivalent to the L_c value of the particular fibre–matrix combination.

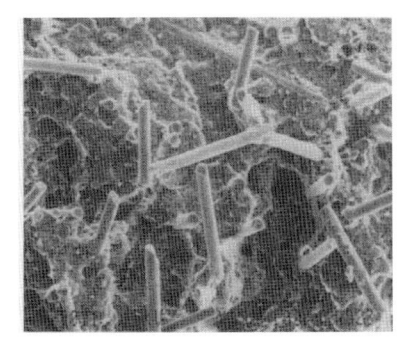

Fracture surface of a thermoplastic composite containing glass fibres with length $L > L_c$. Source: Mascia (1974).

For fibre lengths shorter than L_c, the fibres cannot reach the maximum possible value and, therefore, failure occurs at the interface between fibre and matrix by a shear mechanism, as indicated in the second micrograph. In this case the composite undergoes yielding deformation, causing the rotation of non-aligned fibres along the direction of the applied stress.

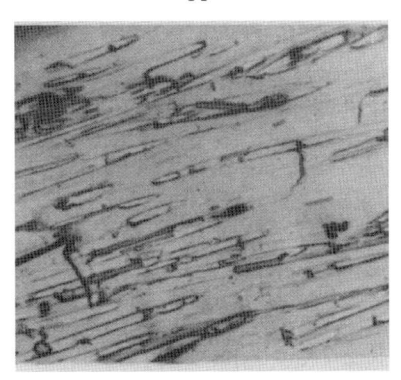

Fracture surface of a thermoplastic composite containing glass fibres with length $L < L_c$. Source: Mascia (1974).

Note that, if the fibre–matrix interfacial bond is very strong, failure will take place within the matrix near the interface at the point when the interfacial shear stresses reach the shear strength of the matrix. This makes it possible to obtain an estimate of the value of the critical fibre length by taking the balance of the tensile force acting within the fibre and the shear force acting at the interface, as shown.

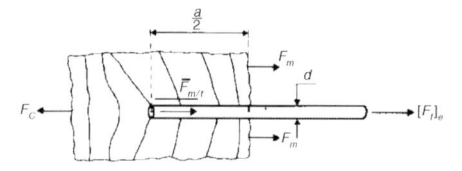

Force balance between tensile stress on a fibre and the shear force at the fibre–matrix interface when $L = L_c$.

When the length of the fibre (L) is equal to L_c, the average tensile force on the fibres ($\overline{F_f}$) is equal to the average shear force at the interface ($\overline{F_{mf}}$), where

$$\overline{F_f} = \sigma_f^* \pi d^2 / 4$$

and

$$\overline{F_{mf}} = \bar{\tau} \pi d \frac{L_c}{2}.$$

Here d is the diameter of the fibre and $\bar{\tau}$ is the interfacial shear strength or the shear strength of the matrix when the interfacial bond is sufficiently strong to cause failure within the matrix rather than at the interface. Making the appropriate substitutions in the above equation, it is possible to obtain an estimated value of the critical fibre length from the expression

$$L_c = \sigma_f^* d / 2\bar{\tau}.$$

A strong interfacial bond and a high matrix strength are, therefore, required to obtain a low value for L_c, so that the tensile strength of the composite remains well above the strength of the matrix. On the other hand, because the propagation of the fracture from the fibres to the matrix takes place through a pull-out of the fibre from the matrix over a length equal to L_c, a large value for L_c becomes beneficial for toughness enhancement by providing an increase in the energy absorbed during fracture.

The estimated pull-out energy U_f is related to L_c and $\bar{\tau}$ by the expression

$$U_f = \phi_f \bar{\tau} L_c^2 / 12d$$

The reduction in tensile strength resulting from shortening of the fibres in unidirectional composites can be estimated through a modification of the law of mixtures for isostrain conditions based on the average value of stress carried by the fibres at the point of fracture. When $L \gg L_c$ the average stress can be estimated by approximating the variation of the stress along the fibre to the geometry of a trapezium, as seen in the diagram illustrating the concept of L_c. Since fracture originates in the middle section of the fibre at the point where the fibre reaches its ultimate tensile strength, it follows that the average stress, $\bar{\sigma}$, carried by the fibres at this point can be related to the strength of the fibre, σ_f^*, by the expression

$$\bar{\sigma} = \sigma_f^* (1 - L_c/2L).$$

This can be used in the law of mixtures for the strength of the composite, σ_c^*, which becomes

$$\sigma_c^* = \phi_f \sigma_f^* (1 - L_c/2L) + (1 - \phi_f) \sigma_m^*,$$

where σ_m^* is the stress carried by the matrix at the point of the initiation of the fracture in the fibres.

155. Critical Strain Energy Release Rate (G_c)

(See Fracture mechanics.)

156. Critical Stress Intensity Factor (K_c)

(See Fracture mechanics.)

157. Critical Wetting Tension

Also known as the critical surface energy, corresponds to the minimum value of the surface tension (or surface energy) that is required for a liquid to obtain spontaneous wetting of the surface of a solid. (See Zisman plot.)

158. Cross-Linked Thermoplastic

Linear polymers containing a small number of cross-links between polymer chains, introduced through second-stage reactions after a product has been manufactured, or as an intermediate stage for a final process operation. An example of the latter is the cross-linking of thermoplastic tubes, usually by electron beam radiation, before these are expanded to produce heat-shrinkable tubes. The polymers that are most widely cross-linked are: polyethylene (cross-linked polyethylene, XPE), both low- and high-density types; poly(vinyl chloride), mostly as a plasticized polymer; poly(vinylidene difluoride); and thermoplastic elastomers. Cross-linking of thermoplastics is usually carried out either directly through free-radical reactions, using peroxides or high-energy radiation techniques, or indirectly via functional groups grafted onto the polymer chains. One widely used technique for cables is to graft alkoxysilane functional groups onto the polymer chains, allowing the cross-linking reactions to take place through the formation of siloxane links, which are obtained by hydrolysis and condensation reactions when the product is exposed to natural atmospheric environments.

159. Cross-Linking

A term used to denote the formation of chemical bonds (cross-links) between polymer or oligomer chains, leading to the formation of a network. (See Curing.)

160. Cross-Ply

The variable angle arrangement of unidirectional fibres, or fabrics, in the production of composites to obtain planar isotropy. A typical arrangement for cross-plies obtained in the production of pipes or cylindrical vessels by filament winding is shown.

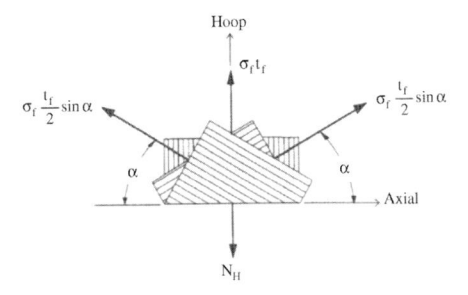

Cross-ply layout and stresses acting on the fibres of different layers in filament winding of tubular products. The stress on the fibre, σ_f, is normalized with respect to the total thickness t_f.

161. Crow Feet

Defects on the surface of calendered sheets, particularly in PVC, resulting from the presence of rigid polymer particles, known as gels, dragged by the rotation of rolls, forming wakes on the surface of the sheets that resemble the feet of a crow.

162. Crystalline Polymer

A polymer with a molecular structure consisting of a mixture of aligned molecular chains (crystals) and random chains (amorphous regions). During crystallization, polymer chains will fold over at some specific chain length, forming a lamella. Individual polymer chains enter into the adjacent lamellas via amorphous regions. An illustration of the structure of crystalline polymers and the formation of lamellae is shown.

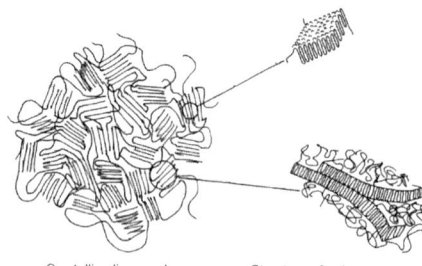

Crystalline linear polymer   Structure of polymer crystals

The sketch on the left depicts the early interpretation of the structure of polymer crystals, known as a 'fringed micelle', consisting of domains containing aligned segments of polymer chains and others in which sections of the polymer chains are coiled, forming amorphous regions. The sketches on the right show the actual structure of polymer crystals, consisting of polymer chains folding over a well-defined distance to form lamellae of crystals, surrounded by amorphous regions consisting of randomly coiled chains. The same polymer molecule can participate in the chain folding process for the formation of lamellae as well as being part of the amorphous domains. The latter are often referred to as 'tie molecules'. (See Spherulite.)

163. Crystallinity

Denotes the crystalline state of a polymer. The 'degree of crystallinity' of a polymer, on the other hand, defines the percentage of crystalline domain present in a particular sample. This depends on the nature of the polymer and on the processing conditions used to produce a given sample or product. The degree of crystallinity of polymers varies from about 10% (low-crystallinity polymers) to about 60–70% for polymers with a high degree of crystallinity.

The degree of crystallinity of polymers has a major influence on most properties, and therefore it is often adjusted to obtain the best balance of properties for a particular application. (See Ethylene polymer.)

164. Crystallization

Denotes the formation of crystals through cooling of a polymer melt or a polymer solution. It takes place in two steps, known as nucleation and growth. Nucleation can be divided into homogeneous nucleation and heterogeneous nucleation, according to whether the nuclei are formed either spontaneously or on the surface of impurities or existing heterogeneities present. Nucleation and growth of crystals take place only when the temperature is higher than the glass transition temperature (T_g). The maximum crystallization rate takes place at some temperature between the melting point (T_m) and the T_g of the polymer. The growth of polymer crystals takes place through the formation of lamellae by the folding of molecular chains in the direction perpendicular to the basal plane. (See Crystalline polymer.)

The thickness of the lamellae is between 5 and 50 nm, and is inversely related to the degree of supercooling from the melt. Supercooling is defined as the difference between the equilibrium melting point and the ambient temperature at which crystallization takes place. In the case of solvent crystallization, the extent of supercooling is assessed with respect to the 'equilibrium' dissolution temperature. Crystallization can also take place when the polymer is heated from the quenched amorphous state to a temperature above the glass transition temperature. This is often referred to as 'cold crystallization', which can be facilitated by the application of external stresses, as in the injection blow moulding of poly(ethylene terephthalate) (PET). Stress-induced crystallization can take place also above the melting point of the polymer, as in the case of stretched natural rubber, and also within the die of an extruder when some crystalline polymers are processed at temperatures just above their T_m.

165. Curable Adhesive

Systems that will produce cross-links in the final structure. (See Adhesive.)

166. Curative

Cross-linking agent used to induce curing reactions. An alternative term for 'hardener' for thermosetting resins or 'vulcanizing agent' for elastomeric materials.

167. Cure-Meter

A device or an apparatus to evaluate the curing characteristics of a rubber mix (compound). The most widely used commercial system is the Mooney viscometer, which consists of a rotating bi-conical disc oscillating in a cavity containing the sample. Radial ribs on the surface of the disc allow the contacting rubber sample to oscillate and, therefore, to shear between the static surface of the cavity and the moving surface of the disc. The basic construction is shown.

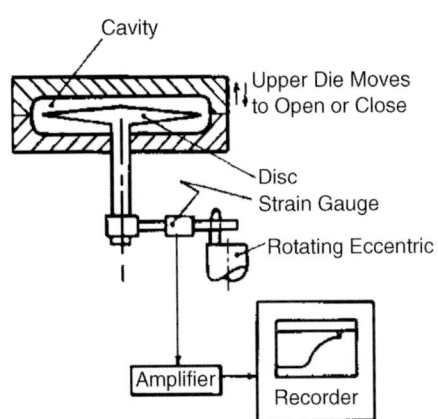

Schematic diagram of a rubber cure-meter. Source: Eirich (1978).

The cure-meter records the variation of the torque exerted by the disc on the sample while it is cured. The temperature of the chamber and the amplitude of the oscillations can be changed at will to suit the nature of the sample examined and to study the effects of related processing parameters. (See Cure time.)

168. Cure Time

A term denoting the time required by a thermosetting resin, or vulcanizable elastomer, to develop the required network density for optimum physical properties. Typical curves for the variation of viscosity or torque to maintain constant the amplitude of the oscillations of the disc in the rheometer used in the measurements is shown.

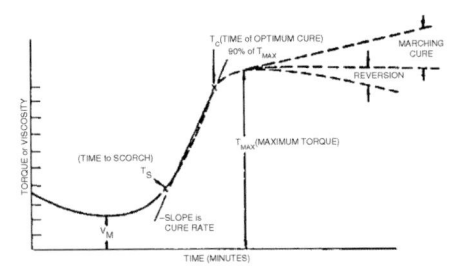

Cure curves recorded in measurements with an oscillating disc rheometer.

In the diagram the following parameters are identified:

- V_M – The minimum torque as an indicator of the viscosity of the mix before curing. (Note that the initial decrease in torque is an artefact arising from the warming up of the sample fed into the rheometer at room temperature.)
- T_S – The 'scorch' time, corresponding to the time required to reach an arbitrary increase in viscosity, above which the material will no longer be workable.
- T_C – The 'optimum cure' time, arbitrarily defined as the time required to reach 90%

of the maximum achievable cross-linking density. It is assumed that a moulded product will continue to cure to the maximum as it cools down naturally after being removed from the mould.
- Slope before T_C – A parameter related to the rate of cross-linking reactions (cure rate).
- T_{MAX} – An indicator of the degree of cross-links (cross-linking density of the network).
- Slope after T_{MAX} – An indicator of the stability of the network produced. A negative slope, known as 'reversion', results from the breakdown of cross-links. A positive slope, known as 'marching cure', indicates that there is a continual increase in the degree of cross-linking, albeit at much slower rate due to the depletion of reactive species present.

The optimum curing conditions for thermosets are established *ad hoc* through measurements of the mechanical properties of moulded specimens. Typical changes in mechanical properties for thermoset polymers are shown. The indicator line at point A shows the 'optimum cure time' determined in terms of the best balance of properties achievable for a particular system.

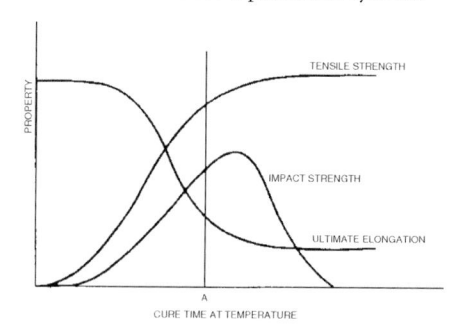

Change in mechanical properties with reaction time at a given processing temperature for a rigid thermoset polymer. Source: Goodman (1998).

169. Curing

A term originally used to describe the reduction in compression set in rubber with the addition of sulfur to the formulation. Nowadays it is generally used to denote cross-linking reactions for both thermosetting resins and elastomers. In the latter case, the process is also known as vulcanization. Similarly to polymerization, curing reactions can take place by addition or condensation reactions. The latter is generally avoided, if at all possible, in order to avoid the formation of volatiles, which may produce bubbles or have other undesirable side-effects.

170. Current Density

For conduction through the bulk of a medium, the current density corresponds to the intensity of the current flowing through the medium divided by the cross-sectional area. For surface conduction, the current density is defined as the intensity of the current divided by the width, perimeter or circumference of the medium. (See Conductivity, Volume resistivity and Surface resistivity.)

171. Cyclic Butylene Terephthalate

Cyclic oligomers containing a small number of butylene terephthalate units, used for *in situ* polymerization manufacturing operations.

172. Cycloaliphatic Epoxy Resin

Type of epoxy resins used to obtain products with higher transparency and better water-like clarity than aromatic epoxy resins. Widely used also for electrical applications because of the better tracking resistance, arising from the absence of aromatic rings in the network, which would give rise to the formation of graphitic polynuclear rings under the influence of arcs formed by high electrical stresses. (See Epoxy resin.)

173. Cyanate Ester

Type of thermosetting resin used in adhesives or composites for the high T_g values achievable through curing reactions involving trimerization of isocyanate groups with the formation of triazine rings. The reaction scheme shows how these reactions can take place.

Cyanate esters are often used as mixtures with epoxy resins to reduce the cure temperature and to improve the wettability with graphite fibres. (See Urethane polymer and resin.)

Scheme for trimerization reactions of isocyanate groups.

D

1. DABCO

A tertiary cyclic diamine (2,4-diazobicyclo [2.2.2]octane) used as a catalyst for isocyanate reactions in the production of polyurethane or as a catalyst for esterification reactions in paints. (See Urethane polymer and resin.)

2. Damage Tolerance

A term used to denote the resistance of a material to damage caused by loads at levels that would cause catastrophic fracture. It is normally assessed by measuring the deterioration in mechanical properties after a panel has been subjected to impact loads. (See Composite.)

3. Damping

The attenuation of oscillatory excitations, mechanical vibrations or sound waves applied to a structure. The phenomenon is also known as attenuation. (See Loss angle.)

4. Damping Factor

The extent of damping or attenuation that a material undergoes is expressed in terms of $\tan \delta$, which is known as the damping or attenuation factor, where δ is known as the loss angle. (See Dynamic mechanical thermal analysis and Loss angle.)

5. Debonding

A term used for removing or loosening the bond (adhesion) of adhesives towards the adherend (substrate). Widely used also to denote the loss of adhesion between matrix and fibres or fillers in composites.

6. Deborah Number

A dimensionless number used in rheology, defined as the ratio of the characteristic time of the material (usually the relaxation time or retardation time) to the time over which the phenomenon examined is considered.

7. Decomposition

Denotes the breakdown of the molecular structure of a polymer by heat. Widely used term in thermogravimetric analysis, where various reference temperatures are identified as relevant features of the decomposition process. For instance the 'onset decomposition temperature' is the temperature at which the incipient loss of matter is observed in a thermal scan of the analysis.

8. Decompression Zone

The zone of the screw of a compounding extruder, usually within the region where vacuum is applied to remove volatiles. This is illustrated in the diagram, which shows also the rapid drop in pressure on the melt on approaching the devolatilization zone of the screw/barrel assembly.

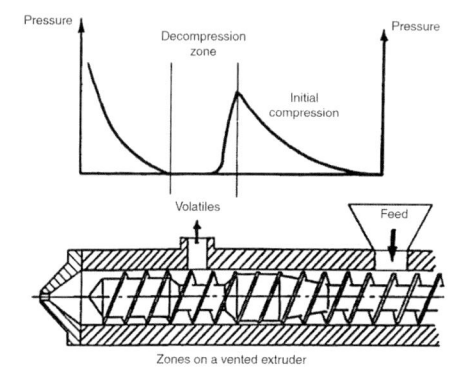

Extruder with devolatilization port and decompression zone in screw.

9. Deconvolution

A procedure that allows one to distinguish clearly between two merging absorption peaks in spectra obtained by spectroscopic analysis, such as IR or UV spectroscopy.

10. Deflection Under Load Temperature (DULT)

The temperature at which a specimen subjected to a specified flexural load (three-point bending mode), when heated at a specified rate of temperature rise, reaches a specified central deflection. This is also known as the heat distortion temperature. (See Heat distortion temperature.)

11. Deformation

A generic term for the response of a specimen or structure to a mechanical force. This takes place in the form of a change in length when a tensile or compressive force is applied, or as a geometric or torsional distortion when the applied forces are shear or torsional types.

12. Deformational Behaviour

Refers to the different states of a polymer over a wide range of temperatures. The changes that take place with increasing temperature presented as force versus extension curves up to fracture are shown.

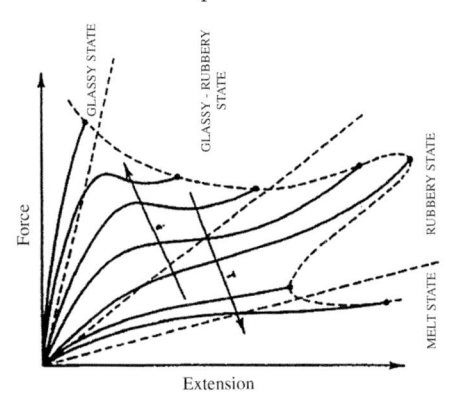

The deformational behaviour can also be interpreted in terms of the deformational states of a polymer with changes in temperature, illustrated as nominal plots of a parameter representing the deformability, or compliance, as a function of temperature. It is important to note that the transition from the 'glassy' state to the 'rubbery' state for the case of a linear polymer, or one with a very low cross-link density (as in the case of elastomers), brings about an increase in deformability by three or four orders of magnitude.

It is noted that the transition from the rubbery state to a nominal viscous state takes place via an intermediate 'melt' state. In the melt state a polymer acquires the ability to undergo flow while it retains some of the characteristics of the rubbery state, known as melt elasticity (see diagram).

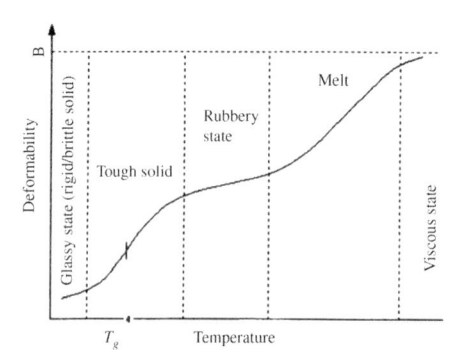

Various deformational states of polymers.

13. Degradation

The deterioration of properties or optical discoloration resulting from oxidative reactions taking place during processing or environmental ageing of a polymer sample or product. (See Thermal degradation and UV degradation.)

14. Degree of Cross-Linking

Refers to the quantity of cross-links present in network polymers, such as rubbers (elastomers) or thermoset plastics. It is normally expressed in terms of the molecular weight of chains between cross-links. For vulcanized rubber, the molecular weight between cross-links is in the region of 2000–3000; whereas for thermoset plastics, the value can vary much more widely but can be estimated to be within the range of 300–1000. It has to be borne in mind that the molecular constituents of thermoset plastics are much more bulky than those of elastomers.

15. Degree of Crystallinity

Represents the volume fraction of crystalline domains in a polymer, expressed as a percentage. Can be obtained from thermal analysis or X-ray diffraction methods. (See Crystallinity.)

16. Degree of Polymerization

Represents the number of monomeric units present in a polymer chain. The diagram shows the length of a free polymer chain with a degree of polymerization equal to about 1000. Although this is a fairly low degree of polymerization for most commercial polymers, the actual length of the chains is very large, as indicated in the diagram.

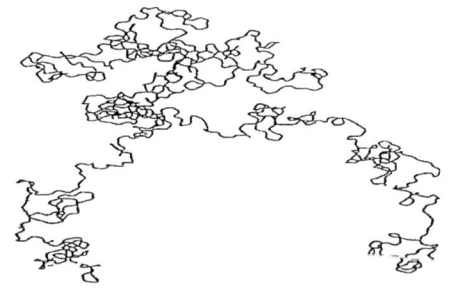

A free polymer chain with a degree of polymerization in the region of 1000.

For polyethylene, for instance, a degree of polymerization of 1000 corresponds to a molecular weight of 28 000. This is equivalent to a low-molecular-weight grade used in injection moulding. In comparison, an ultra-high-molecular-weight polyethylene (UHMWPE) grade has a degree of polymerization in the region of 40 000. However, for polymers containing aromatic units in the backbone chains, such as polycarbonates or polysulfones, the degree of polymerization is even lower than 1000.

17. Delamination

The breaking up of the interlayers in laminar composites or other multilayer constructions.

18. Demoulding

The removal of a moulded part from a moulding process at the end of a cycle.

19. Dendrimer

Can be defined as a branched molecule with a perfect structural and molar mass regularity. The principle used to build up this type of 'perfect' structure through various stages is illustrated. The concept has been used to produce industrially useful products known 'hyperbranched polymers'. (See Hyperbranched polymer.)

23. Devolatilization

The removal of volatiles, particularly water, from a compound. The term also refers to the section of a compounding extruder where a vacuum port in the barrel is used to remove volatiles from the decompression zone of the screw. (See Decompression zone.)

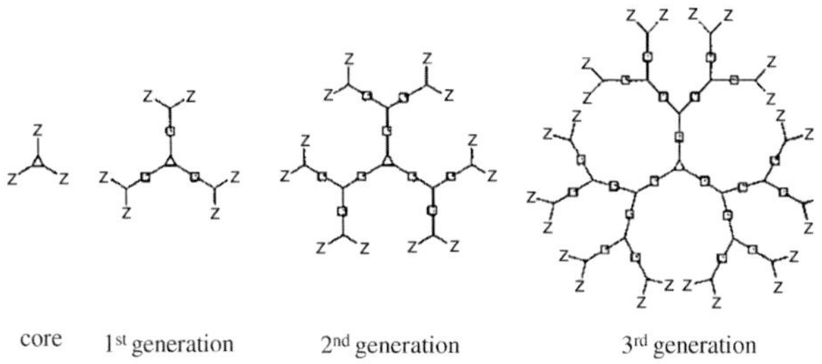

core   1st generation   2nd generation   3rd generation

Formation of dendritic molecular structures through condensation reactions. Source: Teegarden (2004).

20. Denier

A unit of measurement of the diameter of fibres or textile yarns, expressed in weight in grams per 9000 metre length. (See Tex.)

21. Depolymerization

The reverse process to polymerization. Some polymers, such as poly(tetrafluoroethylene), poly(methyl methacrylate) and polyoxymethylene (POM), will depolymerize back to the original monomer through unzipping reactions.

22. Desorption

The loss of volatiles from a substance when is exposed to an environment where the concentration of the previously absorbed volatiles is either zero or very small. (See Diffusion.)

24. Diallyl Phthalate and Isophthalate (DAP and DIAP)

A tetrafunctional unsaturated monomer used for the production of thermosetting resin moulding compositions. The chemical formula is:

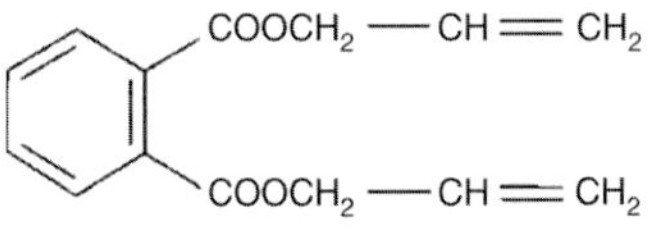

DAP and DIAP are primarily used as cross-linking agents for unsaturated polyesters and as polymerizable plasticizer in PVC formulations. The high reactivity of the allyl groups gives rise to networks with a high cross-link density and products with a very high glass transition temperature (T_g).

25. Die

The part of an extruder that shapes the melt into the desired geometry. Fitted near the screw tip, a die receives the melt from the metering zone of the extruder and delivers the extrudate to cooling devices before arriving at the take-off units. The term 'die', however, is sometimes used synonymously with the term mould. (See Extrusion.)

26. Die Analysis

Refers to the flow analysis within the die of an extruder to predict the flow rate (Q) as a function of the pressure drop (ΔP) from the die entry to the screw tip. This can be generalized by the equation

$$Q = K\Delta P/\eta,$$

where K is a constant related to the geometry of the die and η is the viscosity of the melt. Because the die is composed of several sections, the analysis has to be carried out for each section from knowledge of the pressure profile. In general, the highest pressure drop takes place in the flow through the die lips or die gap. Therefore, approximate estimates can be obtained by neglecting the pressure drops arising before the melts enter the die lips or die gap. The calculation of K for the most common types of channels in extrusion dies are shown below for Newtonian flow.

- **Circular sections** (cylindrical rods):

$$K = \frac{\pi R^4}{8L},$$

where R is the radius and L is the length (for $L/R \gg 1$).
- **Rectangular sections** (slit dies):

$$K = \frac{SWH^3}{12L},$$

where $S = 1 - 0.65(H/W)$ is the shape factor, H is the height of the channel and W is the width of the channel.
- **Axial annular sections** (thin-wall tubular dies):

$$K = \frac{\pi R_o^4 [1 - R^4 - (1-R^2)^2/\ln(1/R)]}{8L},$$

where R_o is the external radius, R_i is the internal radius, with $R = R_i/R_o$, and L is the length of the channel.

Note that for the case of polymer melts the viscosity is dependent on the shear rate and temperature and it is, therefore, difficult to obtain accurate estimates from the above equations without a detailed knowledge of the variation of viscosity. (See Extrusion and Screw–die interaction.)

27. Die Drool

The build-up of solid deposits at the die exit of an extruder, resulting from the removal of 'loose' matter that has accumulated on the surface during the flow path.

28. Die Exit Phenomena

(See Sharkskin.)

29. Die Gap and Die Lip

The upper and lower walls of the exit channel of a die used in polymer extrusion. These are often adjustable to change the thickness of the extrudate. (See Extrusion.)

30. Die Swell Ratio (*B*)

The ratio of the diameter of the extrudate to the diameter of the die for circular capillary dies, that is,

$$B = \frac{D(\text{extrudate})}{D(\text{die})}.$$

It is frequently quoted as an empirical parameter characterizing the elasticity of polymer melts. (See Melt elasticity and

Normal stress difference.) The value of B depends on the shear rate ($\dot{\gamma}$), temperature (T) and the length/diameter ratio (L/D) of the die, as shown in the diagrams.

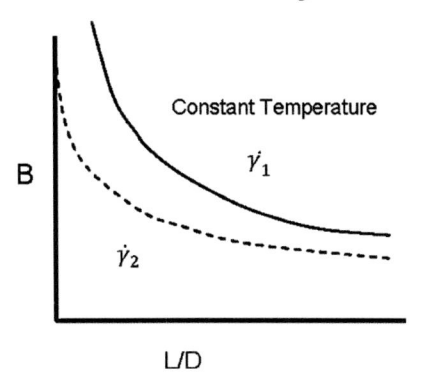

Variation of swell ratio (B) with increasing L/D of a capillary die at two shear rates, $\dot{\gamma}_1 > \dot{\gamma}_2$.

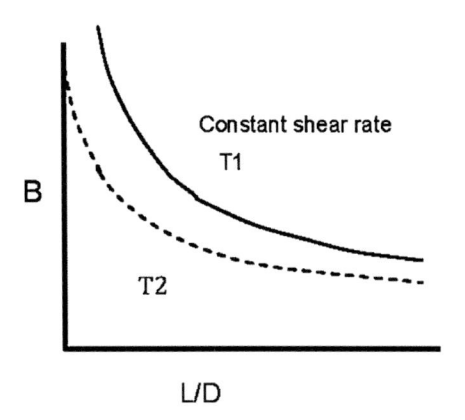

Variation of swell ratio (B) with increasing L/D of a capillary die at two temperatures, $T_2 > T_1$.

The increase in die swell ratio with increasing shear rate is related to the higher degree of molecular orientation of the polymer chains during the flow down the capillary. The reduction in swell ratio with increasing L/D is thought to be due to the gradual relaxation of the molecular orientation acquired in the die entry region. Similarly, the reduction in die swell ratio with increasing temperature results from the more rapid relaxation of molecular orientation before the melt reaches the die exit. For the same type of polymer, an increase in molecular weight results in an increase in die swell ratio. Although the die swell ratio curves suggest that a limiting constant value of B is reached at high L/D ratios, this value is always greater than 1 due to the elastic characteristics of polymer melts. At the same time, the limiting value of B cannot in itself be considered a fundamental property insofar as it cannot predict the extent of swelling for dies of different geometry, such as slit dies.

31. Dielectric

A generic term for an electrical insulating material, normally for low-voltage circuits.

32. Dielectric Constant

Known also as the relative permittivity, it is defined as the ratio of the permittivity of a dielectric to the permittivity of air. It is, therefore, a dimensionless number. The permittivity of air is close to 1, which is the lowest value achievable insofar as it corresponds to the value for a perfect dielectric, that is, vacuum. (See Permittivity and Dielectric properties.)

33. Dielectric Properties

Properties that define the relationship between an applied electrical stress (voltage gradient along the line of current flow) and the resulting current density (current per unit cross-sectional area).

For DC situations, the relationship can be described in terms of 'volume resistivity' (ρ) and 'bulk conductivity' ($1/\rho$), where

$$\rho = (V/d)(A/I).$$

A perfect dielectric is one whose volume resistivity or bulk conductivity is independent of the duration of the applied voltage (linear behaviour).

For AC situations, the dielectric properties are described by the 'specific impedance' (Z_p) and 'permittivity' (ε). For a perfect dielectric, the specific impedance and permittivity are independent of the frequency of the current in the circuit. The deviation from ideal behaviour can be described by models (analogues) consisting of combinations of a 'resistor' (perfect dielectric) and a 'capacitor' (the component responsible for the decrease in volume resistivity with time, and associated with loss of electrical energy in AC circuits). The approach is similar to that used for the viscoelastic behaviour, using combinations of a spring (elastic solid) and a dashpot (viscous component).

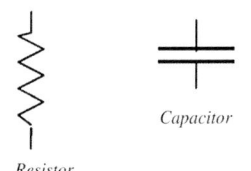

Resistor *Capacitor*

In this respect, appropriate combinations of the two elements can be used to model the reduction in volume resistivity with time (the opposite for conductivity) and the increase in permittivity with frequency (the opposite for specific impedance). The ratio of the constants for the elements determines the time- or frequency-dependent behaviour in terms of a characteristic time, known as the 'relaxation' or 'retardation' time, λ.

For AC circuits the deviation from perfect dielectric behaviour is defined by the loss components Z_p'' and ε'', or the loss tangent of the phase angle

$$\tan\delta = Z_p''/Z_p' = \varepsilon''/\varepsilon'.$$

Similarly to the viscoelastic behaviour for mechanical properties, the variation of the dielectric loss component ε'' and $\tan\delta$ at constant frequency goes through a maximum with respect to temperature, corresponding to the glass transition temperature, T_g. The corresponding variations with respect to frequency have a maximum when the frequency is equal to $1/\lambda$.

34. Dielectric Strength

The value of the electrical stress at which failure takes place. Failure denotes the loss of dielectric characteristics through the formation of carbonaceous channels or the formation of holes resulting from depolymerization reactions, as for PTFE. The typical measurement set-up for dielectric strength is shown in the diagram.

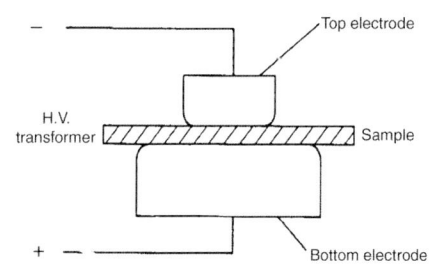

Electrode set-up for measuring dielectric strength.

Note that often the specimen is immersed in transformer oil to prevent arcing over the edges of the specimen caused by the breakdown of air.

35. Dielectric Thermal Analysis (DETA)

Known also as dielectric relaxation spectroscopy, is a characterization technique based on the measurement of the permittivity (ε) and loss factor ($\tan\delta$) over a wide range of frequencies (f) and temperatures (T). (See Permittivity.) A sample is subjected to an AC voltage to induce polarization of the dipoles associated with the constituent chemical groups. The response is analysed in terms of the charges generated and loss of input energy, from which are calculated the two components of the permittivity, respectively the in-phase (ε') and out-of-phase (ε'') components. In DETA measurements there are two possible transitions that polymers can undergo, varying according to the excitation frequency. These are associated with interfacial polarization for frequencies around 100 Hz and with dipolar polarization at frequencies between 100 and 10^6 Hz. (See Polarization and Dynamic mechanical thermal analysis.) The treatment is similar to that developed for the viscoelasticity of polymers, using models that can be used to define fundamental parameters that can be correlated with the chemical and physical structure. The models consist of a combination of resistors and capacitors to represent the actual behaviour between the two extremes exhibited by each individual element. The vector diagram for the two components of the current resulting from the applied voltage, respectively I_c (out-of-phase component, representing the charging current) and I_e (in-phase component, representing the loss current), gives the phase angle δ, known as the loss angle.

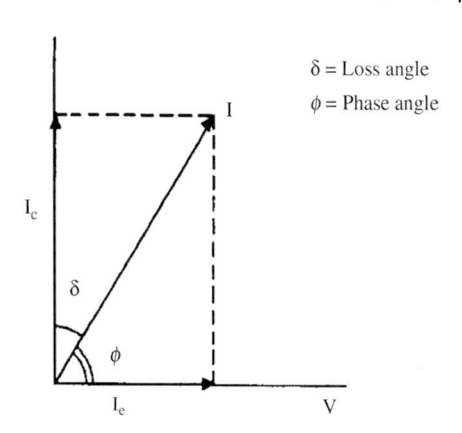

Vector diagram of electric current resulting from an AC voltage applied to a dielectric.

The input and response signals are used to generate the data required to calculate the parameters representing the frequency-related dielectric characteristics of polymers in terms of the components of the complex permittivity ε^*, that is, $\varepsilon^* = \varepsilon' - i\varepsilon''$, where i is the imaginary number ($\sqrt{-1}$), ε' is the in-phase component and ε'' is the out-of-phase (loss) component, so that $\tan \delta = I_e/I_c$ and also $\tan \delta = \varepsilon''/\varepsilon'$. A transition is manifested as a rapid decrease in the ε^* or ε' values and a peak in the $\tan \delta$ or ε'' values with increasing excitation frequency – see diagram for poly(methyl methacrylate) (PMMA).

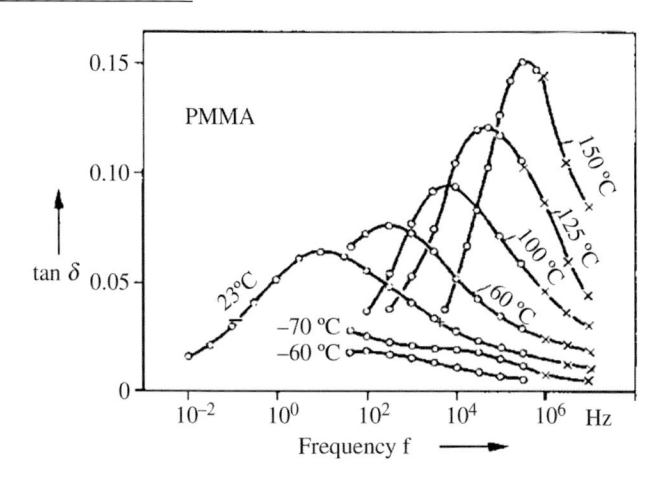

Variation of dielectric tan δ with frequency at different temperatures for a PMMA sample. Source: Koppelmann (1979).

36. Diene Elastomer

A homopolymer or copolymer based on a diene monomer. The latter are available as random copolymers requiring conventional sulfur vulcanization for curing, or as block copolymers in the form of thermoplastic elastomers. The more widely used systems are described below.

36.1 Polybutadiene (Butadiene Rubber, BR)

Polymer chains with mostly cis-1,4-, some trans-1,4- conformations and small amounts of vinyl-1,2- type units, as depicted. Produced by a number of polymerization techniques, giving a variety of grades with different molecular structure, molecular weight and polydispersity. The grades with the highest cis-1,4- conformation gives the lowest T_g, around $-90\,°C$ and a melting point of approximately $+1\,°C$.

$$\cdots\!+\!CH_2\!-\!CH\!=\!CH\!-\!CH_2\!+\!CH_2\!-\!CH\!=\!CH\!-\!CH_2\!+\!CH_2\!-\!CH\!-\!\cdots$$
$$\qquad\qquad\qquad\qquad\qquad\qquad\qquad\qquad\qquad\qquad |$$
$$\qquad\qquad\qquad\qquad\qquad\qquad\qquad\qquad\qquad\qquad CH$$
$$\qquad\qquad\qquad\qquad\qquad\qquad\qquad\qquad\qquad\qquad ||$$
$$\qquad\qquad\qquad\qquad\qquad\qquad\qquad\qquad\qquad\qquad CH_2$$

1,4-Butadiene | 1,2-Butadiene Units

The polymer does not exhibit stress-induced crystallization at room temperature, although the tendency to stress crystallization increases with increasing amount of 1,2- units in the polymer chains, resulting in a decrease in wear resistance. The average molecular weight of commercial BR is in the region of 250 000–300 000. Such BRs are mainly used in blends with natural rubber (NR) and styrene–butadiene rubber (SBR) in sulfur vulcanization compounds, primarily for use in tyre manufacture.

36.2 Polyisoprene

The polymer constituent of natural rubber (NR) and synthetic polyisoprene (isoprene rubber, IR). Both have chains with a cis-1,4- configuration content greater than 92%. The chemical formula is:

$$-\!\!\left[CH_2 - \underset{\underset{CH_3}{|}}{C} = CH - CH_2\right]_n\!\!-$$

The molecular weight of both synthetic IR and NR is very high, in the region of $(1-1.5)\times 10^6$, and they have similar properties. The density is $0.934\,g/cm^3$ and IR undergoes crystallization if cooled slowly from about 10 to $-35\,°C$. Stress-induced crystallization, on the other hand, can take place at higher temperatures. NR has to be masticated in order to reduce the molecular weight and render it compoundable and processable. Its main use is for the production of tyres.

36.3 Polychloroprene Rubber (Chloroprene Rubber, CR)

Sometimes also known as 'neoprene', is essentially poly(2-chlorobutadiene), which can be represented by the formula:

$$-\!\!\left[CH_2 - \underset{\underset{Cl}{|}}{C} = CH - CH_2\right]_n\!\!-$$

Some commercial grades may contain small amounts of 2,3-dichlorobutadiene, acrylonitrile or styrene, as a means of controlling the crystallization characteristics of the polymer. Contrary to all other diene elastomers, the vulcanization of CR is not carried out using sulfur but through reactions with metal oxides involving the scission of the C–Cl bond. Typical oxides are MgO, ZnO and PbO or Pb_2O_4, even though the cross-linking reactions are assisted by the addition of conventional accelerators, such as thiuram or mercaptans. The main characteristics of CR are flame retardancy and oil resistance due to the presence of large quantities of chlorine in the structure.

36.4 Styrene–Butadiene Rubber (SBR)

A general-purpose rubber containing 23–40% styrene and 60–77% butadiene as random copolymers, represented by the formula:

$$-\!\!\left[-CH_2CH=CHCH_2\ CH_2CH(\text{Ph})-\right]_n-$$

There are many grades available commercially, varying according to the method of polymerization used, and broadly known as E-SBR (emulsion-polymerized) and L-SBR (solution-polymerized). The difference in structure can be quite substantial, owing mainly to variations in *trans*-1,4- conformation and *vinyl*-1,2-butadiene content. The glass transition temperature, T_g, is usually in the region of $-50\,°C$ and increases with the styrene content of the polymer chains. E-SBR has a molecular weight in the range 250 000–800 000 and is used mainly in blends with polybutadiene (BR) in tyre manufacture. This combination brings about a substantial decrease in the T_g value. L-SBR is a purer grade with a less broad molecular-weight distribution and is less branched than E-SBR. It is characterized by a higher abrasion resistance and a lower heat build-up than E-SBR. Owing to the random nature of the chain structure, SBR does not exhibit stress-induced crystallization and, therefore, the products have inferior mechanical properties than natural rubber or synthetic polyisoprene rubber.

36.5 Acrylonitrile–Butadiene Rubber (NBR)

Also known as nitrile rubber, is a based on random copolymers with acrylonitrile content in the range 18–51%, represented by the formula:

$$-\!\!\left[-CH_2CH\!\underset{CN}{|}\!CH_2CH=CHCH_2-\right]_n-$$

There are also many variations in grades available due to differences in *trans*-1,4- conformation and *vinyl*-1,2-butadiene content. The T_g varies from about -10 to $-40\,°C$ with increasing butadiene content. Owing to the random structure and the lack of uniformity of polymer chains NBR does not exhibit stress-induced crystallization and, therefore, relies on the use of carbon black reinforcement to achieve adequate strength characteristics. They are widely used in blends with natural rubber, polybutadiene and EPDM rubbers. An important characteristic is the resistance to mineral oils due to the high polarity of the chains, resulting from the presence of the acrylonitrile units. The polar nature of the chains makes NBR miscible with poly(vinyl chloride) (PVC) and the phthalate plasticizers, which are widely used for low-voltage insulation in wire and cable applications.

37. Differential Scanning Calorimetry (DSC)

A technique that measures changes in heat flow to and from a sample contained in a cell while it is undergoing chemical or physical changes. This is done by comparison to a reference sample, which remains stable during the same scan at a constant rate of temperature rise (see diagram).

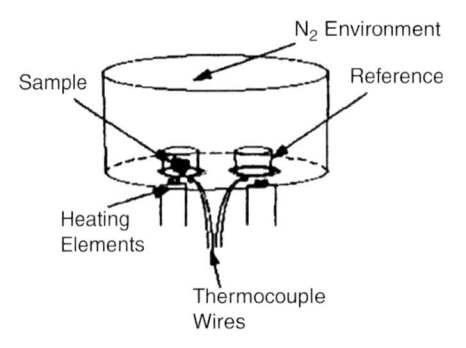

Schematic diagram of a DSC cell.

The measurements are usually made in a nitrogen atmosphere, unless it is intended to examine the effects of an oxidative environment. Heats of reaction are more usually measured with isothermal scans, while physical transitions, such as the glass transition temperature and the melting point, are determined with a ramp scan. The measured enthalpy can be used to measure the degree of crystallinity of a polymer from knowledge of the corresponding heat of fusion for a fully crystalline sample, normally estimated from theoretical calculations. The weight fraction of crystalline content is equal to the ratio of the measured enthalpy to that of the fully crystalline sample. The thermogram shows the glass transition (T_g) in the form of a sharp increase in the baseline of the heat flow, corresponding to the increase in specific heat ΔC_p, and a peak associated with the endothermic heat flow, which results from the melting of crystals.

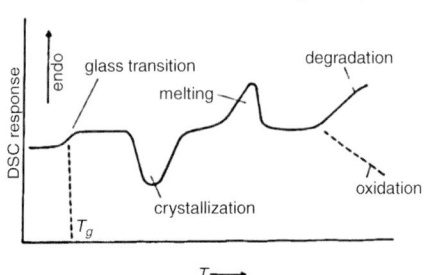

Thermogram of a polymer examined by DSC.

Melting takes place over a range of temperatures owing to the non-homogeneity of the crystals present and is completed at the peak temperature, taken as the thermodynamic melting point (T_m). The diagram also shows the endothermic oxidation of the polymer, which takes place at higher temperatures.

38. Differential Thermal Analysis (DTA)

A technique that measures the difference in temperature between an inert reference material and the sample arising from heat changes in experiments carried out in a heating scan in the measuring apparatus. (See Differential scanning calorimetry.) Endothermic events will bring about a reduction in temperature of the sample examined relative to the reference sample. The opposite changes are recorded if the thermal process is exothermic.

39. Diffusion

A term used to denote the transport of gases or liquids through a solid medium. There are various types of diffusion. The more common are the Fickian and the case II types. Fickian diffusion takes place through a random movement of the diffusant molecules, driven by concentration gradients through the thickness. Case II diffusion takes place via an advancing front layer, which gradually increases in thickness until it traverses the entire cross-section of the medium, driven by the hydrostatic tension ahead of the swollen advancing front layer.

40. Diffusion Coefficient (or Diffusivity)

The coefficient (D) that relates the rate of diffusion (F) of gases or liquids through a solid medium under steady-state conditions, generally known as Fick's law, that is, $F = D(dc/dt)$, where dc/dt is the concentration gradient. The diffusion coefficient can be measured from experiments carried out on thin sheets or films either by monitoring the amount of diffusant that has been transported through the medium and plotted against time, or by measuring the mass uptake of the sample through absorption when immersed in the diffusant.

For the first case, the mass (M_t or F_t) transported through the medium is approximately related to D by the expression

$$M_t = \frac{DC_0}{x(t - x^2/6D)},$$

where C_0 is the concentration of diffusant at the surface and x is the thickness of the film. The quantity $x^2/6D$, known as the 'time lag' (L), can be used to calculate the value of D. A typical graph obtained from the above plot is shown.

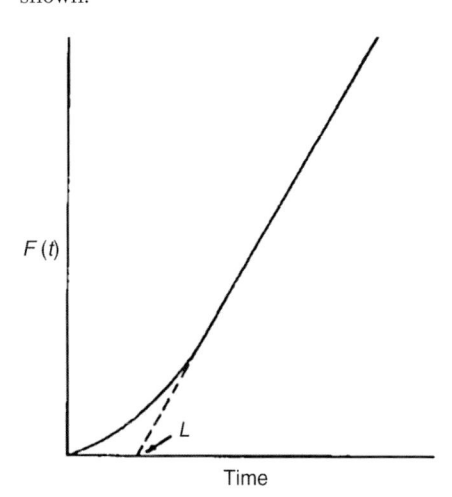

Concept of 'time lag' (L in diffusion phenomena).

For absorption experiments, the mass uptake of diffusant (M_t or F_t) is approximately related to D by the expression

$$\frac{M_t}{M_\infty} = \frac{4}{\pi^{1/2}(Dt/x^2)^{1/2}},$$

from which the value of D can be obtained by measuring the mass uptake with time (t) up to equilibrium to obtain the value of M_∞ (i.e. the maximum amount absorbed by the sample). This equation is valid, however, only for M_t/M_∞ ratios up to 0.5, so that the value of D can be estimated from the gradient of the initial part of the curve obtained from the plot of M_t/M_∞ against $(t/x^2)^{1/2}$, where x is the thickness (see diagram). The time for $M_t/M_\infty = 0.5$ is known as the 'half-time'.

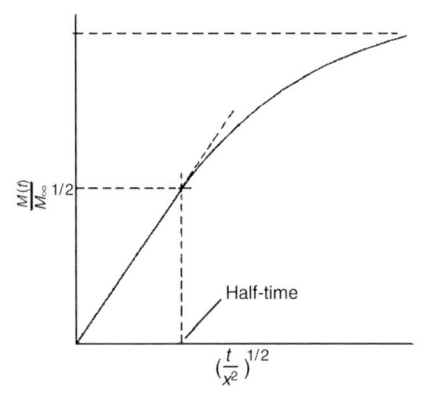

Concept of 'half-time' in diffusion phenomena.

A linear diffusion behaviour implies that the value of D does not depend on the concentration of diffusant C_0 used in the measurements. The validity of the assumption of Fickian behaviour of diffusion through polymer samples can be checked by carrying out experiments with samples of different thickness. All the data on the plot should fall on the same curve up to the 'half-time'.

The variation of the coefficient of diffusion with temperature obeys the Arrhenius equation on account of the nature of the phenomenon as a physical rate process. (See Arrhenius equation and Activation energy.)

41. Dilatant Fluid

A fluid whose viscosity increases with shear rate. This is the opposite of pseudoplastic behaviour. (See Rheology.)

42. Dilatation Coefficient

Equivalent to the volumetric expansion coefficient.

43. Dilatometer

An apparatus used to measure the change in linear dimensions of a sample resulting from a change in temperature. (See Thermomechanical analysis.)

44. Dimensional Stability

A term used to describe the resistance of a polymer product to changes in dimension as a result of the action of external environmental agents, such as moisture or heat.

45. Dip Coating

A technique used to produce coatings on articles through immersion into the coating fluid or dispersion, which can be carried out as a continuous operation.

46. Dipole

A concept used to describe the imbalance of electron density along a chemical bond, arising from the difference in electronegativity between the atoms involved in the bond considered. Dipoles can be divided into 'permanent dipoles' and 'induced dipoles'. The latter refer to the polarization of a chemical bond arising from neighbouring species. Dipoles are indicated as $\delta-$ if there is an excess of electrons on a particular atom or $\delta+$ if there is a defect of electrons.

47. Direct Current (DC)

A current resulting from the application of a constant voltage.

48. Direct Reflection Factor

Defined as the ratio of the light flux reflected from a surface to the incident light flux, it characterizes the ability of a product to reflect light.

49. Direct Transmission Factor

Known also as the specular transmittance, it is defined as the ratio of the light flux transmitted to the incident light flux. It denotes the light transmission characteristics of a medium.

50. Dispersion

A term used to denote the state of distribution of an additive of filler into the bulk of the polymer.

51. Dispersity

Normally refers to the breath of the distribution of the molecular weight of a polymer.

52. Dispersity Index

Also known as polydispersity, denotes the ratio of the weight-average to the number-average molecular weight and provides a measure of the spread of the distribution of molecular weight, based on the assumption that the distribution is Gaussian (i.e. statistically symmetrical).

53. Dispersive Mixing

A term used to describe the mixing of additives through a dispersive action. (See Mixing.)

54. Distortion and Warping (Moulding)

Planar deviations of a large-area moulding (moulded part) from the geometry of the mould cavity, which arise from the effect of cooling rate on the density of a crystalline polymer or a rigid thermosetting 'compound'.

A higher cooling rate produces a larger amount of cross-links in a thermosetting resin and a lower degree of crystallinity in a crystalline polymer. In both cases, this brings about a lower density of the moulded part.

The diagram shows that a flat geometry is maintained if the cooling rate at the two surfaces, S_1 and S_2, is the same (left). On the other hand, if the cooling rate at surface S_1 is lower than that at surface S_2, the moulded part has to bow inwards in order to accommodate the difference in density.

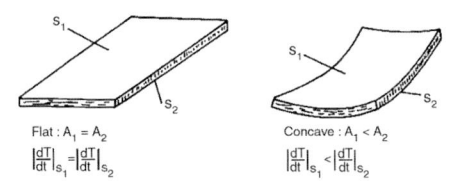

Effect of differential cooling between the opposite surfaces of a flat moulded part. Source: Mascia (1989).

Warping in moulded parts, on the other hand, results primarily from the orientation of polymer molecules, or fibre reinforcement in the case of composites, even though the two types of morphological orientation have the opposite effect. In the diagram, depicting a comparison of side-feed gates with central feeding, it is shown that the direction of the orientation coincides with the direction of flow of the melt while feeding the cavity.

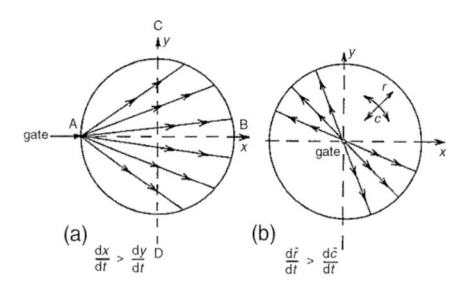

Orientation direction of polymer molecules or reinforcing fibres in filling a mould cavity. Source: Mascia (1989).

For the case of molecular orientation, the shrinkage that takes place during cooling of the moulded part is greater along the orientation direction (owing to molecular relaxations taking place during cooling) than in the transverse direction. For the case of a disc-shaped cavity, a side-gate melt inlet has a main flow path along the diameter A–B of the disc. Owing to the differential shrinkage in the two perpendicular directions, after cooling, the diameter A–B will be smaller than the diameter C–D. This situation is accommodated by an inward bowing (distortion) of the moulded disc. For the case of centre-fed cavities, the molecular orientation and the associated linear shrinkage are greater in the radial direction than in the transverse direction, which corresponds to the circumferential direction. The implications are that, in the absence of radial orientation, the geometrical relationship $C = 2\pi R$ (C is the circumference and R is the radius) holds both before and after cooling. When radial orientation occurs, on the other hand, the extent of shrinkage in the radial direction becomes greater than that along the circumferential direction. This situation results in the relationship $C > 2\pi R$, which can only be accommodated if the disc assumes an out-of-plane hyperbolic–parabolic (figure-of-eight) 'warped' geometry.

The situation is completely reversed for both types of gates if the orientation is caused by the alignment of fibres, as in the case of short-fibre composites with a matrix that does not bring about molecular orientation (e.g. thermosetting resins) and, therefore, is not susceptible to differential shrinkage. This is because the shrinkage along the fibre orientation direction is now less than along the circumferential direction. The case of glass-reinforced thermoplastics represents an ideal solution to distortions and warping problems, insofar as the shrinkage associated with fibre orientation and that resulting from molecular orientation tend to take place in opposite directions and therefore will minimize or even eliminate the differential shrinkage. In practice, the differential shrinkage is completely eliminated with the incorporation of small quantities of mica flakes whose plate-like geometry reduces the extent of orientation of the reinforcing fibres by creating greater turbulence along the flow direction.

55. Distribution of Molecular Weight

(See Molecular weight.)

56. Distribution of Relaxation Times

Defines the spread of relaxation times for a large number of models acting in series, either Maxwell or standard linear solid (SLS) models, as a means of describing more accurately the properties of polymers with respect to their linear viscoelastic behaviour. A single model, characterized by a single value of the relaxation time, predicts a rapid change in the relaxation modulus when the duration of the experiment approaches the value of the relaxation time of the model (i.e. $t = \lambda$). Since this is not sufficiently close to the true behaviour of polymers, a large number of elements (connected in series), each exhibiting a different relaxation time, are used in the model to produce a bland decrease in relaxation modulus, which provides a more realistic description of the true behaviour of polymers. The mathematical equation for the variation of the relaxation modulus with time, $E(t)$, for an infinite number of SLS models acting in series becomes

$$E(t) = E_\infty + \int_{-\infty}^{\infty} H(\ln \lambda)\, e^{-t/\lambda}\, d(\ln \lambda),$$

where $H(\ln \lambda)$ is a parameter known as the distribution of relaxation times and E_∞ is the modulus at infinite time, known also as the relaxed modulus.

57. Distribution of Retardation Times

Defines the spread of retardation times for a large number of models acting in series, either Kelvin–Voigt or standard linear solid (SLS) models, as a means of describing more accurately the viscoelastic behaviour of polymers under creep conditions. A simple model characterized by a single value of the retardation time results in a very rapid increase in the creep compliance when the duration of the experiment approaches the value of the retardation time of the model (i.e. $t = \lambda$). Since this is not sufficiently realistic, a large number of elements (connected in series), each exhibiting a different retardation time, are used in the model to obtain a more bland change in the creep compliance in order to bring the predictions of the model closer to the actual behaviour of polymers. The mathematical equation for the variation of the creep compliance with time, for an infinite number of SLS models acting in series, becomes

$$D(t) = D_0 + \int_{-\infty}^{\infty} L(\ln \lambda)[1 - e^{-t/\lambda}]\, d(\ln \lambda),$$

where $L(\ln \lambda)$ is a parameter known as the distribution of retardation times and D_0 is the compliance at time zero, known also as the instantaneous compliance.

58. DMP-30

Tradename for a tertiary amine used as a curing agent for epoxy resins, that is, 2,4,6-tris(dimethylaminomethyl)phenol. The phenolic hydroxyl group enhances the reactivity. The corresponding DMP-30 tri-(2-ethyl)-hexoate salt has been used in resins for electrical applications in order to improve adhesion to metal components.

59. Dog-Bone Specimen

Also known as dumb-bell, is a type of specimen used in tensile tests. A typical example is shown. The waist section ensures that failure takes place in an area subjected to uniaxial tension. When rectangular specimens are used in tensile tests, failure is likely to take place at the clamps, where spurious triaxial stresses arise from the combination of the clamping forces and the applied tension.

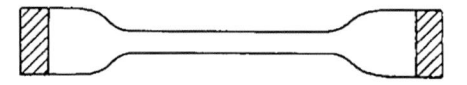

Dog-bone shaped specimen for tensile tests.

60. Doping

A term used to denote the enhancement of specific properties by the addition of small amounts an additive. A typical example is the addition of very small amounts of strong acids to an intrinsically conductive polymer, such as polyaniline, to increase the electron transfer efficiency along the conjugated double bonds.

61. Dough Moulding Compound (DMC)

(See Bulk moulding compound.)

62. Drag Flow

Also known as Couette flow, is a type of flow produced between two surfaces, one of which is stationary while the other is moving. When the two surfaces are sufficiently close, the flow produces a linear velocity gradient, corresponding to a state of constant shear rate across the gap. This type of flow occurs very often in polymer processing and melt mixing operations. For instance, the flow in the channels of the metering zone of an extruder takes place by the rotation action of the screw against the stationary wall of the barrel. (See Extrusion theory.)

63. Drape Forming

A technique used in vacuum forming (thermoforming) using a male mould. The article is formed by the application of vacuum to draw the sheet against the contours of the male mould after the sheet is 'draped' (pushed into shape). (See Thermoforming.)

64. Draw Ratio (or Extension Ratio)

The ratio of the final length after drawing a sample or product (usually in the form of filaments, tapes or films) to the original length. In manufacturing processes, this corresponds to the ratio of the velocity of the two sets of rolls used to draw the product.

65. Drawdown Ratio

The ratio of the velocity of an extrudate at the exit of the tensioning rolls to the velocity of the melt through the die. It is normally estimated from the ratio of the final cross-sectional area of the extrudate to the corresponding area at the die lips.

66. Drawdown Resonance

The formation of regular diameter fluctuations in an extrudate arising by the drawing of the polymer melt from the die. This phenomenon occurs under conditions in which the drawdown ratio exceeds a critical value. This phenomenon occurs primarily in the production of fibres, filaments or cast films. (See Melt strength.)

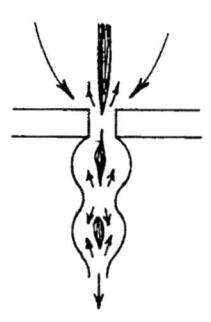

Undulations in diameter of filaments or thickness of films resulting from drawdown resonance.

67. Dry Band

A term applied to the formation of a dry zone on the surface of high-voltage insulators subjected to surface contamination with salt solutions. (See Tracking.)

68. Dry Blend

A mixture of PVC powder (suspension type) and plasticizer, which is forced to occupy the internal pores of the PVC particles through high-speed mixing operations.

69. Drying Resin

An alkyd resin containing large amounts of an unsaturated vegetable oil. The term 'drying' derives from their ability to become increasingly less tacky during curing, due to cross-linking reactions within the oil component.

70. Dual Sorption

Phenomenon related to the dissolution of gases in glassy polymers, consisting of two sorption components, hence the term 'dual sorption'. One corresponds to the amount of gases occupying free volumes within polymer chains, known as the Henry solubility; while the other corresponds to the amount of gas residing in microcavities, known also as Langmuir absorption. Both terms are related to pressure according to the equation

$$C_t = S_D p + \frac{C'_H bp}{1+bp},$$

where C_t is the total concentration of gas or vapour absorbed, S_D is the solubility coefficient, C'_H is the void absorption constant, b is the void affinity constant and p is the pressure.

71. Ductile

A term used to describe the ability of a material to undergo large deformation through yielding, known also as plastic deformations. (See Brittle fracture.)

72. Ductile–Brittle Transition

(See Brittle point and Brittle–tough transition.)

73. Ductile Failure

Failures occurring through yielding or plastic deformations. (See Load–extension curve.)

74. Dumb-Bell Specimen

(See Dog-bone specimen.)

75. Dye

A colouring agent that is molecularly miscible or dispersible on the nanometre scale, which permits the polymer to develop a colour without losing its transparency. (See Colorant and Colour matching.) The chemical structure of some typical dyes used for the coloration of transparent polymers is shown below.

- **Spirit-soluble**: Salts of organic bases, for example, Zapon Fast Scarlet GG:

[structure: diazo compound with OH and SO$_3^-$ groups, $[(C_8H_{17})_3NH_3]_3$]

- **Oil-soluble**: Various non-ionic aromatic compounds containing alkyl groups to improve compatibility, for example, Oil Orange C.1.24:

[structure: phenyl-N=N-naphthol]

76. Dynamic Mechanical Thermal Analysis (Dynamic Mechanical Analysis)

Abbreviated to DMTA or DMA, denote tests carried out under cyclic loading conditions, where the stress applied to a specimen and the resulting strain vary in a sinusoidal manner with time at a certain frequency. DMTA and DMA tests measure also the phase angle resulting from the delayed response to the applied stress, which arises from the viscoelastic behaviour of polymers. This makes it possible to calculate the components of the complex modulus over a wide range of frequencies and temperatures. (See Viscoelasticity.) The plots of these parameters against frequency or temperature provide fingerprints for the dynamic mechanical behaviour of polymers, known as 'mechanical spectra'. From these it is possible to obtain valuable information about the chemical and morphological structure.

77. Dynamic Mechanical Spectra

Plots for the variation of the loss tangent ($\tan \delta$), or loss modulus (E'' or G''), with temperature (T) and oscillation frequency (ω). Typical dynamic mechanical spectra are depicted, which show the variation of the elastic modulus (E') and the loss modulus (E'') with angular velocity at two different temperatures (T_2 and T_1).

Note that, in cyclic phenomena, the angular velocity ω is related to the frequency f by the expression $\omega = 2\pi f$. The glass transition temperature of the polymer (T_g) is obtained from the maximum of the E'' plot against temperature. Often the $\tan \delta$ value ($\tan \delta = E''/E'$) is plotted against temperature. The maximum, in this case, occurs at a

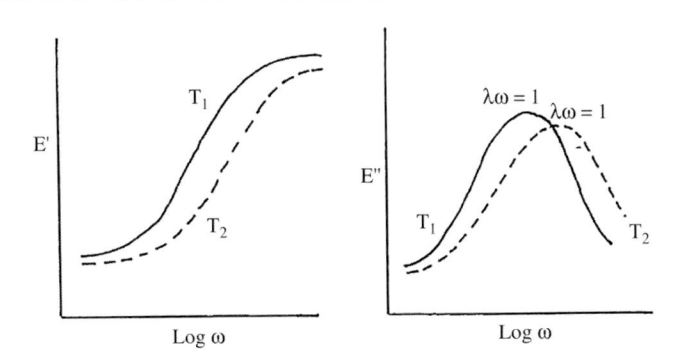

Variation of elastic and loss moduli with angular velocity at two different temperatures ($T_2 > T_1$).

slightly higher temperature. The modulus and compliance functions are obtained through mathematical operations on the relationships for stress and strain, that is, the function

$$\sigma = \sigma_0 \sin(\omega t + \delta)$$

can be expanded to

$$\sigma = \sigma_0 \cos\delta \sin(\omega t) + \sigma_0 \sin\delta \cos(\omega t),$$

which means that the associated vector has been decomposed into two vectors forming a phase angle of 90° with respect to each other. Relative to the strain vector,

$$\varepsilon = \varepsilon_0 \sin(\omega t),$$

the first term represents the in-phase component and the second term is the out-of-phase component. This equation can be written in 'complex' notation, that is,

$$\sigma = \sigma' \sin(\omega t) + \sigma'' \cos(\omega t),$$

where σ' corresponds to the stress component that is in-phase with the 'complex' strain,

$$\varepsilon* = \varepsilon_0 \sin(\omega t),$$

and σ'' corresponds to the stress component out-of-phase with respect to the strain.

78. Dynamic (Oscillatory) Flow

A drag flow that takes place in cone-and-plate and parallel-plate rheometers operated in an oscillatory mode. The results are obtained in terms of the complex viscosity η^* and the melt elasticity parameter G'. (See Viscosity.)

79. Dynamic Vulcanization

A procedure whereby one of the components of a blend is an elastomer that is vulcanized, or partially vulcanized, during the mixing process, while the other component remains non-cross-linked. This procedure is often used for the production of thermoplastic elastomers where the non-cross-linked component is a thermoplastic polymer. The procedure is also used as a means of improving the dispersion of fillers in the blend and also to restrict the location of the filler primarily to one phase. The phase contrast micrograph, taken on a dynamic vulcanized blend of natural rubber (NR) and an ethylene–propylene copolymer (ethylene–propylene rubber, EPR) shows that the phases are co-continuous.

Phase contrast micrograph of a dynamically vulcanized NR–EPR blend. Source: Tinker and Jones (1998).

E

1. Effective Modulus

The modulus value (E_{eff}) used for fracture analysis under plane strain conditions, that is,

$$E_{\text{eff}} = E/(1-\nu^2),$$

where E is Young's modulus and ν is Poisson's ratio.

2. Efficient Vulcanizing Cure (EV Cure)

An elastomer formulation containing a high accelerator-to-sulfur ratio in order to provide large amounts of monosulfide cross-links. Semi-efficient vulcanizing cure systems have an accelerator-to-sulfur ratio that is intermediate between EV cure systems and ordinary sulfur cure systems.

3. Ejection Mechanism

The assembly of a mould that provides a mechanism for the ejection of moulded parts from the cavities, consisting of ejector pins fitted onto the ejector plate. A simple example of the operation of ejector pins is shown in the diagram with reference to compression moulding of thermosetting polymers.

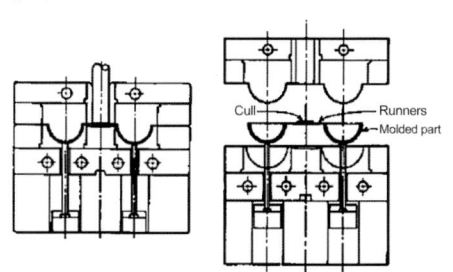

Pin ejection of moulded parts in transfer moulding operations. Source: (N. M. Bikales, Ed., Molding of plastics: Encyclopedia reprints, 1971, Wiley - Interscience).

For the removal of a thin moulding, where there is a risk that the pins could damage the moulded part, ejection is carried out with the use of a 'stripper plate'. The principle of operation of a stripper plate ejection is shown.

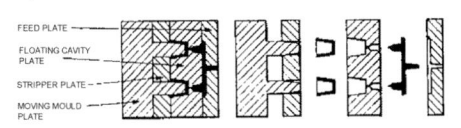

Stripper plate ejection mechanism. Source: (Unidentified original source).

4. Elastic Behaviour

Indicates that the Young's modulus and the shear modulus of a material are independent of both the loading history and the magnitude of the applied load. This means that the load–deformation relationship (see diagram) is linear and the gradient of the plot of the two variables is invariant irrespective of the duration of the applied load (force) or whether it acts in a static or dynamic manner.

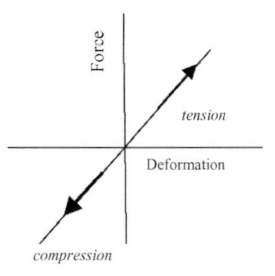

Force–deformation diagram for an elastic material.

5. Elastic Memory

A term that is sometimes used to describe the partial recovery of the deformation imposed on a polymer melt upon removal

Polymers in Industry from A–Z: A Concise Encyclopedia, First Edition. Leno Mascia.
© 2012 Wiley-VCH Verlag GmbH & Co. KGaA. Published 2012 by Wiley-VCH Verlag GmbH & Co. KGaA.

of the stress. This manifests itself in the form of swelling when a polymer melt emerges from the die of an extruder or capillary rheometer. (See Die swell ratio, Barus effect and Melt elasticity.)

6. **Elastic Recovery**

The instantaneous recovery component of a linear viscoelastic material. (See Viscoelasticity and Recovery.)

7. **Elasticity**

A characteristic of materials denoting the ability to recover an imposed deformation when the applied force or stress is removed. In theoretical mechanics, the term is used also to denote conditions in which the imposed strains are small and the deformations are directly proportional to the applied forces. (See Elastic behaviour and Rubber elasticity.)

8. **Elastomer**

A semi-scientific term for rubber, that is, polymers with a glass transition temperature (T_g) well below room temperature and containing physical or chemical cross-links. Accordingly, they are known as either thermoplastic elastomers or vulcanized elastomers, depending on whether the main component is a linear polymer with reinforcing crystalline domains (physical cross-links) or a polymer containing a small number of chemical cross-links.

9. **Electric Strength**

The value of the electric stress at which electrical failure occurs. (See Dielectric strength.)

10. **Electrical Failure**

Failure of a dielectric or insulating component via the formation of conductive channels either through the bulk or over the surface. (See Tracking.) For polymers that do not carbonize as a result of thermal degradation, failure can take place through the formation of holes resulting from localized depolymerization. A typical example is polytetrafluoroethylene (PTFE).

11. **Electrical Properties**

Properties describing the behaviour of a polymer under the influence of electrical stresses. Electrical stress is defined as the voltage gradient in the direction of the current, which is also known as the 'electric field'. (See Dielectric.)

12. **Electrolytic Deposition**

A method for depositing polymer coatings on highly conductive substrates from waterborne solutions, emulsions or dispersions. The outer surface layers of the polymer particles contain ions. On the application of a voltage, those ions anchored on the surface of the polymer particle, or attached to the polymer molecules in solution, migrate to one electrode, transporting the polymer, while the counter-ion migrates to the other electrode. The surface of the polymer may contain either positive ions (cations), usually protonated amine types, or negative ions, normally carboxylate anions. Accordingly, the deposition process is known as cationic electrolytic deposition (CED) or anionic electrolytic deposition (AED). In addition to the deposition of the ions, there are also electrolytic chemical reactions taking place at the respective electrodes.

- For CED:

  $2H_2O + 2e^- \rightarrow H_2 + 2OH^-$ (cathodic reaction)

  polymer–NR_2H^+ + $OH^- \rightarrow$ polymer–NR_2 + H_2O

- For AED:

  $2H_2O \rightarrow 4H^+ + O_2 + 4e^-$ (anodic reaction)

  polymer–COO^- + $H^+ \rightarrow$ polymer–COOH

A schematic diagram for the cathodic electrodeposition is shown. Note that the oxygen and hydrogen gases formed during the electrolysis play an important role in determining the thickness of the final coating. These gases produce a cellular structure in the deposited wet polymer coating, which contributes to building up a resistance to the deposition of additional polymer through the formation of an isolating layer for the electrode where the polymer is deposited.

13. Electromagnetic Radiation

Radiation that includes visible light, UV and IR, with wavelengths ranging from 200 to 760 nm. The amount of energy (E) carried by radiation is related to the frequency (ν) by Planck's law, that is, $E = h\nu$, where h is Planck's constant.

14. Electron Beaming

A method used for hardening thermosetting unsaturated oligomers, such as vinyl esters, and to produce cross-links in polymeric or elastomeric products obtained by extrusion, using high-energy electrons 'beamed' onto the product. Electron beaming is a low-temperature curing method, sometimes referred to as radiation processing. The bombarding of aliphatic polymers, or oligomers, with highly energetic electrons introduces very reactive free radicals in the molecular chains, which can interact with similar radicals from a different chain to form new covalent bonds and ultimately lead to the formation of cross-links.

The presence of unsaturated groups, as either functional groups or structural defects (cf. linear polyethylene), reduces the energy levels required to produce the free radicals, while the flexibility of the polymer chains provides the favourable conditions for the chain-to-chain interactions required for the formation of cross-links via radical

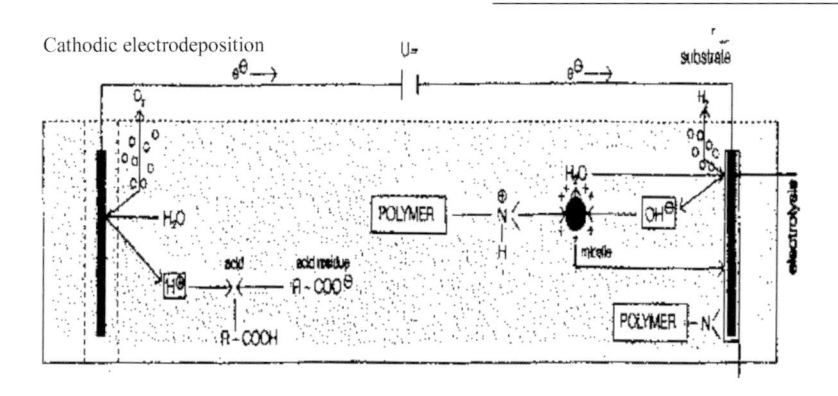

Schematic illustration of cathodic deposition.

recombination reactions. For the cross-linking of linear polymers, a co-agent (also known as 'pro-rad'), consisting of a multi-functional unsaturated monomer, such as triallyl isocyanurate, is invariably used to reduce the energy levels to promote the formation of free radicals and to speed up the rate of cross-linking.

Aromatic groups in the polymer chains, on the other hand, have a high capacity to dissipate the absorbed energy through electron delocalization. At the same time, the restriction that aromatic groups impose on molecular relaxations, owing to the rigidity of the polymer chains, provides obstacles for inter-chain interactions and prevents the formation of cross-links. The latter applies to all polymers with a high T_g, even if they are capable of producing free radicals by chain scission. Good examples are poly(methyl methacrylate) (PMMA) and polystyrene (PS). For polymers such as polypropylene (PP), despite the low T_g of the amorphous phase, the rate of chain scission reactions is always much faster than the rate of intermolecular interactions, so that they ultimately undergo degradation reactions with deterioration of mechanical properties.

15. Electron Microscopy

(See Scanning electron microscopy and Transmission electron microscopy.)

16. Electron Spectroscopy for Chemical Analysis (ESCA)

A technique that identifies the elements present on the surface of a sample, except for H and He. A sample is subjected to soft X-rays under high-vacuum conditions, and the emitted electrons are collected by an electrostatic energy analyser and detected as a function of the kinetic energy E_K, producing a spectrum. The intensity of the peaks is proportional to the number of atoms within the sampled volume, from which the surface composition can be deduced.

17. Electron Spin Resonance (ESR)

A spectroscopic technique that detects the transitions induced by electromagnetic radiation between the energy levels of electron spins in the presence of a static magnetic field. The method can be used to detect species containing one or more unpaired electrons, such as the free radicals that are generated by the scission of polymer chains.

18. Electronic Conductivity

Conductivity arising through the movement of electrons in the direction of the applied electrical stress.

19. Electrostatic Charge

Charges accumulated on the surface of a polymer as a result of contact with a non-conductive surface. Positive charges denote an electron donor characteristic of the surface, while negative charges indicates an electron acceptor behaviour. The electrostatic charging characteristics of polymers are related to their dielectric constant. Generally a polymer with a high dielectric constant acts as an electron donor (positive charging behaviour). Most thermoplastics show high charging tendencies owing to their high surface resistivity, which is above 10^{14} $\Omega/\square$, which prevents the acquired charges from leaking to earth. On the other hand, polymer formulations containing antistatic agents, or carbon black, have negligible charging characteristics in view of their relatively low surface resistivity, which is below 10^{10} $\Omega/\square$.

20. Electrostatic Fluidized Bed Coating

(See Powder coating and Electrostatic spraying.)

21. Electrostatic Spraying

A process by which polymer powder is charged electrostatically in its passage through the nozzle of a gun and, as a consequence, becomes attracted onto a metallic object connected to earth. When a specific coating thickness has built up, the excess powder falls off the coating through repulsion from the similarly charged outer particles of the coating. This is caused by the insulation created by the coating, which prevents them from being discharged to earth. After this stage the coated object is transferred to an oven for curing (thermosets) or sintering (thermoplastics).

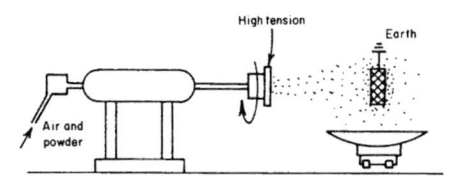

Schematic diagram of electrostatic spray coating. Source: Mascia (1989).

22. Ellis Equation

An equation that describes the variation of the viscosity (η) of a polymer melt with the imposed shear stress (τ), that is,

$$\eta/\eta_0 = 1 + (\tau/\tau_*)^{n-1},$$

where η_0, τ_* and n are empirical constants.

23. Elongational Flow

Known also as extensional flow, arises as a result of tensile stresses acting on a melt, as in the case of fibres or films drawn from the die by the take-off rolls. Elongational flow arises also in converging flow situations owing to the increase in velocity of the melt as it flows towards the exit. In this case the melt undergoes both shear flow, manifested as a velocity gradient through the cross-section of the channel, and elongational flow, resulting from the axial velocity gradient.

24. Elongational Viscosity

Defined as the coefficient that relates an elongational (tensile) stress to the resulting elongational rate (axial velocity gradient), that is,

$$\sigma = \eta_E (d\varepsilon/dt),$$

where σ is the elongational stress, η_E is the elongational viscosity and $d\varepsilon/dt$ is the elongation rate.

For ideal (Newtonian and incompressible) liquids, the elongational viscosity is three times greater than the shear viscosity. In polymer melts, the elongational viscosity is many times greater than the shear viscosity. The discrepancy diminishes with increasing temperature. An apparatus for measuring the elongational viscosity of polymer melts is shown in the section on melt strength. (see Melt strength.)

25. Empirical

An adjective usually used in relation to a test, methodology or property. An empirical property is one that is not derived from fundamental considerations and, therefore, it varies according to the test method or the geometry of the specimen used in the test. Typical empirical mechanical properties are scratch resistance, hardness, impact strength and abrasion resistance.

26. Emulsion

A sub-micrometre suspension of monomer droplets or polymer particles in a liquid, usually water. In the case of a monomer emulsion, the suspended droplets are stabilized against coalescence by the micelle formed by the surfactant, which entraps the monomer (hydrophobic component) in the centre layers of a multitude of vertically oriented surfactant molecules. The individual particles of an emulsion are very small (about 30–100 nm) but they tend to aggregate into larger microscopic secondary particles, which are responsible for the milky appearance, resulting from the scattering of the impinging visible light. The TEM micrograph, showing an OsO_4-stained diene elastomer emulsion, indicates that the size of discrete particles is in the region of 50–100 nm.

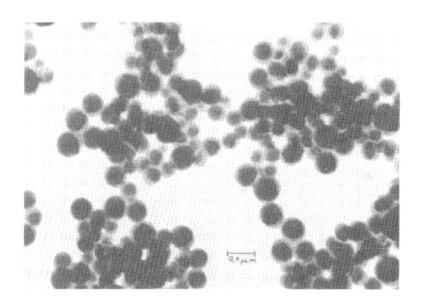

TEM micrograph of OsO_4-stained particles of a diene elastomer emulsion. Source: Kampf (1986).

27. Emulsion Polymerization

A very widely used polymerization method for vinyl, styrene and acrylic monomers. A liquid monomer is dispersed in water containing amounts of surfactants in excess of the critical concentration required to form micelles to entrap the monomer molecules. A water-soluble initiator, such as potassium persulfate or hydrogen peroxide, is used to initiate the polymerization within the micelles. The initiator is attracted into the micelles by the thermodynamic drive derived from the polymerization reaction, producing polymer particles swollen with monomer. As monomer micelles are converted into polymer particles, new monomer micelles are created at the expense of monomer droplets, because the initiation and growth of polymer chains occur preferentially in the monomer micelles. All the components present during emulsion polymerization are shown in the diagram.

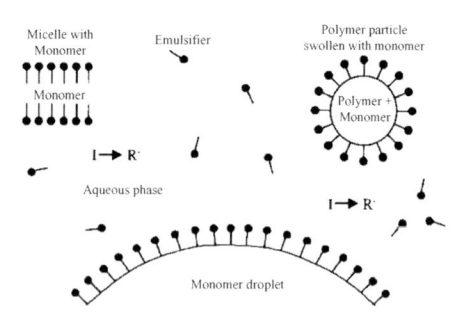

Species present during emulsion polymerization.

28. Encapsulation

A process by which a component is covered completely through a large inlet and entrapped by either a paste or a curing resin in a mould. The process is particularly used for the protection of delicate electronic devices.

29. Endothermic

An adjective for chemical or physical processes that absorb heat from the surrounding atmosphere.

30. Energy-Dispersive Scanning Analysis (EDS or EDX)

An analytical technique for detecting atoms on the surface of a specimen during examinations by scanning electron microscopy

(SEM). (See Scanning electron microscopy.) Characteristic X-rays have well-defined energies from different atoms. Thus analytical information can be obtained from an X-ray spectrum.

31. Engineering Design

A methodology that seeks to determine the dimensions and to optimize the geometry of an article from fundamental principles and/or through rational use of experimental data available for the candidate materials. The most widely used engineering design principle is the prediction of the deflection resulting from an applied load, or the determination of the maximum load that a structure can support. The general design equation is

$$\Delta = (P/E) \times \text{function}(\nu, \text{geometric parameters}),$$

where Δ is the deflection, P is the load, E is Young's modulus and ν is Poisson's ratio of the material. For instance, the deflection of a thin centrally loaded circular plate supported at the rim is calculated by the equation

$$\Delta = (Pr^2/Ed^3)[3(3+\nu)(1-n^2)/4\pi(1+\nu)],$$

where r is the radius and d is the thickness.

31.1 Estimation of Deflection

The above design equations indicate that, for elastic materials, it is necessary to know only two constants, E and ν, to be able to perform the design calculations. The nonlinear viscoelastic behaviour of polymers, coupled with their sensitivity to temperature changes, means that a large amount of data is needed to solve the equations for the wide range of possible conditions that the designed article is likely to encounter in service. While the variations of Poisson's ratio are not large enough to make a substantial difference to the calculated load–deflection values, this is not the case with respect to changes in Young's modulus. The nonlinear viscoelastic behaviour makes it necessary to know the value of the modulus for the particular loading history. (See Nonlinear viscoelastic behaviour.) Particularly important is to know whether the product is likely to be subjected to continuous or intermittent loads.

For the latter case it is also necessary to know the length of the recovery period relative to that of the application of the load (creep period). The plot of strain versus log (total creep time) at two different stress levels ($\sigma_2 > \sigma_1$) shows that the accumulated strain from creep periods is reduced substantially if there are intermittent long recovery periods (dashed curves). The lower dashed curves for a given stress level indicate that, the longer the recovery time relative to the creep time, the lower the accumulated strain. (See Creep curve.)

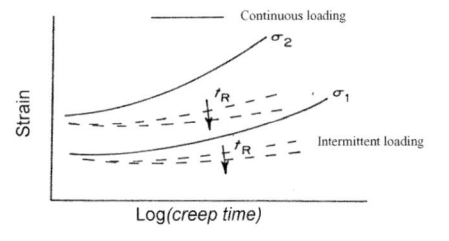

Schematic representation of the creep behaviour of polymers under continuous and intermittent loading conditions. Here, t_R is the ratio of recovery time/creep time, known as the reduced time.

The availability of these creep curves at a range of temperatures of interest makes it possible to determine the value of the creep modulus, E, as the reciprocal of the compliance, which is the ratio of accumulated strain to applied stress. Any discrepancy arising for the calculated creep modulus at different stress levels results from the nonlinear viscoelastic behaviour. Reasonable estimates can be obtained for different conditions through interpolations and extrapolation procedures.

31.2 Estimation of Maximum Allowable Load

The maximum load that a structure can tolerate is determined either by the stipulation of the maximum allowable deflection or by the strain level that would lead to failure through yield deformations or brittle fracture. The maximum deflection can be determined by external factors or materials limitations, such as the occurrence of crazing phenomena. In this case, knowing the values of critical crazing strain for the particular service conditions, the corresponding values for the deflection of the article can be estimated and set as the maximum deflection. (See Crazing.) The load that would bring about failure by yielding, on the other hand, can be determined from available data on strength as a function of loading time at different temperatures. (See Creep rupture.)

32. Engineering Polymer

A polymer that has characteristics suitable for the production of components expected to be subjected to high stress levels. An engineering polymer generally has a high modulus and high strength, as well as a high resistance to wear, and is capable of tolerating changes in temperature without loss of performance. Typical engineering polymers are polycarbonate, polyamides, acetals and aromatic polymers in general.

Examples of components made from engineering polymer. Source: ICI (1965).

33. Enthalpy

The heat component (H) of the total internal energy (E) of a system, that is,

$$E = H + pV,$$

where p is the pressure acting on the system and V is the volume. Very often the change in the pV term during a process, particularly those involving solids or liquids, is very small, so that the change in internal energy becomes equal to the change in enthalpy (heat content) of the system.

34. Entropy

A theoretical concept related to the degree of disorder in a system. It is used in the second law of thermodynamics to determine the conditions that have to be satisfied for the occurrence of chemical and physical phenomena, written as

$$\Delta G = \Delta E - T\Delta S,$$

where ΔG is the change in Gibbs free energy, ΔE is the change in internal energy, T is the absolute temperature (K) and ΔS is the change in entropy; ΔE is equal to the change in enthalpy (ΔH) if the pressure is low and there is no change in volume. (See Enthalpy.) Since an event will occur only if it leads to a decrease in free energy, the conditions are satisfied if there is a decrease in enthalpy, corresponding to an exothermic process, and an increase in entropy through a higher level of disorder.

Note that polymerization is a favourable process despite the reduction in entropy that results from the much higher order, and lower degrees of freedom, of the monomeric units in a polymer chain relative to the free monomer. The favourable thermodynamic conditions, therefore, arise from the dominance of the decrease (negative change) in enthalpy due to the exothermic nature of the polymerization reactions. The melting of polymer crystals, on the other hand, is favoured by the large increase in entropy resulting from the high disorder acquired by the polymer chains in its liquid (melt) state, despite the endothermic nature of the melting process, which requires heat (an increase in enthalpy) to overcome the intermolecular forces holding the chains within the lamella of the crystals. In thermodynamics terms, therefore, melting takes place because the term $T\Delta S$ is larger than ΔH. Note, in any case, that 'entropy' is a purely theoretical parameter and cannot be measured experimentally.

where X = CN, Cl, COOEt, etc.

Cross-linking and chain scission resulting from UV degradation of vinyl polymers.

35. Entry Effect

An expression used to describe the pressure drop occurring in the flow of a polymer melt at the entry of a die. This may be the cause of errors in the calculation of the shear stresses at the walls of a die. For this reason, a specific procedure has to be used to estimate the pressure drop resulting from 'entry effects'. (See Bagley correction.)

Oxygen-assisted UV degradation of polystyrene.

36. Environmental Ageing

A term that refers to the deterioration of the properties of a product resulting from the action of environmental agents, such as temperature, oxygen, UV radiation, moisture and pollutants in the atmosphere. Polyolefins and diene elastomers are particularly prone to degradation reactions caused by the combined action of oxygen and UV light. Condensation polymers, such as polyesters, polyamides and polyurethanes, are also prone to degradation reactions caused by hydrolysis, which is accelerated by acidic or basic environments. Two brief schemes of degradation reactions induced by environmental ageing are shown.

37. Environmental Stress Cracking

A phenomenon related to fracture induced by environmental agents, which is entirely due to physical factors. The term was originally used to describe the fracture of polyethylene bottles induced by detergents. It is now more generally used to denote the adverse effect of some environmental agents on the fracture toughness of polymers. The standard test, known also as the Bell test, was originally devised specifically to assess the environmental stress cracking resistance (ESCR) of low-density polyethylene. ESCR is an empirical test that can give misleading results, particularly when used to test the more rigid grades of polyethylene. In the latter cases, the bending of the specimens into an arch produces very large strains in the outer layers, which can easily exceed the levels that bring about 'yielding' (left-hand diagram). This will reduce the elastic (stored) energy in the notched region of the specimen. Another oddity of the Bell test in terms of fundamental principles is the insertion of a central cut (notch) along the bending direction, which creates a very complex state of stresses that cannot be rationalized in a manner that would allow for variations in the geometry of the specimen. A more rational approach has been used in devising a test at low strains with notches across the width of the specimen, which can be carried out with a simple jig (right-hand diagram).

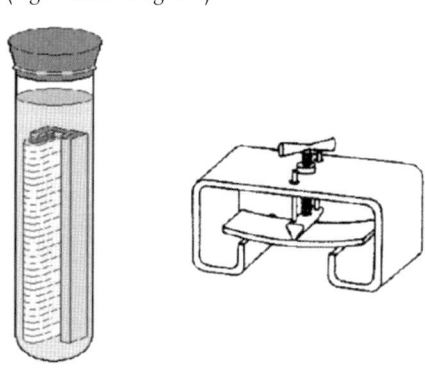

Typical methods for measuring ESCR: (left) high-strain standard test; (right) low-strain test for rigid polymers.

Fundamentally, ESCR has to be evaluated in terms of the decrease in the value of fracture toughness parameters of a polymer, K_c or G_c, brought about by a particular environment. (See Fracture mechanics.) For glassy linear polymers, stress cracking is aggravated by the presence of solvents in the atmosphere, which can bring about a reduction in the strain required to induce crazing and also accelerate the conversion of crazes into cracks. (See Crazing.) For glassy cross-linked polymers, solvents can also have a detrimental effect on the fracture resistance, even though the phenomenon will manifest itself without the visible formation of crazes. Environmental stress cracking phenomena observed in glassy polymers may also be connected to localized antiplasticization effects caused by the solvent, which may be considered as the underlying cause of the decrease in the critical strain for crazing, or the general reduction in fracture toughness due to localized restrictions on molecular relaxations at the crack tip.

38. Epichlorohydrin Rubber (Epichlorohydrin–Ethylene Oxide Copolymer, ECO)

Consists essentially of epichlorohydrin and ethylene oxide copolymers, having the repeating units:

$$\begin{array}{c} -CH_2-CH-O-CH_2-CH_2-O- \\ CH_2Cl \end{array}$$

Cross-linking is induced via reactions with some of the pendent $-CH_2Cl$ groups, using thiourea and mixtures with metal oxide acid acceptors, such as MgO and Pb_3O_4. Some terpolymers (epichlorohydrin–ethylene oxide–diene terpolymer, ETER) containing small amounts of diene units are also available, which can be vulcanized with sulfur-based curatives and peroxides. The polyether constituent units confer a high polarity to the polymer chains, which provides a high resistance to oils. The random distribution of the components prevents the formation of crystalline structure, either thermally or stress-induced. The T_g value is in the region of -30 to $-40\,^\circ C$, depending on the ratio of the two main structural components.

39. Epikote

Tradename for a variety of epoxy resins, known also as Epon.

40. Epoxidation

A chemical reaction that introduces epoxy groups into a polymer or oligomer. An example of epoxidation is the conversion of unsaturated groups in a diene elastomer to epoxy groups through controlled oxidation reactions.

41. Epoxide

An synonym for epoxy resins.

42. Epoxidized Soya Bean Oil

An epoxidized oil used as a reactive internal plasticizer for epoxy resin formulations and as a co-stabilizer for PVC.

43. Epoxy Equivalent (or Epoxide Equivalent)

The weight of resin in grams that contains one gram equivalent (mole) of epoxy groups. For an unbranched epoxy resin, this corresponds to half the number-average molecular weight.

44. Epoxy Group

Also known as an oxirane ring, corresponds to a strained ether ring between either two adjacent CH groups or consecutive CH and CH_2 groups. (See Epoxy resin.)

45. Epoxy Resin

A thermosetting resin containing two or more epoxy groups, respectively referred to as bifunctional or multifunctional epoxy resins. Monofunctional epoxy compounds are also available but are used only as reactive diluents for conventional epoxy resins in order to reduce the cross-linking density of the cured products. This has the effect of reducing the glass transition temperature (T_g), which also results in an increase in the ductility of the final product. A typical monofunctional epoxy diluent (PGE) is shown.

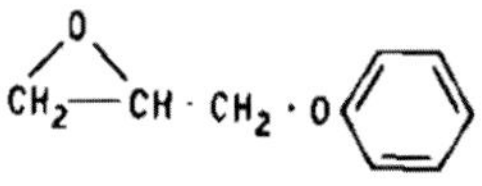

Phenyl glycidyl ether (PGE).

The most common type of epoxy resin is the diglycidyl ether of bisphenol A (DGEBA), represented by:

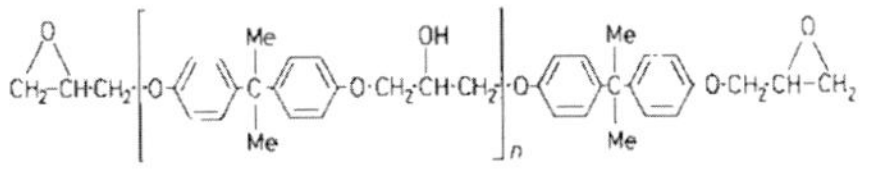

where the value of n varies from about 0.2 for liquid resins to values between 5 and about 20 for solid resins. Better thermal oxidative stability can be achieved by replacing the propylidene group, CH_3-C-CH_3, of the bisphenol A unit with the CF_3-C-CF_3 group contained in bisphenol F.

Among the multifunctional resins, the most widely known are the epoxidized novolacs and tetraglycidoxylmethylene p,p'-diphenylene diamine (TGDM), which are both shown.

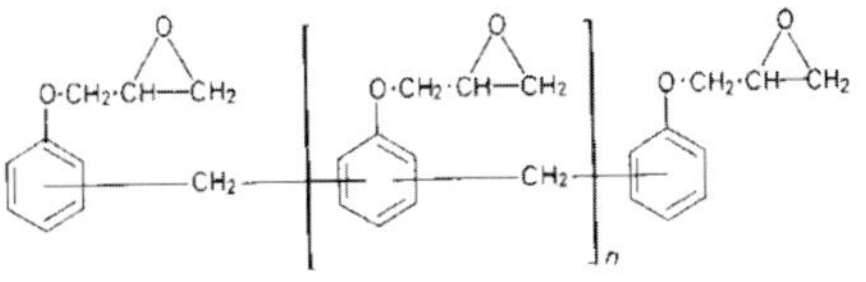

Structure a typical epoxidized novolac resin.

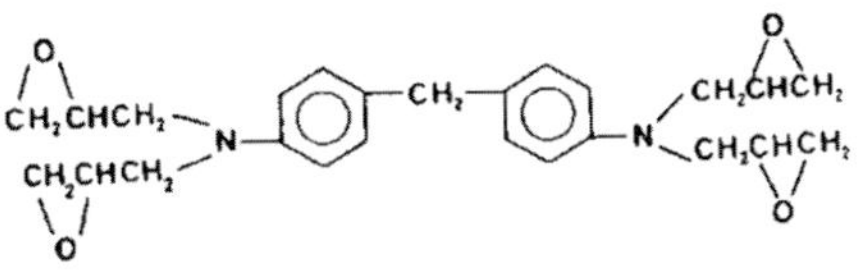

Structure of TGDM.

Multifunctional epoxy resins give a higher cross-linking density than conventional bifunctional resins. The use of an aromatic hardener as a curing agent produces networks with a high glass transition temperature.

Other types of epoxy resins are the aliphatic and cycloaliphatic resins. These structures give cured products with rubbery or high flexibility characteristics. The structures of one typical aliphatic and two typical cycloaliphatic epoxy resins are shown.

Typical aliphatic epoxy resin produced from poly(propylene glycol).

Typical cycloaliphatic epoxy resin: vinyl cyclohex-3-ene.

Typical cycloaliphatic epoxy resin: 3,4-epoxycyclohexylmethyl-3,4- epoxycyclohexane carboxylate.

These cycloaliphatic resins are widely used for high-voltage electrical applications owing to their high tracking resistance in comparison to aromatic-type resins. The absence of aromatic rings reduces the possibility of forming polynuclear graphitic structures, which are responsible for the formation of conductive 'tracks'. (See Tracking.)

45.1 Hardener for Epoxy Resin

The main types of hardener can be divided into anhydrides and multifunctional amines. Other hardeners are used for very specific reasons.

45.2 Anhydride

These hardeners exert a 'curing' function by reacting with epoxy resins first through ring opening of the epoxy groups either via the reaction of carboxylic acid groups (present as impurities) and/or via the reaction with carboxylate ions (formed by the interaction of the anhydride with a tertiary amine used as catalyst). The opening of the epoxy ring is followed by esterification reactions between the anhydride groups and the newly formed hydroxyl groups, which is also catalysed by the tertiary amine catalyst (see diagram).

Reaction scheme for the ring opening of the anhydride groups by tertiary amines and subsequent reaction with epoxy groups.

The most widely used anhydride hardeners are shown. Note that aromatic anhydrides are not normally used as sole hardeners owing to their low reactivity with the resin.

Typical anhydride curing agents: (left) tetrahydrophthalic anhydride; (right) methylnadic anhydride.

Dodecylsuccinic anhydride (DDSA) is sometimes used when a higher flexibility is required. This results from the internal plasticization effect caused by the dodecyl groups.

$$Me \cdot (CH_2)_2 \cdot CHMe \cdot CH_2 \cdot CMe : CH \cdot CMe_2 \cdot CH \cdot CO$$
$$|\rangle O$$
$$CH_2CO$$

Dodecylsuccinic anhydride.

45.3 Amine Hardener

These hardeners react directly with the epoxy groups but do not react with the hydroxyl groups. So while amines react much faster than anhydrides, the resulting cross-linking density is lower. The reaction scheme of primary and secondary amines with epoxy groups is shown.

$$R \cdot NH_2 + CH_2\text{—}CH\text{—} \longrightarrow R \cdot NH \cdot CH_2 \cdot CH\text{—}$$
(with O bridge on left; OH on right)

$$R \cdot NH \cdot CH_2 \cdot CH\text{—} + CH_2\text{—}CH\text{—} \longrightarrow R \cdot N \begin{matrix} CH_2 \cdot CH\text{—} \\ CH_2 \cdot CH\text{—} \\ OH \end{matrix}$$

Reaction scheme for the curing of epoxy resins with amine hardeners.

45.4 Aliphatic Amine Hardener

The more common types are diethylenetriamine (DETA), H_2N–CH_2CH_2–NH–CH_2CH_2–NH_2, and triethylenetetramine (TETA), H_2N–CH_2CH_2–NH–CH_2CH_2–NH–CH_2–NH_2, used primarily for cold-curing adhesives. Another type that could enter this classification is dicyanodiamide (DICY), which exists in three tautomeric forms, as shown.

DICY is used in quantities much smaller than the amounts expected from cross-linking reactions alone, which suggests that the reactions take place primarily through homopolymerization, involving ring opening of the epoxy groups to produce long sequences of linear polyethers, also known as phenoxy structures.

45.5 Cycloaliphatic Amine Hardener

These are very reactive and therefore are suitable for cold-curing systems. The cyclic structure confers a higher rigidity to the network than linear aliphatic hardeners, thereby imparting a higher T_g value for the cured products than those obtained with the use of DETA or TETA. A widely used cycloaliphatic amine hardener is 4,4'-methylene-bis-cyclohexanamine (PACM), whose structure is shown.

$$H_2N\text{—}\bigcirc\text{—}CH_2\text{—}\bigcirc\text{—}NH_2$$

Structure of PACM.

45.6 Aromatic Amine

These are used to obtain cured products with high T_g values, resulting from the rigidity of the aromatic rings in the network. These are slow curing systems and therefore require high-temperature conditions for the curing reactions to go to completion. Typical aromatic amine hardeners are m-phenylenediamine (MPD) and 4,4'-diaminodiphenylsulfone (DDS). The structures of these hardeners are shown.

$$H_2N \cdot C : N \cdot CN \rightleftharpoons H_2N \cdot C \cdot NH \cdot CN \rightleftharpoons H_2N \cdot \overset{\oplus}{C} \cdot NH : C : \overset{\ominus}{N}$$
$$| \| \|$$
$$NH_2 NH NH$$
$$SOLID IN\ SOLUTION$$

Tautomeric forms of DICY.

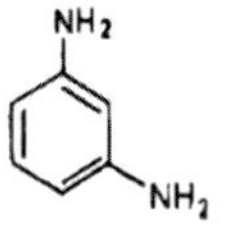

MPD

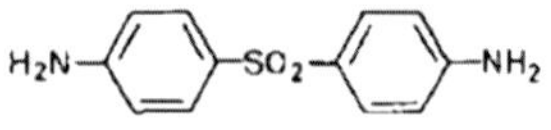

DDS

Note that MPD is often used in the form of eutectic mixtures in order to reduce the melting point so that it can be more easily handled in manufacturing. DDS, on the other hand, is widely used for curing multifunctional epoxy resins in matrices for carbon fibre composites for aerospace applications, owing to the very high T_g values achievable as a result of the chain stiffening effect of the $-SO_2-$ groups.

45.7 Ketimine

Chemically, ketimines can be regarded as 'blocked' amines. Since the C=N bond is readily hydrolysed to the original ketone and amine, they can be regarded as 'latent' curing agents. These are useful for one-pot resin systems in view of the long pot life achievable, resulting from the delay of the curing reactions, which can only take place after the hydrolysis step required to generate the amine. A typical ketimine, as a 'blocked amine hardener, is shown.

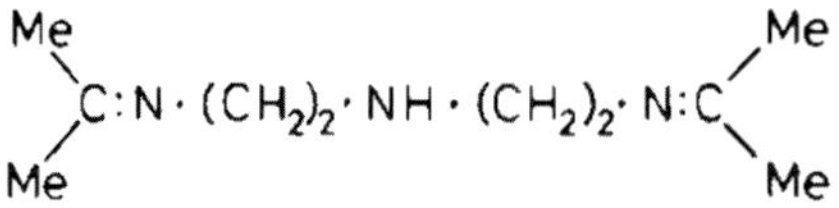

45.8 Polyamide Hardener

These are amines based on polyamides, produced from the reaction of dimerized or trimerized fatty acids from vegetable oils with polyamines. For the example shown, the reaction takes place by a Diels–Alder mechanism between 9,12- and 9,11-linoleic acids.

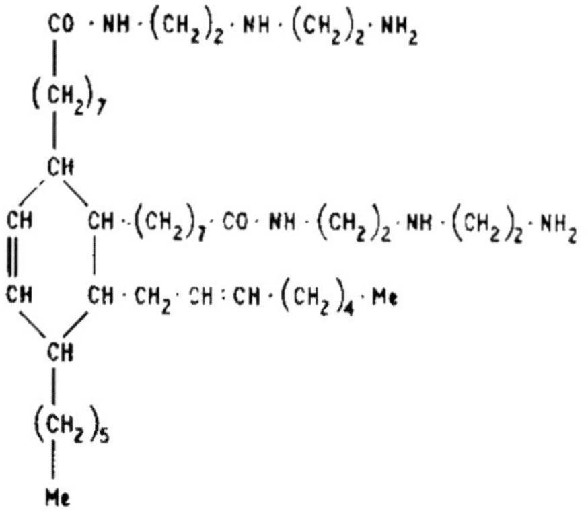

Typical amine-functionalized polyamide.

Although they can act as curing agents in their own right, polyamide hardeners are to be regarded primarily as flexibilizers for use in mixtures with other amine hardeners.

45.9 Catalytic Curing Agent

These consist of Lewis bases and Lewis acids. Lewis base curatives for epoxy resins include tertiary amines, such as o-(dimethylaminomethyl)phenol and tris-(dimethylaminomethyl)phenol and its salt, as well as imidazole compounds. Although these are referred to as 'catalysts', they are true hardeners insofar as they react with the resin and remain part of the structure of the cured products. The reaction mechanism for the curing of epoxy resins with Lewis bases is illustrated.

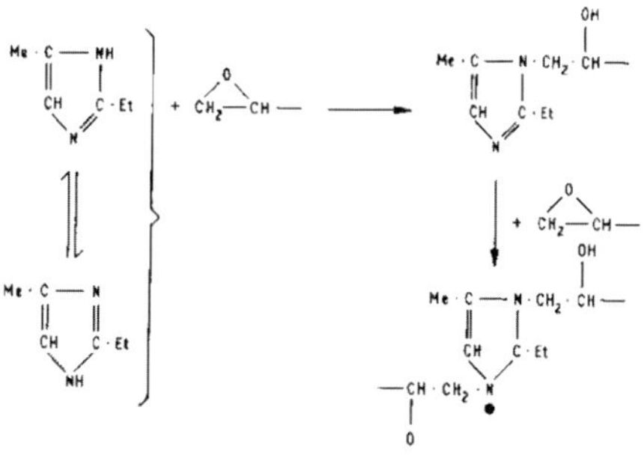

Mechanism of curing reactions with imidazoles.

Lewis acid catalysts function as curing agents by forming coordinated structures with the oxygen atom in the epoxy group, which facilitates the transfer of a proton for the opening of the epoxy ring. These curatives are usually complexes of boron trifluoride with amines, such as monomethylamines. Owing to their extremely high reactivity, they are often used to speed up the curing rate in conjunction with anhydride hardeners.

45.10 Polysulfide

These are end-terminated thiol oligomers used particularly as flexibilizing curing agents. A typical polysulfide curing agent is shown. Curing takes place through the epoxy ring opening reaction by mercaptan $-SH$ groups on the end of chains, as shown.

$$HS\cdot[(CH_2)_2\cdot O\cdot CH_2\cdot O\cdot (CH_2)_2\cdot S\cdot S]_n\cdot (CH_2)_2\cdot O\cdot CH_2\cdot O\cdot (CH_2)_2\cdot SH$$

Typical polysulfide, where $n = 2-26$.

$$R\cdot SH + CH_2\overset{O}{-}CH- \longrightarrow R\cdot S\cdot CH_2\overset{OH}{CH}-$$

Curing by a mercaptan group.

45.11 Reactive Toughening Modifier

These are low-molecular-weight polymers with terminal functionalities consisting of $-COOH$ or $-NH_2$ groups, often referred to as 'liquid rubbers' to indicate that they are in the liquid state, which makes them easy to dissolve in an epoxy resin. The most widely used systems are random copolymers of butadiene and acrylonitrile, with molecular weight in the region of 2000–10 000, known as carboxyl-terminated butadiene–acrylonitrile oligomer (CTBN) for carboxylic acid functionalized systems and as amine-terminated butadiene–acrylonitrile oligomer (ATBN) for systems with amine functionality. The acrylonitrile content is in the region of 25–40% for both systems. The procedure that has to be used to induce the precipitation of rubbery particles during curing is to react the functionalized modifier with a large excess of epoxy resin in the presence of a selective catalyst, such as triphenylphosphine (TPP), before the hardener is added for the subsequent curing stage. This reaction produces telechelic extension to the molecular chains of the modifier, as shown in the scheme.

Reaction scheme for the telechelic extension of a liquid elastomer modifier. Source: Riew (1989).

Many other systems have been studied. The most interesting among these are the telechelic extended polydimethylsiloxanes and perfluoroether oligomers, owing to their very thigh thermal and UV stability, which would overcome the deficiency of diene-based liquid rubbers. These ABA-type oligomers are very immiscible and therefore have to be telechelically extended by reaction with functionalized miscible segments in order to acquire the necessary 'compatibility', evidenced by the formation of a transparent epoxy resin mixture prior to curing. Since the central segments of theses ABA oligomers are not molecularly miscible with the epoxy resin, they provide ready-made nuclei for the precipitation of the rubbery particles, which can take place at much lower levels of addition than for CTBN or ATBN systems. These systems, however, are not available commercially. The multicomponent morphological structure observed in precipitated rubbery particles in a DGEBA resin containing about 4–5% of a hydroxyl-terminated perfluoroether, cured with methylnadic anhydride (MNA) is shown.

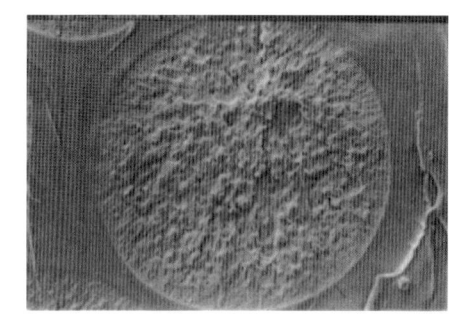

Structure of 'soft' particles derived from a telechelically extended perfluoroether oligomer in an epoxy matrix. Source: Zitouni (1994).

Note that the morphology of the precipitated particles of these telechelic perfluoroether systems resembles the structure obtained through phase inversion of acrylonitrile–butadiene–styrene (ABS) and high-impact polystyrene (HIPS) polymers produced from mass polymerization, exhibiting also the characteristic broad distribution of particle size. (See Phase inversion.)

45.12 Epoxy–Silica Hybrid

These are epoxy resins containing fine three-dimensional domains of organo-modified silica produced *in situ* via the sol–gel process. These can be produced by functionalizing the epoxy resin with a trialkoxysilane to about the 10% level, as shown, and then mixing with a siloxane precursor, containing tetraethoxysilane (TEOS).

Reaction of an amine bis(propyltrimethoxysilane) with an epoxy resin.

Small amounts of water and γ-glycidoxylpropyltrimethoxysilane are added to the TEOS precursor, respectively, to induce hydrolysis of the ethoxysilane groups and to compatibilize the organic and inorganic domains. The latter is assisted by the use of a rapidly reacting amine hardener, such as PACM, for the curing of the epoxy resin domains. (See Organic–inorganic hybrid.)

46. Ester–Amide Polymer

A class of biodegradable polymers produced from condensation reactions of 1,4-butanediol, adipic acid and hexamethylenediamine, with varying ester/amide ratios. These have a melting point in the range of 120–180 °C and a T_g varying from −40 to −5 °C. Accordingly, the mechanical properties are related to the degree of crystallinity, which is determined by the ester/amide ratio. In general, they are quite flexible, intermediate between low-density polyethylene (LDPE) and medium-density polyethylene (MDPE).

47. Esterification

Reaction resulting in the formation of esters, usually from reaction between a carboxylic acid, a glycol and a hydroxyl-containing oligomer.

48. Ethenoid Polymer

Polymers derived from alkylene monomers, such as ethylene, propylene, butylene and isobutylene. These polymers are also known as polyolefins. Their main feature is the hydrocarbon nature of the polymer chains, which confers to them a highly

49. Ethylene Polymer

Polymers with ethylene units as the main components of the polymer chains. They are either homopolymers or copolymers of ethylene and small amounts of a higher olefin (butene, hexene or dodecene), which produce short branches along the backbone chains. These branches are introduced as a means of controlling the density through variations in the degree of crystallinity. Accordingly, these polymers are divided into low-density polyethylene (LDPE), known also as high-pressure polyethylene (HP-PE), very low-density linear polyethylene (VLLDPE), linear low-density polyethylene (LLDPE), medium-density polyethylene (MDPE), high-density polyethylene (HDPE) and ultra-high-molecular-weight polyethylene (UHMWPE). The first polyethylene that was made available commercially was LDPE produced by a high-pressure method, containing irregular long branches attached on the main chains. The LLDPE, MDPE and HDPE types are produced by Ziegler–Natta or 'metallocene' polymerization methods. HDPE can also be produced by the Phillips process. The typical characteristics of polyethylenes are summarized in the table.

Polyethylene type	Average density (g/cm^3)	Degree of crystallinity (%)	Melting point (°C)
LDPE	0.925	>30	110
VLLDPE	0.915	>30	115
LLDPE	0.925	30–45	125
MDPE	0.9330	45–55	128
HDPE	0.945	55–70	138
UHMWPE	0.940	55–65	137

Note that, while the melting point varies considerably with the degree of crystallinity for the various types of polyethylene, the T_g value remains fairly constant at around $-100\,°C$. The increase in density, resulting from the higher degree of crystallinity, also brings about an increase in modulus (rigidity) and yield strength, as well as in increase in opacity of the products. The wide range of types of polyethylene available commercially is complemented by an even wider range of different grades varying in molecular weight and molecular-weight distribution. The number-average molecular weight can vary by a factor of 100 across the whole range, from as low as 30 000 for some easy-flow injection moulding grades to 3 000 000. The different grades are identified by the density (degree of crystallinity), melt flow index (MFI, an empirical parameter related to the average molecular weight) and sometimes also melt flow ratio (i.e. the ratio of two MFI values obtained with loads of 5.00 and 2.16 kg, respectively). The melt flow ratio is related to the molecular-weight distribution. (See Melt flow index.)

The increase in density, from LDPE to HDPE, brings about large changes in properties such as modulus, strength, solvent resistance and barrier properties. Increasing molecular weight results in improvements in environmental stress cracking resistance and toughness at low temperatures. One of the attractive properties of UHMWPE, for instance, is the extremely high toughness at cryogenic temperatures (the highest of all known materials), and the generally high wear resistance at temperatures up to about 60–70 °C, owing to a combination of low coefficient of friction and toughness. The grades with the highest molecular weight, that is, above 1 000 000, cannot be processed by the conventional methods used for thermoplastics in view of the very high melt viscosity.

One notable difference between the original LDPE and the relatively new LLDPE is with respect to the rheological properties, which require some adjustments in the processing conditions and equipment features in switching from one type to another. LLDPE has a viscosity with a lower shear-rate sensitivity than LDPE (higher power-law index). On the other hand, the elongational viscosity of LLPE increases to a lesser extent with increasing extension ratio than LDPE. Both aspects are relevant for the production of films, where both types of polyethylene are extensively used. Frequently, different grades of polyethylene, particularly LDPE and LLDPE, are blended to tailor the properties to specific end-use (e.g. better tear strength) or process requirements (e.g. improved melt drawdown characteristics). All types and grades of polyethylene are characterized by excellent dielectric properties, due to the complete absence of polar groups or polarizable structural features. The additives used in grades for cable applications have to be carefully chosen to avoid deterioration in dielectric losses even when used at very low levels of addition. Furthermore, the lack of polar groups in the polymer chains confers an extremely high resistance to water absorption for all grades of polyethylene, along with other polyolefins.

49.1 Ethylene–Carbon Monoxide Copolymer (ECO)

These usually contain less than 5% CO. They have been produced for applications requiring rapid degradation reactions when exposed to UV radiation. These can be utilized for the production of biodegradable packaging films. (See Biopolymers.)

49.2 Ethylene–Vinyl Alcohol Copolymer (EVOH)

These copolymers are obtained by the hydrolysis of ethylene–vinyl acetate (EVA) copolymers. Widely used in packaging films as a barrier layer and/or as binding layer for sandwich constructions of polar and non-polar polymers, such as polyethylene and a polyamide. The ethylene content of commercial EVOH polymers is usually within the range 30–50 mol%. The melting point decreases with increasing ethylene content within the region of about 165–185 °C.

49.3 Ethylene–Ethyl Acrylate Copolymer (EEA)

Similar characteristics to EVA but with better thermal oxidative stability. (See later subsection in this entry on ethylene–vinyl acetate copolymers.)

49.4 Ethylene–Ethyl Acrylate Terpolymer Elastomer (EAM)

Copolymers of ethylene containing large quantities of an acrylate comonomer and a small amount of an acidic monomer, such as acrylic acid, as a means of introducing reactive sites for cross-linking with the addition of a diamine. (See Ionomer.)

49.5 Ethylene–Methyl Methacrylate Copolymer (EMA)

Very stable copolymer widely used to produce master batches and for coatings by extrusion due to its capability of resisting very high processing temperatures.

49.6 Ethylene–Propylene Rubber (EPR)

Random copolymers with very low level of crystallinity and rubber-like characteristics. The chemical structure is shown, where the molar ratio of ethylene to propylene is around 1–1.5 : 1.

$$\left[CH_2\ CH_2\ CH_2\ \underset{\underset{CH_3}{|}}{CH} \right]_n$$

Chemical structure of an EPR rubber.

The absence of unsaturated groups in the molecular chains prevents EPR from being cross-linked by conventional sulfur vulcanization methods, requiring the use of peroxide curative systems. Cross-linking takes place via the formation of free radicals resulting from abstraction of the hydrogen atom of the *tert* CH group.

49.7 Ethylene–Propylene–Diene Monomer Rubber (EPDM)

Produced from a terpolymer consisting primarily of ethylene and propylene units with minor amounts of a diene monomer to introduce unsaturation in the chains, so that it can be vulcanized by conventional sulfur-based rubber curatives as well as by peroxide initiators. The diene monomer is usually either hexane-1,4-diene (type I), dicyclopentadiene (type II) or 4-ethylidenenorbornene (type III). The schematic chemical structure of these diene monomeric units in an EPDM chain is shown.

Various types of unsaturated groups along EPDM chains for cross-links by a free radical mechanism.

49.8 Ethylene–Vinyl Acetate Copolymer (EVA)

Random copolymers containing 10–30% vinyl acetate (VA) units along with the predominant ethylene units. Designed to reduce the level of crystallinity and to increase the polarity in the polymer chains, thereby producing a more rubbery material with a lower melting point, as well as exhibiting a higher resistance to hydrocarbon solvents. At the upper end of the range of VA content, these copolymers are very rubbery, comparable to plasticized PVC. Grades with higher levels of VA are also available and are used primarily as adhesives and coatings, due to the higher affinity for polar substrates, through hydrogen-bonding interactions, making them particularly suitable for the production of hot-melt adhesives. They can also be readily cross-linked with peroxides and high-energy radiation, making them suitable for low-voltage cable insulation. Copolymers with very high VA contents, up to about 90%, are used for the manufacture of emulsion paints and soft adhesives. (See Vinyl polymer.)

50. Excited-State Quencher

An additive that interacts with photo-excited polymer molecules and brings them back to their ground state by dissipating the acquired energy through harmless infrared radiation. These are typically nickel complexes, normally used as co-stabilizers for UV protection. (See UV stabilizer.)

51. Exfoliated Nanocomposite

Composites formed when the silicate nanolayers of a clay are individually dispersed within a polymer or resin matrix through two consecutive processes, known as intercalation and exfoliation. The original layered silica clay contains aggregates of primary layers, also known as tactoids. The layer thickness is around 1 nm, while the lateral dimensions of the platelets can vary from a 30 nm to several micrometres. The tactoids are characterized by moderate surface charges, known as the cation exchange capacity (CEC) expressed as mequiv/100 g (mmol/100 g). The structure of a typical interlayered nanoclay is shown.

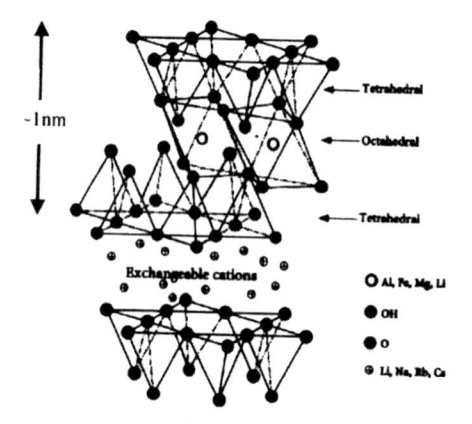

A cationic surfactant is used to exchange the alkali or alkaline-earth cations on the surface of individual silicate layers with quaternary ammonium cations of the surfactant, which provides the driving force for the intercalation of the silicate layers with the long aliphatic chains of the ammonium cation. After mixing with a resin or a polymer, the intercalated layers are separated and become dispersed in the matrix to produce a nanocomposite. The various steps and structures of the nanofiller are illustrated in the diagram.

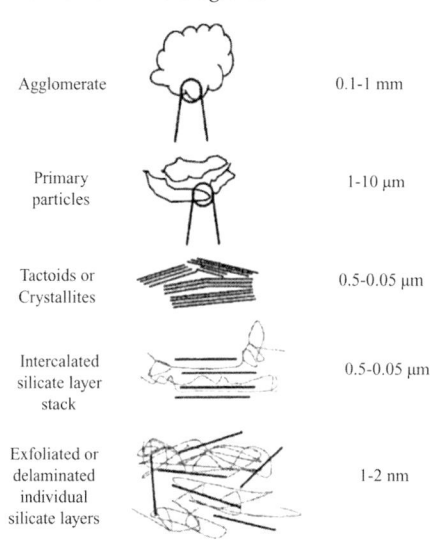

The micrographs show an example of the morphology of exfoliated nanoclays in a polymer matrix nanocomposite.

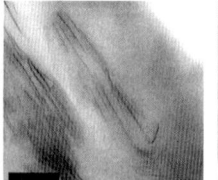

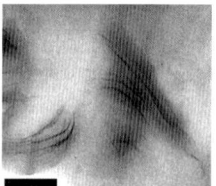

Bright-field TEM micrographs of 6 wt% Cloisite 15A* dispersed in polyamide 6 by intermeshing twin-screw extrusion [*montmorillonite intercalated with dimethyl-dihydrogenated tallow–quaternary ammonium cations (125 mequiv/100 g clay), Southern Clay Product]. Source: Courtesy of D. Acierno, University of Naples, Italy.

52. Exothermic

An adjective for a chemical or physical process that gives out heat to the surroundings, as a result of the temperature rise caused by events of the process.

53. Expanded Polystyrene

Foams usually produced from physically blown polystyrene beads.

54. Extender

A liquid containing both aliphatic and aromatic components, often referred to as 'oil', used in rubbers to improve the processing characteristics, as well as the dispersion of additives and reinforcing filler. At the same time, an extender acts also as a plasticizer and/or diluent insofar as it decreases both the T_g and the modulus of the cured elastomer.

55. Extension Ratio

Known also as the draw ratio, corresponds to the ratio of the final length of a specimen divided by the initial length, after being stretched. (See Draw ratio.)

56. Extensional Flow

Also known as elongational flow, represents the flow arising from the action of tensile stresses. It can take place as a unidirectional or planar flow depending on whether the tensile stresses are monoaxial or biaxial. (See Elongational flow.)

57. Extensional Viscosity

(See Elongational viscosity.)

58. Extensometer

A device that it is attached to a tensile specimen to record the localized extension during a tensile test.

59. External Lubricant

Additive used to reduce the friction and adhesion of a polymer melt with the surface of the processing equipment. (See Lubricant.)

60. External Plasticization

The lowering of the glass transition temperature of a glassy polymer by the addition of a plasticizer. (See Plasticization and Internal plasticization.)

61. Extinction Coefficient

The constant, ε, in the Beer–Lambert law, which relates the intensity of absorbed light (I) to the concentration of absorbing species (c), that is,

$$\log(I/I_0) = \varepsilon Z c,$$

where I_0 is the intensity of the incident light and Z is the thickness of the medium receiving the light.

62. Extinction Index

A coefficient, n'', related to the light transmission characteristics of a polymer via the equation

$$I/I_0 = \exp(-4\pi n'' Z / \lambda),$$

where I is the intensity of the transmitted light, I_0 is the intensity of the incident light, Z is the sample thickness and λ is the wavelength. The extinction index n'' corresponds to the imaginary component of the refractive index n^* ($n^* = n' - i n''$), where i is the complex number ($\sqrt{-1}$).

63. Extrudate

The product that emerges from the die of an extruder.

64. Extruder

Known also as single-screw extruder, is the equipment used to extrude a polymer melt through a die. It consists of an Archimedean screw rotating inside a heated barrel, performing the following operations: (a) 'feeds' the cold polymer granules or powder arriving from the hopper, (b) 'melts' the solid polymer, and (c) 'delivers' it to the die, as well as acting as a pump to force the melt through the die, where it acquires the shape of the extruded product.

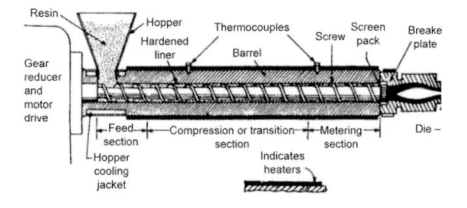

Example of a single-screw extruder. Source: Unidentified original source.

65. Extruder Screw

The screw of a single-screw extruder consists of three zones, respectively, the 'feed zone' (near the feeding hopper), the 'compression or transition zone' (central section) and the 'metering zone' (near the die). There is also a reliance on the metering zone to stabilize the flow stream arriving from the transition zone and to enhance the mixing of additives and fillers. This is particularly important for operations where additives are incorporated directly into the polymer without passing the mixes through a preceding compounding operation. Special mixing devices (see diagram) are attached to the end of the screw or within the metering zone for this purpose.

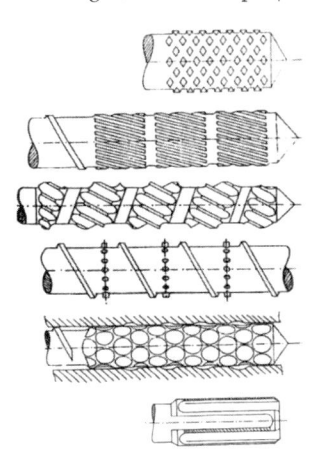

Mixing devices attached to screws of extruders. Source: Unidentified original source.

Some extruders operate with screws characterized by a sudden compression, as in the case of nylon screws, which have a long feed zone, a very short transition zone (over half a turn of the screw) and a relatively short metering zone. (See Nylon screw.) In some other extruders the sudden transition takes place after a relatively short feed section where the barrel contains circumferential grooves to induce rapid melting of the granules through frictional forces (see diagram). This is followed by a long metering zone, which is particularly advantageous for the production of thermoplastic foams using chemical blowing agents. Such a system allows the pressure to build up very rapidly so that, if the blowing agent starts to decompose the gases formed are dissolved into the polymer melt rather than escaping though the spaces between granules and then through the hopper.

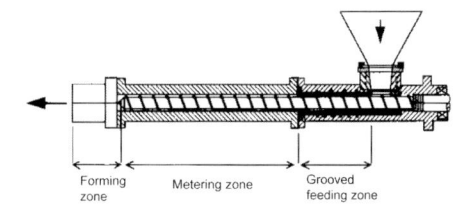

Grooved barrel extruder. Source: Scholtz (2009).

Another variant of screws for polymer extrusion is the type used for rubbers. The feedstock can be in the form of a strip arriving directly from a two-roll mill in line with an internal mixer. Alternatively, the mass of rubber compound can be force-fed into the extruder by a reciprocating hydraulic ram.

66. Extrusion

A continuous process that produces articles of unlimited length and with a constant cross-section, for example, pipes, sheets and profiles. The most common type of extruders for polymer extrusion is the single screw type. (See Extruder.) The layout of a typical extrusion line for the production of sheets is shown, as an example.

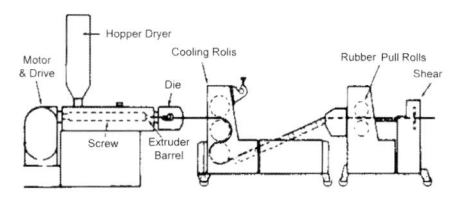

Sheet extrusion line using a unit with three cooling rolls. Source: Rosato (1998).

67. Extrusion Die

The shaping component of an extruder fitted to the exit end of the barrel. Sketches of typical dies used for polymer extrusion are shown.

In dies used for tubular extrusions, there is an air inlet in the centre of the die to prevent collapse of the walls of the tubular melt as it emerges from the die. This feature is not required for systems fitted with a vacuum sizing die.

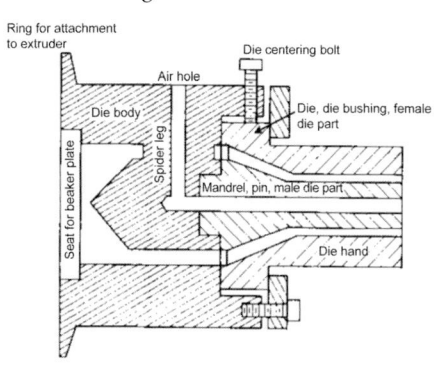

Die used for tubing and pipe extrusion. Source: McCrum et al. (1988).

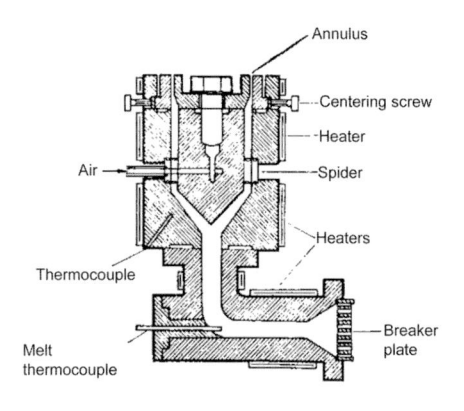

Die used for the production of tubular films. Source: Unidentified original source.

The problem arising from the difficulty of obtaining a uniform flow rate along the circumference of the die has been overcome by means of spiralling channels before the melt reaches the annular die lips at the exit. The rotation of the melt before reaching the die exit also has the function of destroying the weld lines formed by the flow stream after leaving the spider legs section, which is used to fit the central mandrel to the body of the die.

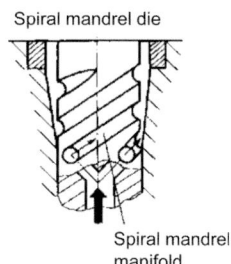

Principle of flow through spiral mandrel dies.

The die for wire coating is also known as a pressure wire die, which indicates that the melt is forced on the wire surface by the pressure inside the die. Vacuum is applied at the wire entry to remove any air that becomes entrapped at the interface with the wire. For dies where the melt tubing is impinged on the wire at the exit of the die, the application of vacuum also assists the drawing of the melt against the surface of the wire.

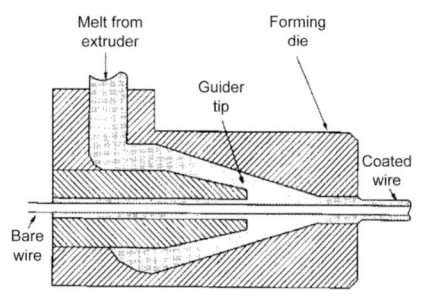

Die used for wire coating. Source: Baird and Collias (1998).

A cross-section of a typical sheet die, featuring a flexible lip adjustment, is shown (where b is a transverse channel for flow equalization). The overall thickness of the extruded sheet can be adjusted by closing or opening the die gap by a push–pull action imposed on the die lips by a series of bolts. The flow rate through the restriction section is controlled by the choke-bar in the same way.

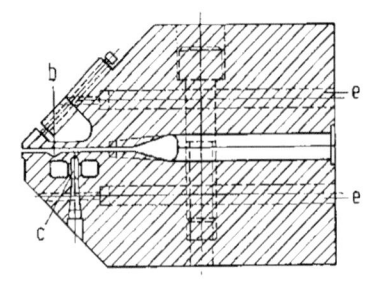

Die with flexible lips for sheet and flat film extrusion. Source: Unidentified original source.

The features for the lateral spreading of a flow of melt delivered by the screw of an extruder to a die via a fish-tail-shaped channel (indicators 3 and 4 in diagram) carrying the melt to the die lips is shown. A larger trough parallel to the die lips (indicator 2 in the diagram) is often provided to accommodate a flexible choke-bar, which is used as a means of controlling the flow rate in zones across the width. This is to ensure that the melt at the die lip is subjected to a constant pressure across the width so that the thickness of the sheet at the die exit is uniform. The main equalization of pressure along the width of the die, however, has to be achieved by an appropriate design of the angle of the manifold channel feeding the melt to the die lips.

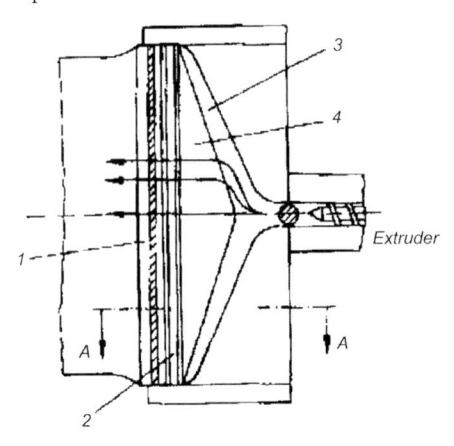

Sketch of spreading flow in sheet dies via a fish-tail-shaped channel. Source: Michaeli (1992).

Both tubular dies and sheet dies can be used to produce multilayer films using different polymers delivered to a common die from separate extruders. (See Co-extrusion.)

68. Extrusion Theory

Refers to the analysis of the feeding, melting and pumping action of the screw in delivering the melt to the die entry of an extruder.

68.1 Analysis of Events in the Feed Zone

The theoretical analysis of the feed zone of a single-screw extruder assumes that the granules or powder particles are compacted together within one or two turns of the screw flights to form a solid plug that slides forward by virtue of the difference in friction forces between the root surface of the screw and the inner surface of the barrel. From an analysis of the forces developed, the average velocity of the plug (V_z) is related to the velocity of the screw surface (V_b) by the expression

$$V_z = V_b \tan \theta / (\tan \beta + \tan \theta),$$

where θ is the helix angle of the screw channel and β is an angle defining the axial movement of the plug along the channel in terms of changes in position of the plug as it moves forwards.

The volumetric conveying rate of the feed section of the screw (G) is calculated by multiplying the velocity V_z of the plug by the cross-sectional area of the channel, from which an expression for the solid conveying rate (G_s) can be obtained, for example,

$$G_s = \rho_s \pi^2 N H_f D_0 (D_0 - H_f) \tan \theta \tan \beta / (\tan \theta + \tan \beta),$$

where ρ_s is the density of the solid plug, N is the screw rotation frequency, H_f is the channel depth and D_0 is the screw diameter.

The solid conveying angle β can be estimated from the expression

$$\cos(\theta + \beta) = (f_s/f_b)[2(H_f/W) + 1] + (H_f/f_b)d(\ln P)/dz,$$

where f_s is the coefficient of friction against the screw, f_b is the coefficient of friction against the barrel, W is the width of the channel, and $d(\ln P)/dz$ is the slope of the plot of logarithmic pressure against distance from the entry.

68.2 Melting and Flow within the Transition Zone

Before the polymer reaches the metering zone, it has to be fully melted to avoid the entrapment of air, which would cause the formation of bubbles in the extrudate. The melting of the polymer plug arriving from the feed zone takes place via the formation of a molten film at the barrel surface. This is scraped by the ploughing action of the screw and forced to flow into a melt pool formed at the side wall at the upstream side channel of the screw. In the diagram is shown a schematic cross-section of the channels in the transition zone. Note that the material flows along the channel from right to left in the axial direction of the screw.

An expression for the melting rate is obtained from the mass balance:.

$$\frac{\text{rate of melting}}{\text{per unit distance}} = \frac{\text{rate of flow}}{\text{into melt pool}}$$

$$= \frac{\text{rate of melt pool}}{\text{flowing downstream}}$$

This involves also a heat transfer analysis to determine the rate of melting of the polymer at the barrel surface. The final expression for the reduction in the size of the solid plug as it moves downstream becomes

$$\frac{X}{W} = \left\{ \frac{\psi}{A} = \frac{A/\psi - 1}{1 - [(Z/Z_T)(A/\psi)(2 - A/\psi)^{1/2}]} \right\}^2,$$

where X/W is the ratio of solid bed width to channel width, ψ is a parameter containing the heat transfer characteristics and solid plug conveying rate, and Z/Z_T is a dimensionless ratio, corresponding to the fraction of the melting distance considered, Z, relative to the total melt distance, Z_T. The melting of the solid bed, however, starts in the feed zone and extends into the metering zone, as shown by the data in the graphs.

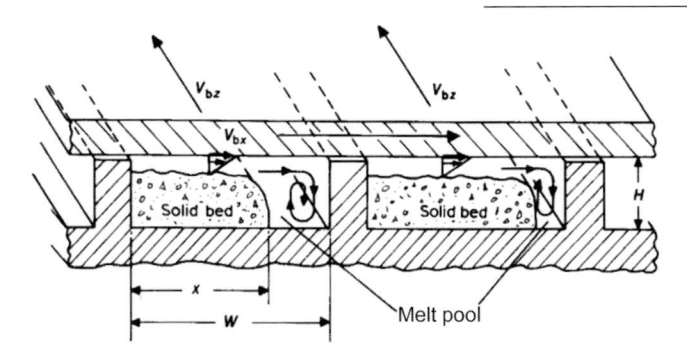

Melting mechanism of polymer powder or granules in the transition zone of an extruder. Melt flow takes place from right (feed zone) to left (metering zone). Source: Adapted from Maddock (1959).

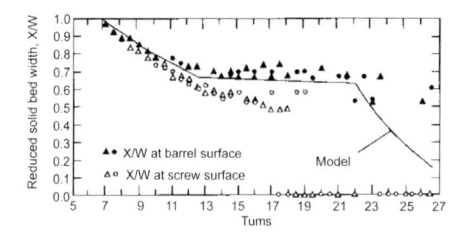

Plots of X/W versus number of turns of the flights along the screw. Source: Osswald (1998).

68.3 Flow Analysis in the Metering Zone

The metering zone has two functions. One function is to dampen any pressure oscillations that arise from pressure fluctuations created by the breaking up of the solid bed in the transition zone. This is necessary in order to avoid surging of the melt (unsteady flow) through the die. The flow rate through the metering zone, hence the pumping capacity of the extruder, is calculated on the basis of a model that assumes that the channel of the rotating screw and the inner surface of the barrel can be unravelled and flattened. Flow is assumed to result from the sliding of the barrel surface over the screw channel, at an angle corresponding to the helix angle of the flight of the screw (see diagram).

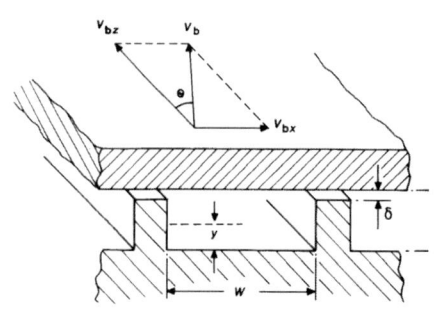

Velocity components of the melt at the inner surface of the barrel sliding over the screw channel and symbols. Source: Mascia (1989).

The analysis consists in setting the appropriate momentum equation and obtaining a solution for steady-state conditions, that is, the velocity does not change with time and, therefore, remains constant along the length of the channel. (See Momentum equation.) The analysis also assumes that the velocity is constant across the width of the channel and that the only velocity gradient along the flow path is the one corresponding to the shear rate through the depth of the channel. Assuming a Newtonian behaviour, the solution leads to an expression for the velocity of a particle within the melt at a distance y from the bottom of the channel in the form of

$$V_z = \frac{y}{H} V_{bz} - \frac{y(H-y)}{2\eta} \frac{\delta P}{\delta z},$$

where V_{bz} is the component of the melt velocity (V_b) in the Z-direction at the barrel surface. The flow rate (Q) can then be calculated as the product of the velocity and the cross-sectional area of the channel, which yields the expression.

$$Q = V_{bz} \frac{WH}{2} - \frac{WH^3}{12\eta} \frac{\delta P}{\delta z},$$

where η is the viscosity of polymer melt, $\delta P/\delta z$ is the pressure gradient along the channel length and H is the height of the channel.

The first term of the flow rate equation corresponds to the 'drag flow rate' (Q_D), which represents the maximum pumping capacity of the screw. The second term is known as the 'pressure flow rate' (Q_P), which represents the reduction in flow rate along the channel of the screw resulting from the building of the pressure due to the restrictions caused by the die. The pressure flow equation can be elaborated to take into account back-flow taking place through the clearance δ between the flights and the barrel (see diagram) and the V_{bz} velocity can be related to the screw revolution frequency,

N (revolutions per second). The flow rate equation can be written as

$$Q = Q_D - Q_P,$$

where

$$Q_D = \alpha N \delta / H \text{ and } Q_P = \beta \Delta P / \eta (1 - \delta / H).$$

The two constants α and β are known as the 'screw characteristics' insofar as they describe the geometric features of the screw. To take into account the non-Newtonian behaviour of polymer melts, the viscosity term η has to be replaced with a function of the shear rate. (See Power law and Carreau model.)

Quite often the viscosity term is replaced by a more complex polynomial equation, obtained from experimental data, which takes into account also the effect of temperature. Note that the flow restrictions imposed by the die can be quantified by a parameter φ, sometimes referred to as the throttle ratio, which is defined as the ratio of the pressure flow (Q_P) to the drag flow (Q_D). The theoretical value of φ varies from 0 for $Q_P = 0$ (open discharge conditions, that is, conditions in the absence of a die) to 1 for $Q_P = Q_D$ (closed discharge conditions, that is, conditions when the die is completely blocked). In the latter case, the only net possible flow is that resulting from leakage flow over the flights of the screw. This is a situation analogous to batch rotary mixers. Hence, the mixing efficiency of an extrusion operation is often defined in terms of the throttle ratio. (See Screw–die interaction.)

69. Eyring Equation

Relates the change in yield strength of a material to strain rate and temperature. The Eyring equation is written as

$$d\varepsilon/dt = B \exp[(\Delta H - v\sigma_Y)/RT] = \dot{\varepsilon}_Y,$$

where $d\varepsilon/dt$ is the applied strain rate, which is equal to the strain rate $\dot{\varepsilon}_Y$ experienced by the material during yielding, B is a material constant, ΔH is the activation energy, v is the activation volume, σ_Y is the yield strength of the material, R is the universal gas constant and T is the absolute temperature. The data are presented as plots of σ_Y/T versus $\log(\dot{\varepsilon}_Y)$, from which the values for the activation volume and activation energy are obtained, as shown in the diagram. (See Activation volume and Arrhenius equation.)

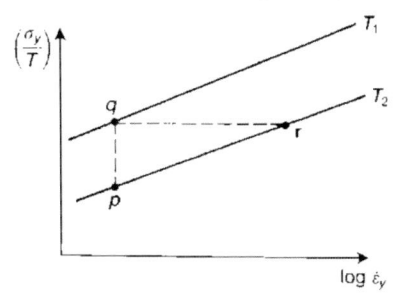

Eyring plots for determining the activation energy and activation volume for yielding failures. Source: McCrum et al. (1988).

F

1. Fabrication

A term used to describe the construction of a structure by assembling and joining various components. (See Tyre construction.)

2. Factice

A term used for softeners or processing aids for rubber. Factices consist of unsaturated oils (vegetable and animal types) that have been reacted with sulfur or mercaptans.

3. Falling-Weight Impact Test

Used to measure the impact strength of materials by dropping a mass on to a specimen from a certain height. This is also known as the ball-drop test. (See Impact strength.)

4. Fatigue Life

A term widely used for fatigue tests on rubber, denoting the number of cycles required to fracture a specimen subjected to dynamic loads.

5. Fatigue Test

A term introduced in the early part of the nineteenth century to denote the failure of metal components when subjected to cyclic loads, over a long period of time, at stress levels below the tensile strength values obtained from standard laboratory tests. Often fatigue tests are classified as static fatigue tests, also known as creep rupture tests, when the stress is kept constant, or dynamic fatigue tests, if the magnitude of the applied stress varies periodically with time. For the convenience of equipment construction, the imposed strain is allowed to vary sinusoidally with time, with the upper and lower limits remaining constant during the test. When the upper and lower limits are both either positive (tension) or negative (compression), the stress cycle is called 'fluctuating'. When the stress changes sign during a cycle, the test is known as 'reversed' cycle fatigue.

There are two cases of reversed and fluctuating cycles used in practice. The first is a symmetrical reverse cycle, in which the mean stress is zero, and the upper and lower limits are equal in magnitude but differ in sign. The second is a fluctuating cycle in which the mean stress is half the magnitude of the maximum stress, with the minimum stress equal to zero. This latter type is widely used for testing rubber samples in tension. The results of fatigue tests are usually presented as linear–log or log–log plots of the applied stress versus number of cycles to failure, known as S-N curves, as illustrated in the diagram.

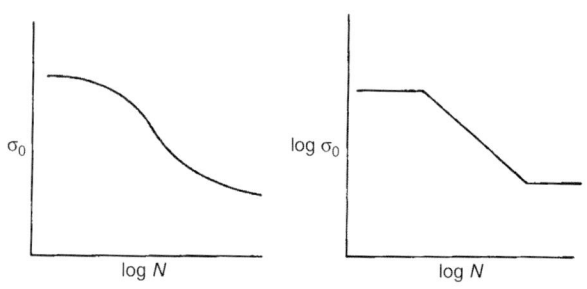

Schematic illustration of fatigue S-N curves: (left) linear–log, (right) log–log.

Fatigue tests are rarely carried out on thermoplastic polymers, unless the oscillation frequency is sufficiently low to prevent large increases in temperature caused by the heat generated through internal friction. In practice, dynamic fatigue tests are carried out on structural adhesives, composites and vulcanized rubber.

6. Fatty Acid

A term for long-chain aliphatic carboxylic acids derived from natural products. For instance, oleic acid is a fatty acid obtained from vegetable oils. However, the term is frequently used more generally and includes aliphatic acids with shorter chain lengths, such as stearic acid and lauric acid.

7. Feed Zone

(See Extruder and Extrusion.)

8. Fibre

Very fine filament normally produced in bundles known as yarns, tows or rovings. Fibres are produced by processes generally known as 'spinning', which can be in the form of a mechanical method, consisting of fibrillation procedures, such as those used for natural fibres and polypropylene, 'wet' spinning and 'melt' spinning. A schematic diagram of a typical melt spinning process is shown.

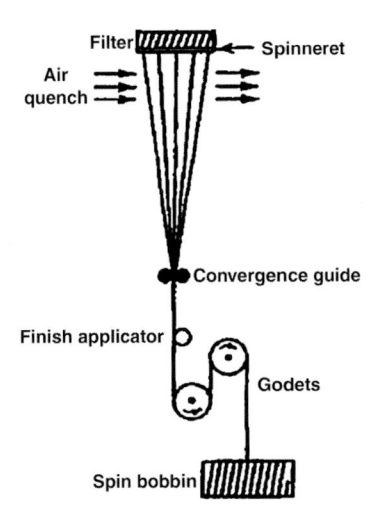

Melt spinning process for linear polymers. Source: Adapted from Teegarden (2004).

Note that the drawdown ratio from the spinneret to the godet rolls is extremely high, so that the orifices of the spinneret can be as a large as a few millimetres in diameter. For the production of inorganic fibres, such as glass fibres, the flow of melt through the spinnerets takes place by gravity feed, possibly assisted by air pressure; while for polymer-based fibres, the melt is delivered to the spinnerets by an extruder via a gear pump. The details of the application of the size (or finish) on the fibres can vary from one process to another depending on both the nature of the fibres and the medium for depositing the coating on the fibres. The process for making carbon fibres is a second-stage operation in which the precursor fibres, acrylic fibres, rayon or pitch fibres, are pyrolysed at very high temperatures in an inert atmosphere.

Note that the high modulus and high strength of inorganic fibres derive from the intrinsically very high internal forces, both within rigid networks and between molecular constituents joined by covalent and ionic bonds. The attainment of high-modulus and high-strength characteristics in organic fibres relies on the molecular chains becoming aligned along the axis of the fibre (monoaxial orientation) so that external forces can be transmitted along the strong covalent bonds of the polymer chains. (See Orientation.) In the case of carbon fibres and carbon nanotubes, the graphene layers are aligned in planes along the fibre axis, so that forces are transmitted along the covalently bonded hexagonal rings, thereby avoiding stresses being transferred between the graphene planes, which are held together by dynamic dipoles created by the free movement of electrons. (See Graphene and Carbon nanotube.)

9. Fibreglass

An alternative name for glass fibre. (See Glass fibre.)

10. Fibre-Reinforced Polymer (FRP)

A composite consisting of a polymer reinforced with fibres exhibiting much higher modulus and strength than the polymer used. (See Composite). Fibres are laid within a polymeric matrix either unidirectionally or in the form of random mats or as woven fabrics.

11. Fibre Reinforcement Theory

A theoretical analysis that relates the properties of composites to the properties of the fibres and those of the matrix, particularly modulus and strength.

11.1 Estimation of Young's Modulus: Unidirectional Continuous Reinforcement

In the analysis it is assumed that (i) both phases (fibres and matrix) behave elastically, (ii) both phases have the same Poisson's ratio, so that no spurious (triaxial) stresses develop from an externally applied axial stress, and (iii) there are no movements at the interface, that is, there is perfect fibre–matrix bonding.

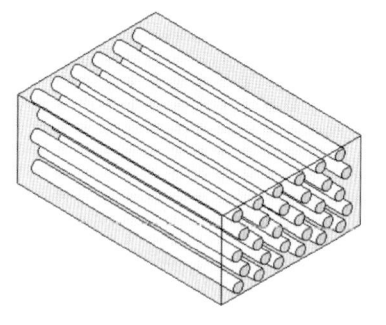

Longitudinal direction External stresses applied along the direction of the fibres produce strains along the fibres of the same magnitude as in the matrix, that is, isostrain conditions ($\varepsilon_c = \varepsilon_f = \varepsilon_m$). (See Law of mixtures.) It follows that the stresses are shared between the fibres and the matrix and, therefore, the stresses (σ) on the composite are equal to the sum of the stresses acting on the two components weighted by their respective volume fractions (ϕ), that is,

$$\sigma_c = \phi_f \sigma_f + (1-\phi_f)\sigma_m,$$

where subscripts c, f and m stand for composite, fibre and matrix, respectively. Dividing by ε_c gives the equation for the Young's modulus, that is,

$$E_c = \phi_f E_f + (1-\phi_f)E_f,$$

which corresponds to the upper limit value predicted from the law of mixtures.

Transverse direction External stresses applied in the direction perpendicular to the fibre direction bring about conditions that vary from isostress at the equator to isostrain at the poles. Between these positions, around the fibre circumference, the conditions are predominantly shear type, as shown in the diagrams.

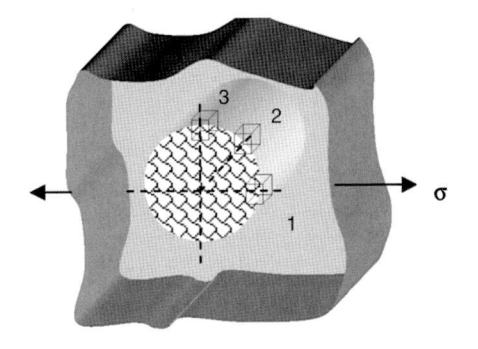

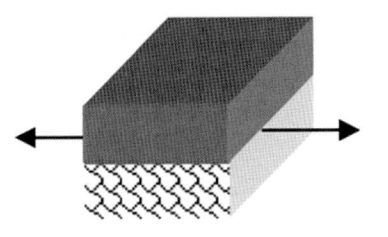

Position 3: $\varepsilon_c = \varepsilon_f = \varepsilon_m$ (isostrain).

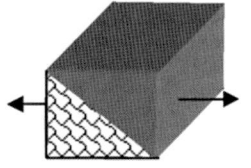

Position 2: $\varepsilon_c \neq \varepsilon_f \neq \varepsilon_m$, $\sigma_c \neq \sigma_f \neq \sigma_m$.

The mixed stress conditions depicted in the last diagram are reflected in the term of the equation for the prediction of the transverse modulus, which is

$$E_c^{(\perp)} = 2[1 - \nu_f + (\nu_f - \nu_m)(1 - \phi_f)]$$
$$\times \left[(1-C) \frac{K_f(2K_m + G_m) - G_m(K_f - K_m)(1 - \phi_f)}{2K_m + G_m + 2(K_f - K_m)(1 - \phi_f)} \right]$$
$$+ C \frac{K_f(2K_m + G_f) + G_f(K_m - K_f)(1 - \phi_f)}{K_m + G_f - 2(K_f - K_m)(1 - \phi f)},$$

where E is Young's modulus, ν is Poisson's ratio, G is the shear modulus,

The stress conditions at three positions around the circumference of the fibres when external stresses are applied in the transverse direction are shown in the three diagrams.

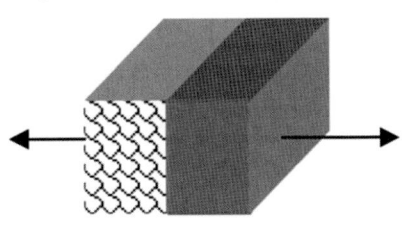

$$K_f = \frac{E_f}{2(1-\nu_f)}, \quad K_m = \frac{E_m}{2(1-\nu_m)},$$
$$G_f = \frac{E_f}{2(1+\nu_f)}, \quad G_m = \frac{E_m}{2(1+\nu_m)},$$

and C is the contiguity factor (distance between fibres, stacking); for isolated fibres, $C = 0$, and for fibres in close contact, $C = 1$.

11.2 Estimation of Young's Modulus: Unidirectional Short-Fibre Reinforcement

When the fibres are not continuous, but are aligned in one direction, the externally

Position 1: $\sigma_c = \sigma_f = \sigma_m$ (isostress).

applied tensile forces in the direction of the fibres produce isostrain conditions only in the central section of the fibres. The region around the fibre ends are subjected to shear deformations, as indicated in the diagram. The fibre ends can be considered as 'ineffective' in the transfer of the stresses from the matrix to the fibres and, therefore, will bring about a reduction in modulus relative to that estimated for continuous fibres.

Halpin–Tsai equation reduces to the lower limit, that is, the 'reciprocal' law of mixtures.

Angular dependence of modulus The variation of the modulus of unidirectional fibre composites between the longitudinal and transverse directions can be derived from considerations of the equilibrium of forces. Converting forces into related stresses and introducing geometrical considerations for the strains, the following expression can be obtained

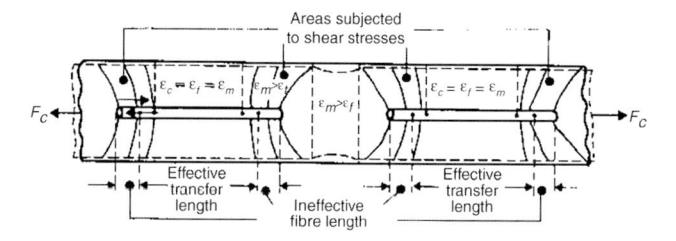

$$E_L/E_\theta = \cos^4\theta + (E_L/E_T)\sin^4\theta + (E_L/G_{LT} - 2\nu_{LT})\cos^2\theta\sin^2\theta,$$

where E_θ, E_L and E_T respectively are the Young's modulus in direction θ and in the longitudinal and transverse directions, G_{LT} and ν_{LT} are the shear modulus and Poisson's ratio related to the planes of the fibre directions, and θ is the angle formed by the direction of the fibres and that of the externally applied stress.

An equation that is more widely used for the estimation of the modulus of short-fibre composites is known as the Halpin–Tsai equation, and it can be used to calculate both the Young's modulus and the shear modulus:

$$\frac{M_c}{M_m} = \frac{1+\xi\eta\phi_f}{1-\eta\phi_f},$$

where

$$\eta = \frac{(M_f/M_m)-1}{(M_f/M_m)+\xi},$$

M_c is the composite modulus (E_c, G_c), M_f is the fibre modulus (E_f, G_f), M_m is the matrix modulus (E_m, G_m), and ξ is a geometric factor that depends on fibre length and loading conditions, with $\xi = 2(l/d)$ for the prediction of the longitudinal Young's modulus and $\xi = 1$ for the prediction of the shear modulus.

It is noted that for $\xi = \infty$, that is, $l/d \to \infty$ (continuous fibres), the Halpin–Tsai equation reduces to the upper limit of the law of mixtures, that is, the longitudinal modulus. For $\xi = 0$, that is, $l/d \to 0$ (fibres in the transverse direction), on the other hand, the

11.3 Estimation of Young's Modulus: Modulus of Random Fibre Reinforcement

When the fibres are arranged in more than one direction as fabric reinforcement and ply laminates, the modulus for each direction can be obtained by averaging procedures taking into account the fraction of fibres present in each of the directions considered. For planar random fibre composites, the modulus will be the same in all directions in the plane and can be calculated directly from the general averaging procedure of a function, that is,

$$\bar{E} = \frac{1}{\pi/2} \int_0^{\pi/2} E(\theta)\, d\theta.$$

11.4 Estimation of Tensile Strength: Unidirectional Continuous Reinforcement

Longitudinal direction Stresses applied in the longitudinal direction produce isometric deformations in the fibres and matrix, that is, $\varepsilon_c = \varepsilon_f = \varepsilon_m$. Under these conditions, the stress acting on the composite is shared between the stress acting on the two phases, which can be calculated from the law of mixtures, that is,

$$\sigma_c = \phi_f \sigma_f + (1-\phi_f)\sigma_m,$$

where ϕ_f is the volume fraction of fibres and $(1-\phi_f)$ is the volume fraction of matrix. If the fracture strain of the fibres is lower than that of the matrix, then the composite breaks through fracture propagation through the matrix at the point when the fibres break. Therefore, the law of mixtures can be applied to calculate the fracture stress for the composite, σ_c^*, from the tensile strength of the fibres, σ_f^*, and the value of the stress on the matrix corresponds to at which the strain is equal to the fracture strain of the fibres, σ'_m, as shown in the diagram.

Therefore, the law of mixtures for fracture conditions can be written as

$$\sigma_c^* = \phi_f \sigma_f^* + (1-\phi_f)\sigma'_m.$$

Assuming elastic behaviour for both components up to fracture conditions, the stresses can be calculated from the respective Young's modulus, that is,

$$\sigma_c^* = \phi_f \varepsilon_f^* E_f + \varepsilon_f^*(1-\phi_f)E_m,$$

where ε_f^* is the strain at which the fibres break.

Transverse direction When the stresses are applied in the transverse direction to the fibres, either failure will take place within the matrix (if the interfacial fibre–matrix bond is very strong), or fracture will start at the interface and propagate through the matrix. For the first case, the strength of the composite is the same as the strength of the matrix; while for the second case, it will be even lower. Therefore, the expression for the strength of the composite in the transverse direction can be written as $\sigma_c^* \leq \sigma_m^*$.

11.5 Estimation of Tensile Strength: Unidirectional Short-Fibre Reinforcement

In short-fibre composites, the stress along the length of the fibre is not constant, but rises from a minimum at the fibre end to a maximum in the centre.

Longitudinal direction When the length of the fibres is greater than the critical fibre length (L_c), the conditions in the central region are isostrain type and, therefore, fracture starts within the fibres at the point when the stress reaches the tensile strength

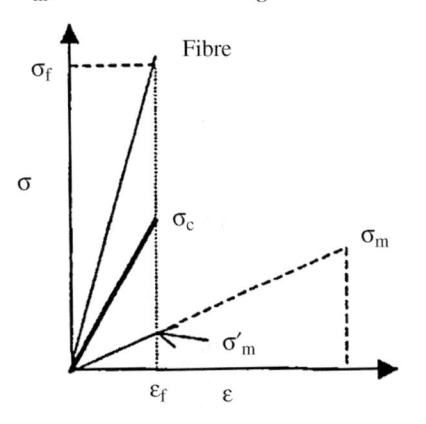

Stress versus strain curves for fibres, matrix and unidirectional continuous fibre composite.

of the fibres. It follows that the average stress, $\bar{\sigma}$, carried by the fibres at the point of fracture can be related to the strength of the fibre, σ_f^*, by the expression

$$\bar{\sigma} = \sigma_f^*(1-L_c/2L),$$

which can be used in the law of mixtures for the strength of the composite, σ_c^*, which becomes

$$\sigma_c^* = \phi_f \sigma_f^*(1-L_c/2L) + (1-\phi_f)\sigma'_m,$$

where σ'_m is the stress carried by the matrix at the point of the initiation of the fracture in the fibres. If the strain at break of the fibres is lower than that of the matrix, the value of σ'_m can be estimated from its Young's modulus, that is, $\sigma'_m = E_m \varepsilon_f$.

When the fibre length is much smaller than L_c, fracture takes place by a shear mechanism initiated either at the interface or within the matrix adjacent to the fibres.

Transverse direction When stresses are applied in the transverse direction, failure takes place within the matrix, even though it may originate by the breaking of the interfacial fibre–matrix bond. Therefore, the strength of the composite is practically equal to the strength of the matrix.

11.6 Angular Tensile Strength of Unidirectional Fibre Composites

When an external tensile force is applied at an angle θ to the direction of the fibres, the internal forces can be resolved into two vectors, in the plane along the directions of the fibres, and in the plane perpendicular to the fibres, respectively, as indicated in the diagram.

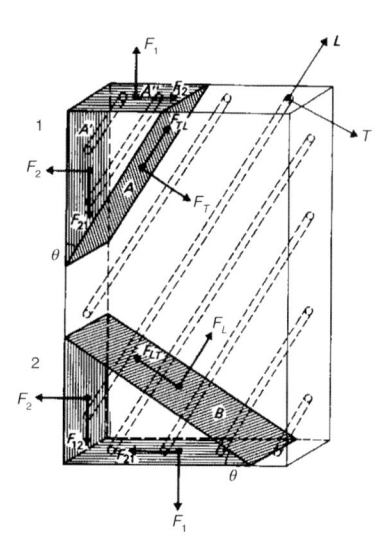

Vector analysis of forces acting on planes parallel and perpendicular to the fibre direction. Source: Mascia (1974).

When the angle is very small, fracture takes place in the plane perpendicular to the fibres, and the strength is practically the same as the longitudinal strength of unidirectional composites. At larger angles, between around 15° and 75°, fracture occurs predominantly through shear slip along the planes between the fibres, as shown.

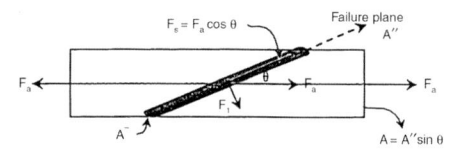

Shear failure of unidirectional fibre composites.

The theoretical analysis assumes that F_s is the shear force responsible for the failure along one of the planes between the fibres. The value of F_s at failure can be estimated from knowledge of the interlaminar shear strength, τ_{m-f}, that is, $F_s = \tau_{m-f} A''$, where A'' is the area of the failure plane. The strength of the composite at angle θ is $\sigma_\theta = F_a/A$. Substituting $F_s = F_a \cos\theta$ and $A = A'' \sin\theta$, one obtains the following expression for the strength of the composite:

$$\sigma_\theta = \frac{\tau_{m-f}}{\sin\theta \cos\theta}.$$

At very large angles, between around 75° and 90°, fracture occurs in the plane between the fibres by a cleavage mechanism, so that, as the angle approaches 90°, the value of the stress that causes fracture becomes equal to the transverse strength of the unidirectional fibre composite. The theoretical equation for the strength, σ_θ, as a function of the angle, θ, becomes

$$\sigma_\theta = \sigma_T/\cos^2\theta,$$

where σ_T is the strength in the transverse direction to the fibres. The overall variation in strength from 0° to 90° is shown schematically in the diagram.

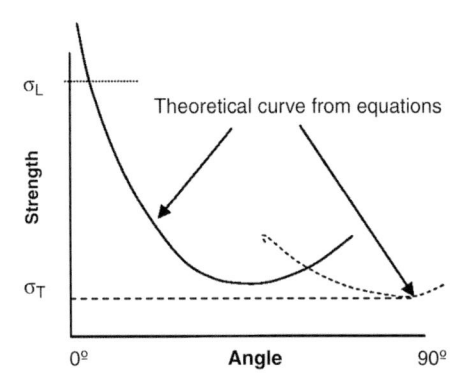

Prediction of the angular strength of unidirectional fibre composites.

In view of the high sensitivity to shear failures of unidirectional fibre composites, products are usually produced with cross-ply configurations or with fabric reinforcement.

12. Fickian Behaviour

A type of diffusion of gases or vapours through a film or sheet that can be described by Fick's law. Accordingly, a plot of the mass M of a sample at time t, under steady-state diffusion conditions (normalized with respect to the mass reached at absorption equilibrium, M_t/M_∞), against the square root of time is linear. The validity of the Fickian behaviour assumption can be ascertained by making measurements on samples of different thickness (L). The plot of M_t/M_∞ against $t^{1/2}/L$ would produce a single curve. (See Diffusion.)

13. Filament Winding

A manufacturing method used to produce hollow products such as pipes and vessels using continuous reinforcing fibres. (See Composite.)

14. Filler

A term for inorganic particles obtained by comminution of low-cost natural minerals. The term derives from their poor reinforcing efficiency so that their main function is to 'fill' a polymer compound in order to reduce the overall cost. Some fillers are also made by a synthetic route. Under this category are included fillers used for their functional properties, such as fire retardancy or electrical characteristics. The main features of fillers used in polymer formulations are shown in the table.

Types	Features
Talc, mica, glass flakes	Platelet geometry
Calcium carbonate, china clay	Quasi-spherical
Carbon black, fumed silica, nanoclays	Very irregular
Aluminium trihydrate, magnesium hydroxide	Quasi-spherical
Silicon carbide, zinc oxide, barites	Quasi-spherical

In terms of mechanical properties, the reinforcement efficiency of fillers is quite low and only elastomers display substantial increases in modulus and strength. The equations normally used for the prediction of shear modulus and viscosity have a certain similarity. For instance, Kerner's equation for the shear modulus of particulate composites is

$$G_c/G_m = 1 + [15\varphi_p(1-\nu_m)/\varphi_m(8-10\nu m)],$$

where G_c/G_m is the shear modulus reinforcement factor (subscripts c and m stand for composite and matrix), φ_p and φ_m are the volume fractions for the particles and matrix, and ν_m is Poisson's ratio of the matrix. One of the more widely used equations for melt viscosity, attributed to Kitano, is

$$\eta_c/\eta_m = [1-\varphi_p/\varphi_{pk}]^{-2},$$

where η_c/η_m is the viscosity 'increment factor' at low shear rates, φ_p is the volume fraction of the particles and φ_{pk} is the maximum volume fraction of the filler particles, known as the packing fraction. (See Reinforcement factor.) The figure shows a comparison of the strength reinforcement efficiency of different fillers in a vulcanized rubber.

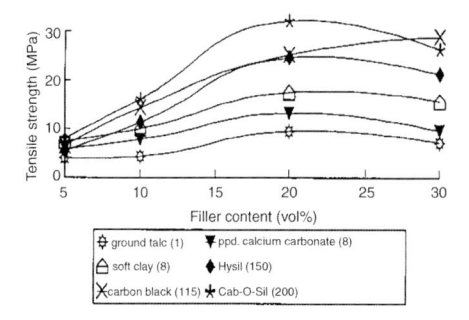

Plot of tensile strength versus filler content of a vulcanized rubber for different types of fillers (numbers in brackets correspond to the surface area of the filler, m^2/g). Source: Wypych (1993).

The data show a good correlation between tensile strength and surface area of the filler and also a substantial increase in strength resulting from the promotion of chemical bonds between filler and matrix, as can be inferred from the data obtained with Hysil (silica) and carbon black. The further data show the effect of filler concentration of different types of fumed silica on the viscosity of an unsaturated polyester resin.

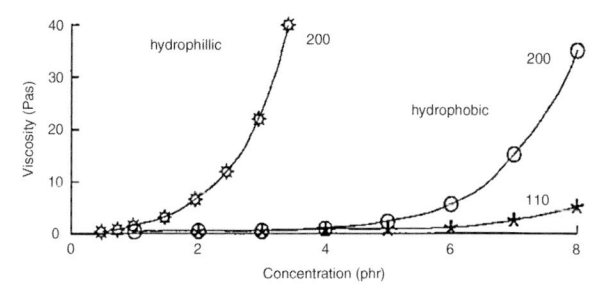

Effect of filler concentration on the viscosity of an unsaturated polyester resin for various grades of fumed silica (numbers on the curves represent the surface area, m^2/g). Source: Wypych (1993).

The data illustrate the strong influence of surface area on viscosity and reveal the important role of surface charges. The presence of OH groups confers a strong hydrophilic character to the surface of the particles, which can immobilize the neighbouring resin molecules through strong hydrogen-bonding interactions. Note that the chemical nature of the surface of the silica particles can be readily changed by treatments with silane coupling agents. (See Coupling agent.)

15. Film Extrusion

(See Extrusion, Blown film and Chill-roll casting.)

16. Film Formation

A term used to describe the formation of a 'solid' coating on a substrate through deposition of a polymer from a solution or dispersion in a liquid medium. The mechanism for the formation of a film from a polymer dispersion is shown schematically.

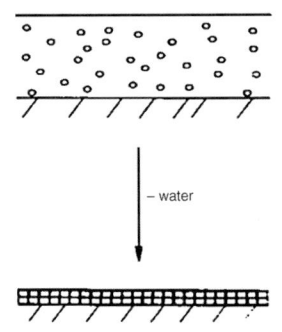

Film formation of coatings from dispersions. Source: Adapted from Denkinger (1996).

After evaporating the water from the deposited coating, the polymer particles fuse with each other to produce a continuous film. For this to take place, the polymer molecules from adjacent particles have to diffuse across the interface and the voids between particles have to disappear. The latter event takes place through a balance between the reduction of surface energy and the rate of energy dissipation by viscous flow, which implies that the ambient temperature has to be higher than the glass transition temperature (T_g) of the polymer to ensure that molecular diffusion can take place at a realistic rate. (See Powder sintering.)

The film formation characteristics of a polymer dispersion are described by an empirical parameter known as the 'film formation temperature' (T_f). The value of T_f is measured experimentally by casting a film on a long plate heated in such a way as to provide a linear temperature gradient along its length. The T_f value is obtained by recording the position, hence the temperature, of the 'frost' line that is observed across the width of the film, which results from the disappearance of voids and brings about the natural transparency of the polymer in the coating. For crystalline polymers, film formation takes place at temperatures T_f above the melting point T_m. In any case, the film formation temperature can be reduced by the incorporation of plasticizers, which remain permanently embedded in the film. Note that, while continuous films can be formed readily from solutions, the removal of residual solvent may be very slow and not complete if the T_g of the polymer is much higher than ambient temperature. For the case of crystalline polymers, the evaporation of solvent may result in the growth of large crystals, which can produce a 'grainy' or powdery coating instead of a continuous film.

17. Film Gate

A channel in the form of a thin slit feeding the flat cavity of an injection mould. (See Gate and Mould.)

18. Fire Retardant

(See Flame retardant.)

19. First-Order Transition

Corresponds to the transition from a highly ordered state of a solid (the crystalline state) to a highly disordered state (the liquid state), that is, the melting transition. Thermodynamically, a first-order transition is defined as the temperature at which a primary variable, such as volume or enthalpy, shows a sudden jump. (See Alpha transition temperature and Beta transition temperature.)

20. Fish Eye

A defect found in sheets due to the presence of 'gel spots', which are dragged by the action of the calendering rolls to form a wake resembling the eye of a fish. (See Gel.)

21. Fish-Tail Die

An extrusion die to produce sheets or flat films. The term derives from its geometric features, which are characterized by the ability to deliver the melt to the die lips at a uniform rate in order to ensure a constant thickness across the width. (See Extrusion die.)

22. Flame Retardant

Additives used to reduce the susceptibility of polymers to be ignited by a flame. There are three main types of additives that can be used to induce flame retardancy in polymer formulations.

22.1 Combustion Inhibitors (or Chemical Flame Retardants)

These are the traditional fire retardant additives, consisting of mixtures of highly halogenated compounds and antimony oxide, with optimum weight ratio at around 2 : 1, derived from the relationship illustrated in the diagram.

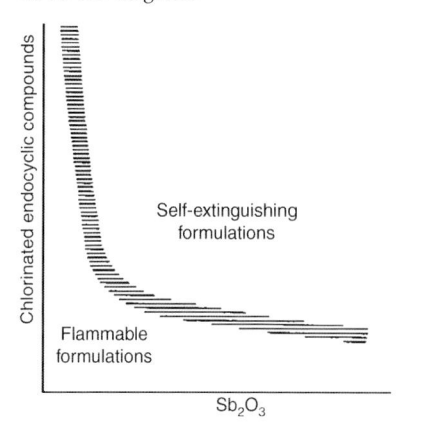

Optimized quantities of synergistic components of flame retardant formulations. Source: Mascia (1974).

At high temperatures, the halogenated compounds decompose into volatile products, consisting of highly reactive species that will intervene in the combustion reactions by quenching the free radicals formed by the pyrolysis of the polymer. These reactions retard the rate of flame propagation and cause the flame to extinguish itself. The addition of antimony oxide (Sb_2O_3) enhances the efficiency of the above intervention through the production of antimony oxychloride 'fumes' capable of adsorbing the very reactive radical species in the combustible gases. The most common types of halogenated flame retardant additives are chlorinated paraffins (C_{10}–C_{30}), derivatives of hexachloro- and hexabromocyclopentadiene, tetrabromododecane and pentabromotoluene.

22.2 Oxygen Diluents (or Physical Flame Retardants)

These are additives that give out large quantities of incombustible gases, such as water, just before the exposed polymer front approaches its ignition conditions in terms of temperature and oxygen concentration. The typical additives that operate on this principle are aluminium trihydrate and magnesium hydroxide. (See Aluminium trihydrate, Magnesium hydroxideAntitracking additive.) These additives very rapidly release 30–35 wt% water at temperatures around 300–400 °C, which reduces the concentration of oxygen available for the combustion of volatile products at the heated front. Since the dehydration of inorganic oxides is an endothermic process, a certain amount of heat will be removed, thereby causing a reduction in temperature in the exposed zone, which also delays the attainment of the conditions required to start ignition. Another additive exerting a similar function is melamine, and its salts, such as melamine cyanurate, which sublime at around 300–400 °C and therefore dilute the mixture of oxygen and combustible gases formed by pyrolysis reactions at the exposed polymer front. It is possible that for some polymers the pyrolysis products may even react with the melamine vapour to form more stable products.

22.3 Heat and Oxygen Barrier Promoters

These are also known as intumescent char-forming additives due to their ability to produce a layer of char as a result of physical and chemical interactions with the decomposing polymer front. (See Intumescent coating.) This can act as a heat and oxygen barrier for the polymer beneath, thereby slowing down the rate of formation of combustible gases from the polymer. Other additives can form a protective barrier layer primarily through the formation of an inorganic glass coating. Typical additives of this type are mixed salts of zinc sulfate, potassium sulfate and sodium sulfate, ammonium perborate and zinc phosphate glasses. Silica gels and nanoclays have also been found to provide some protection by this mechanism.

23. Flash

A thin film of polymer formed at the edge of a moulded article, resulting from the flow of melt between the contacting surfaces of two parts of the mould, consisting of punch and cavity parts.

24. Flash Point

Corresponds to the self-ignition temperature of combustible liquids. It is an important parameter in choosing solvents or plasticizers in situations where there could be fire hazards, 300–400 °C.

25. Flex Cracking

A term denoting the fracture of a flexible sheet by bending.

26. Flex Life

A term to denote the fatigue life of rubber under cyclic loading conditions. (See Fatigue life.)

27. Flexural Modulus

The modulus of a material measured in three-point bending experiments. By applying a load (P) at the centre of a freely supported beam and measuring the resulting deflection (Δ), the Young's modulus (E) is calculated from fundamental considerations. Provided the measurements are made within the applicability of the underlying theory, requiring a large span-to-thickness ratio (i.e. $S/W > 16$) and small deflections (i.e. $S/\Delta > 10$), it is

$$E = PL^3/4\Delta BW^3.$$

The calculated value should be the same as that obtained from tensile or compression tests. However, because the limiting conditions imposed by beam theory cannot be easily imposed in practical measurements on polymers, the E value obtained from a flexural test can be somewhat higher than that obtained from a tensile test. Hence the reason for identifying the value obtained as 'flexural' modulus.

28. Flexural Properties

Mechanical properties measured by methods involving the application of the load in a three-point bending mode, hence the terms flexural modulus and flexural strength.

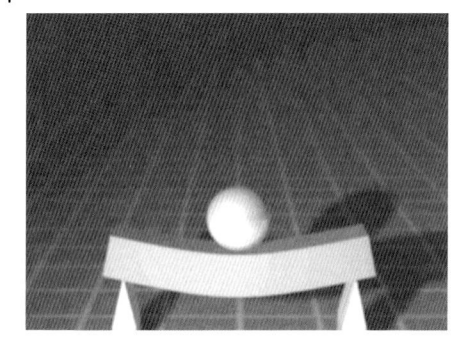

Deflection of a beam loaded in a three-point bending mode. Source: Courtesy of M. Baldwin (2003), EU Mercurio Project.

The three-point bending mode is a convenient way of evaluating the mechanical properties of rigid materials due to the simplicity of the specimens required, typically rectangular bars. Provided that the deflections incurred are small and the span-to-thickness ratio is large, normally greater than 20:1, the recorded forces can be used to calculate the outer skin stresses at the two opposite sides of the loading point (respectively, compression and tension), from which fundamental properties, such as modulus and yield strength, can be estimated. The analysis of a centrally loaded freely supported beam at low levels of deflections makes it possible to calculate the strain, ε, and the stress, σ, at the loading point on both the compression and tension sides of the beam, that is

$$\varepsilon = 6W(\Delta/L^2) \text{ and } \sigma = \tfrac{3}{2}(PL/BW^2),$$

where P is the load, B is the width, W is the thickness and Δ is the central deflection.

29. Flexural Strength

The strength of a material measured in three-point bending experiments. By increasing the load at the centre of a freely supported beam up to the point at which the material fails by yielding (P_Y), the yield strength (σ_Y) can be calculated from fundamental considerations, provided the measurements are made within the applicability of the underlying theory, requiring a large span-to-thickness ratio (i.e. $S/W > 16$) and small deflections (i.e. $S/\Delta > 10$). The equation is

$$\sigma_Y = 3P_Y S/2WB^2,$$

where B is the width of the rectangular specimen. The calculated value should be the same as that obtained from tensile or compression tests. However, because the limiting conditions imposed by beam theory cannot be easily imposed in practical measurements on polymers, the σ_Y value obtained from a flexural test can be higher than that obtained from a tensile test. Furthermore, the tests are widely used to measure the flexural strength even, or especially, when the material fails in a brittle manner or by a complex fracture mechanism, as in composites. In this case, the flexural strength values calculated from the above formula can be considerably higher than the values obtained from tensile tests.

30. Flocculant

(See Flocculation.)

31. Flocculation

A phenomenon or process by which particles are kept in suspension, or prevented from forming agglomerates, when dispersed in a fluid medium. The ability of suspensions to flocculate is determined by considerations based on both the surface characteristics of the particles and the sedimentation velocity. (See Zeta potential and Stokes equation.) The dispersion of particles takes place through repulsions caused by surface charges of equal sign and exhibiting a high zeta potential. This characteristic has to be combined with the stability against sedimentation, through the required balance of particle size, the density of the particles and the viscosity of the medium, as stipulated by Stokes law. The desirable increase in viscosity of waterborne dispersions is often achieved through the addition of water-soluble polymers, known as flocculants, possessing polarity characteristics 'compatible' with the surface charges of the particles.

32. Flow Analysis

Analysis of the flow of polymer melts based on the conservation of momentum. In the analysis of the flow of polymer melts through channels, it is usually assumed that the rate of change in momentum is negligible, even in situations where there is a large change in cross-sectional area along the flow path. This is allowable in view of the very high viscosity of polymer melts, which makes the rate of change in velocity, dV/dt, very small compared to the force required to maintain the flow. (See Momentum equation.) An example of the use of the momentum equation in the flow analysis of polymer melts is to obtain a relationship between pressure and flow rate.

A comparison of the flow though cylindrical and slit dies is shown. From the equations shown, the respective expressions for the flow rate (Q) are obtained as

$$Q = \int_0^R 2\pi r V_z(r)\, dr \quad \text{and}$$

$$Q = \int_0^{H/2} 2W v_z(y)\, dy,$$

where τ_{rz} and τ_{yz} are the shear stresses at distances r and y, and $\delta P/\delta z$ is the pressure gradient along the flow direction.

33. Flow Curves

Plots of shear stress against shear rate, usually on a log–log scale to obtain straight lines, obtained from rheological measurements on polymer melts. (See Power law and Power-law index.)

34. Flow-Induced Crystallization

The formation of crystals taking place in fibre spinning from a melt, a solution or a gel through the combined effect of cooling, or solvent evaporation, and the application of tensile forces to draw the extrudate.

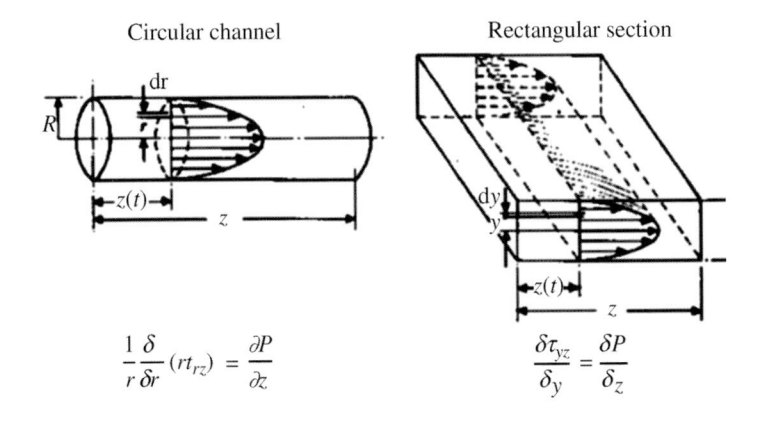

Circular channel

$$\frac{1}{r}\frac{\delta}{\delta r}(r t_{rz}) = \frac{\partial P}{\partial z}$$

Rectangular section

$$\frac{\delta \tau_{yz}}{\delta y} = \frac{\delta P}{\delta z}$$

35. Flow Instability

Fluctuations in the flow rate of a polymer melt emerging from a die. These can arise from 'fractures' at the die entry resulting in a cyclic slip–stick effect at the walls of the die. (See Melt fracture.) In extrusion, flow instability may arise from a phenomenon known as 'surging', which is caused by the breaking up of the solid bed in the transition zone. This event creates pressure fluctuations within the channel of the metering zone that can be transmitted all the way down to the exit of the die, which will be manifested as periodic thickness fluctuation in the extrudate. (See Extrusion theory.)

36. Flow Promoter

A term sometimes used to denote an additive capable of enhancing the flow characteristics of a polymer melt through external lubrication or a reduction in viscosity.

37. Fluid Bed (Fluidized Bed)

A bath consisting of small particles (usually glass Ballotini spheres about 0.15–0.25 mm diameter) suspended in inert hot gas, used primarily for the continuous vulcanization of rubber products, such as tubings or profiles.

38. Fluorescence

A phenomena whereby characteristic chemical groups in molecules absorb incident light and re-emit part of it as radiation at a lower frequency. Fluorescent dyes, for instance, absorb UV light and re-emit it in the visible range. The origin of fluorescence lies within the conjugated double bonds in which the delocalized π–electrons are able to absorb energy with relative ease. This technique is often used to detect degradation in PVC products, which results in the formation of conjugated double bonds.

39. Fluoroelastomer (also known as FKM)

Fluoroelastomers (known under the short form FKM) are produced from copolymerization or terpolymerization of hexafluoropropylene ($CF_2=CFCF_3$) in combination with vinylidene fluoride ($CH_2=CF_2$) and either perfluoro(methyl vinyl ether) ($CF_2=CFOCF_3$) or tetrafluoroethylene ($CF_2=CF_2$). The T_g value varies between -35 and $-50\,°C$ depending on composition. The main attribute of fluoroelastomers is their thermal oxidative stability, which makes them suitable for continuous use up to $200\,°C$. Fluoroelastomers can be cured with diamines in combination with a metal oxide, such as CaO and MgO, as acid acceptor. The amine can abstract a fluorine atom from the molecular chains, producing the desired cross-links, producing HF as by-product. Systems that can be cross-linked with peroxides or by electron beaming are based on terpolymers with minor amounts of vinyl compounds containing bromine groups. Free radicals are produced as a result of the scission of the weak C–Br bond, which produces HBr as by-product and, therefore, these polymers also require a metal oxide to act as acid absorber.

40. Fluoropolymer

A polymer containing fluorine atoms attached to an aliphatic polymer chain. Examples are homopolymers and copolymers of tetrafluoroethylene, vinylidene fluoride and polyhexafluoropropylene.

41. Fluorosilicone (also known as FVMQ)

(See Silicone rubber.)

42. Foaming Agent

(See Blowing agent.)

43. Foam

A product containing large quantities of regularly sized and evenly distributed voids, known as 'cells'. Foams are divided into 'closed cell' foams and 'open cell' foams, according to whether the voids are totally occluded within the polymer matrix or are interconnected to form co-continuous phases of air and polymer. The density of foams ranges from about 300–700 kg/m^3 for structural foams (produced by injection moulding of thermoplastics) down to about 3–5 kg/m^3 for free rising foams from liquid resin systems. The dimensions of the cells can vary from about 0.2 mm for closed cells up to about to 2–3 mm for open cells. The typical structures of cells is shown in the micrographs.

cro-balloons (Ballotini spheres) embedded in a polymer matrix. The mechanism for the formation of foams consists in the nucleation of 'gas clusters' by the blowing agent and the subsequent growth of a large number of cells, through the rapid evolution of gases. These can be generated either through the decomposition of a chemical blowing agent or by the sudden evaporation of fine droplets of a liquid or dissolved gas, known as physical blowing. When foaming is carried out with chemical blowing agents, the formation of clusters and the subsequent growth of cells takes place by the spontaneous chemical decomposition of the blowing agent to produce very rapidly a large quantity of gases. (See Blowing agent.)

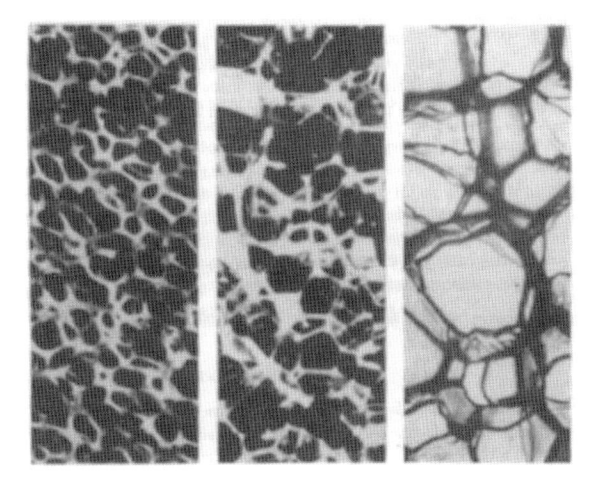

Foam cell structures: (left) expanded polystyrene; (middle) expanded polyethylene; (right) flexible polyurethane. Source: Unidentified original source.

All polymers can be processed to produce cellular structures using similar methods. Two exceptions to this are the manufacture of polystyrene foams by the expandable beads method, and the production of 'syntactic foams', consisting of rigid mi-

44. Foam Density and Properties

An interesting feature of the production of foams is the possibility of estimating the amount of blowing agent that would be required to produce foam with a specified density, on the basis that the contribution of the weight of gas in the foam is negligible and, therefore, the density of the foam, ρ_f, is directly proportional to the volume fraction of the polymer, that is $\rho_f = \varphi_p \rho_p$, where φ_p is the volume fraction of the polymer in the foam and ρ_p is the density of the polymer. (See Law of mixtures.) The volume fraction of gas in the foam, φ_{gas}, therefore, is $(1 - \varphi_p)$, a quantity that can be estimated from the decomposition reaction of a chemical blowing agent or from the solubility–pressure relationship of the foaming gas in the polymer. Alternatively, the density can be calculated from the weight fraction (ω) as

$$1/\rho_f = \omega_{gas}/\rho_{gas} + \omega_p/\rho_p.$$

Solubility diagrams can be used to estimate the pressure required to infuse the minimum amount of gas in the polymer, as well as selecting the right gas for a particular polymer in order to obtain foams of the desired density. Another interesting feature of foams is the possibility to estimate the required foam density in order to achieve a specified change in properties, caused by the replacement of a fraction of polymer with gas. For instance, it is possible to use the Clausius–Mossotti relationship to estimate the required density (ρ) of a cellular structure in order to obtain a certain permittivity value (ε) for a dielectric material, that is,

$$(\varepsilon - 1)/(\varepsilon + 2) = k\rho,$$

where k is a proportionality constant related to the chemical nature of the dielectric.

It is also possible to make also some quite precise predictions for the thermal conductivity of foams for use in thermal insulation. This can be done, for instance, by considering that the heat transfer rate by conduction is given by the equation

$$Q = A(\Delta T/X)[\varphi_{gas} K_{gas} + (1-\varphi_{gas}) K_{solid}],$$

where A is the surface area, ΔT is the temperature difference, X is the thickness, φ_{gas} is the volume fraction of the gas in the cells and K is the thermal conductivity. Knowing that $\rho_{solid} \gg \rho_{gas}$ one obtains

$$Q = A(\Delta T/X)\rho_{solid}[\varphi_{solid} K_{gas} + (K_{solid} - K_{gas})\varphi_{foam}],$$

from which it is predicted that the heat transfer by conduction decreases linearly as the density of the foam decreases.

Although it is more difficult to make accurate estimates for the change in Young's modulus, a reasonable relationship has been found using the expression

$$E_{foam}/E_{solid} = (\rho_{foam}/\rho_{solid})^n,$$

where the exponent n has values between 1 and 2, depending on the cell structure, particularly whether the foam is an open cell or closed cell type. From this it is clearly deduced that the modulus changes more rapidly with increasing gas fraction than the related change in density.

On the other hand, if a bending situation is considered for the case of sandwich or structural foams (i.e. foams with a solid outer skin), there are substantial stiffness advantages to be derived on weight basis arguments. The deflection of a rectangular cantilever beam, y, is given by the expression

$$y = (4PL^3/b)(1/Eh^3),$$

where P is the applied load, L is the length of the beam, E is the Young's modulus, b is the width and h is the thickness of the beam. If the length and width of the beam are kept constant, then the deflection resulting from the applied load P can be written as $y = k/Eh^3$, where k is a proportionality constant. From this it is deduced that the reduction in deflection, y, resulting from an increase in h through foaming is much greater than the increase that results from a decrease in modulus.

45. Foam Formation Mechanism

For foaming operations carried out by means of physical blowing agents, the formation of 'gas clusters' can be nucleated by the incorporation of fine inorganic particles, typically talc or silica, or particles of solid organic compounds, such as citric acid (decomposing at 150–170 °C), as well as inorganic acids, such as boric acid and sodium hydrogencarbonate. The clusters have to reach a critical number and size before they can grow into cells capable of creating the required conditions for the expansion process to occur. These conditions are determined by the equation

$$\frac{\delta r}{\delta t} = \frac{\gamma}{\eta} f\left(\frac{\delta(\Delta P)}{\delta t}\right),$$

where γ is the surface energy and η is the viscosity of the surrounding medium, while $f(\delta(\Delta P)/\delta t)$ denotes a time function of the difference in pressure (ΔP) between the gas in the cells and that of the surroundings. It is noteworthy that the surface energy (γ) is important only for low-viscosity reactive systems. For the case of polymer melts, the predominant factor is the viscosity, η. The size of the cells is controlled, and become dimensionally uniform, through the diffusion of gas from small cells into larger ones, driven by the pressure differential. This sets up a dynamic process of size reduction and size growth and stabilization, till the first leads to vanishing of cells, while the latter is controlled by the conditions set by the equation for $\delta r/\delta t = 0$. These conditions are determined by two possible events: (i) the viscosity becomes infinite through cooling of the melt (thermoplastics) or the formation of cross-links (thermosetting resins), or (ii) the pressure differential between adjacent cells is eliminated when they burst to form interconnected cells.

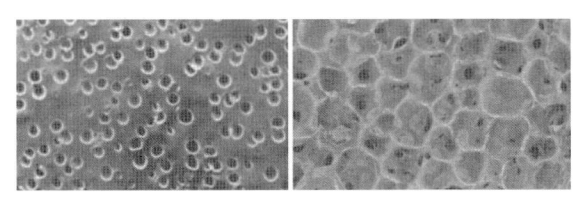

Cell growth (left) and formation of interconnected cells (right) during the production of foams. Source: Unidentified original source.

Images of growing cells and the formation of interconnected cells are shown.

Diffusion of gas from internal cells into the atmosphere is prevented by the formation of a solid skin on the outer layers of the foam. The solid skin is formed by the collapse of local cells caused by the rapid Fickian diffusion of gas in these areas. The outer solid skin helps to maintain at all times a pressure differential between the inner cells and the surroundings. When foaming is induced by the injection of gases in a cross-linked thermoplastic, it is the rubbery plateau modulus that controls the growth of the cells, as the polymer deforms through elastic deformations and not through flow. This has significant implications on the mechanical properties of the foams, insofar as the polymer in the walls of the cells is in a state of biaxial orientation.

46. Foam Manufacture

Four typical methods for the production of foams are illustrated.

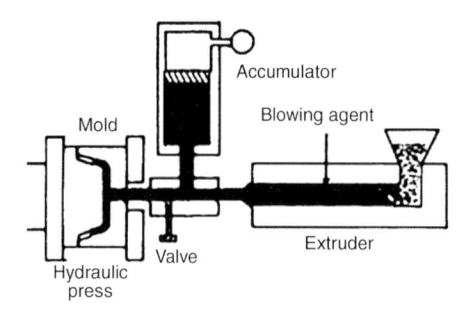

Set-up for the injection moulding of foams with an accumulator to achieve a rapid transfer of the melt into the cavity. Source: Scholtz (2009).

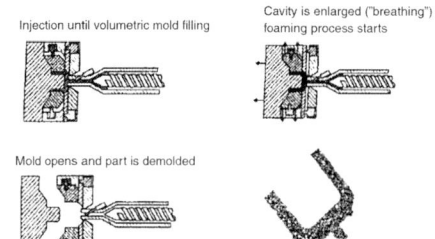

Injection moulding of structural foams with a mould that can be opened to allow for the expansion of cells to take place after injection of the melt. Source: Michaeli et al. (2009).

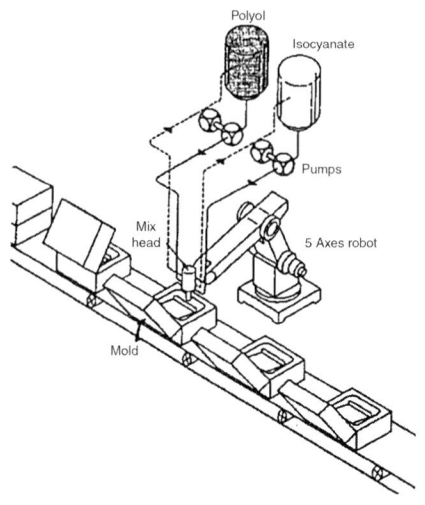

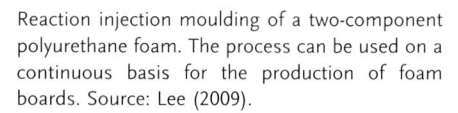

Reaction injection moulding of a two-component polyurethane foam. The process can be used on a continuous basis for the production of foam boards. Source: Lee (2009).

47. Fogging

Haziness of the glass surfaces of the passenger compartment of a motor vehicle. The phenomenon is due to condensation of volatiles, such as plasticizers or residual solvents from polymer products, causing backward light scattering from deposited droplets.

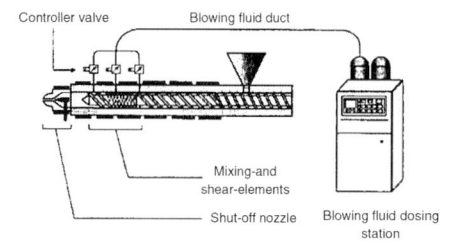

Extrusion of foams with gas injected in the mixing section at the front of the barrel. Source: Michaeli et al. (2009).

48. Force–Deflection Curve

A curve obtained in measurements of the flexural properties of rigid polymers or composites by three-point bending methods.

49. Force–Deformation Curve

A curve obtained in compression tests on polymers and elastomers, or rubber.

50. Force–Extension Curve

A curve obtained in measurements of the tensile properties of polymers or composites in tensile tests.

51. Formica

A tradename for phenolic–paper laminate.

52. Formulation

Detailed list of type and amount of specific ingredients used in a polymer compound or resin mixture. The quantities used are based on parts per hundred polymer, normally abbreviated to phr, where 'r' stands for resin, owing to the traditional reference to thermosetting resin systems.

53. Forward Scatter

Represents the scattering of light from the surface of the medium at the exit side of the incident light flux (see diagram).

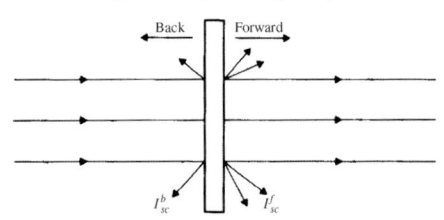

Illustration of forward and backward light scattering.

54. Fractal

A mathematical term that is used to characterize the features of a fractured surface that cannot be described by image analysis techniques in geometrical terms.

55. Fractography

The visualization, description and interpretation of images obtained by examination of fractured surfaces of polymers, adhesives or composites. A very smooth surface denotes the occurrence of a brittle fracture, while an uneven surface with asperities denotes a ductile failure, taking place through localized plastic deformations. (See Fracture mechanics.) For the case of fibre composites, the appearance of 'clean' fibre fragments is a manifestation of poor interfacial bonding.

56. Fracture Mechanics

A branch of science specifically devoted to establishing the principles and conditions for the fracture of products and to measurements of the related fundamental properties of structural materials. The concepts of fracture mechanics have been developed from studies of large plates containing a central crack of dimensions much smaller than the plane dimensions of the plate, that is, $a/W \sim 0$, where a is the length of the crack and W is the width of the plate. Note that the central crack simulates the effects of natural defects in a material, arising during manufacture or during its service life. If the plate is subjected to a tensile stress σ_0, there will be stress intensification at the crack tip in the way shown.

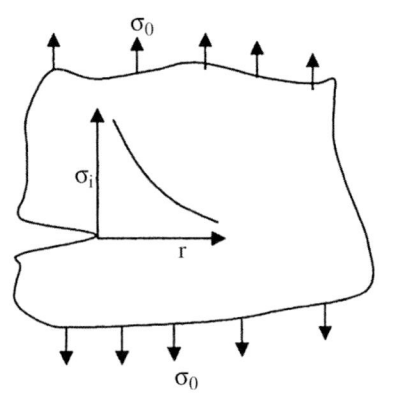

Stress intensification near a crack tip.

56.1 Concept of 'Stress Intensity Factor'

If the plate is very thin, the stresses in the thickness direction, σ_z, are very small and, therefore, do not have any effect on the fracture of the plate. This is known as the plane stress condition, as there are only stresses acting in the x and y directions of the plate. In such a case, the variation of the two stresses, σ_x and σ_y, with distance from the tip of the crack can be related to the dimensions of the crack by

$$\sigma_y = \frac{K}{\sqrt{2\pi x}} \quad \text{and} \quad \sigma_x = \frac{K}{\sqrt{2\pi x}},$$

that is, $\sigma_y = \sigma_x$, where x is the distance from the crack tip and a is half the crack length. Here $K = \sigma_0 \sqrt{\pi a}$ is known as the 'stress intensity factor', and is a factor that combines the effect of the size of the crack (defect) and the magnitude of the externally applied stress (σ_0) on the intensification of the stress ahead of the crack tip. The conditions that give rise to brittle fracture, that is, catastrophic crack propagation, can be expressed in terms of the critical stress intensity factor, K_c. For the testing of realistic specimens, that is, those with finite dimensions, the equation relating K to stress and geometric factors is written as

$$K = Y\sigma_0\sqrt{a},$$

where $Y = f(a/W)$ so that $Y = \sqrt{\pi}$ for $a/W \rightarrow 0$. This is to say, Y is a geometrical factor that depends not only on the a/W ratio but also on the manner in which the external stress are applied, for example, tension or bending.

56.2 Concept of 'Strain Energy Release Rate'

A rectangular specimen of an elastic material containing a single edge notch (SEN), subjected to a load, will store energy (U) equal to the area under the load–deflection (P versus Δ) curve. that is, $U = \frac{1}{2}P\Delta$. From the definition of specimen compliance, $C = \Delta/P$, the equation becomes $U = \frac{1}{2}P^2 C$.

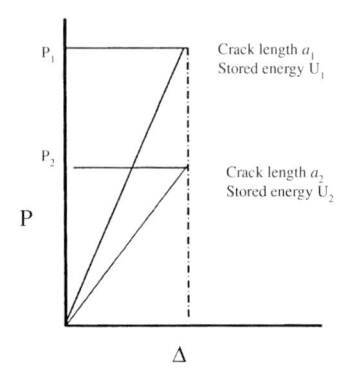

Load and stored energy at constant deformation Δ for specimens at two different crack lengths.

If the crack extends from an original value a_1 to a value a_2, at constant deformation (as in the diagram), there will be release of energy ΔU, which will make the load acting on the specimen drop from the original P_1 value to a new P_2 level. The specific energy released from the specimen as a result of the crack extension is $\Delta U/B(a_1 - a_2)$, where B is the specimen thickness and $B(a_1 - a_2) = B\Delta a$ (crack extended area). If the crack extension $(a_1 - a_2)$ is infinitesimally small, then the associated energy release rate will also be infinitesimally small, and the ratio $dU/B\,da$ is known as the 'strain energy release rate', G. Note that the strain energy release rate, G, can be used instead of the surface energy term 2γ in the Griffith equation, giving $\sigma = \sqrt{GE/\pi a}$. Substituting in this equation the expression $K = Y\sigma\sqrt{a}$, one obtains a relationship between K and G in the form of $K^2 = EG$. The value of the strain energy release rate, G, can be related to the load at the point when crack extension begins by

$$G = \frac{P^2}{2B}\frac{dC}{da},$$

where dC/da is the rate of increase of the specimen compliance with crack length.

This expression can be readily derived by differentiating $CP = \Delta$ (constant) and substituting in $dU/B\,da$. When conditions are reached for fracture to occur at a catastrophic rate, the value of the strain energy release rate is known as the G_c value, that is, the critical value. From the above discussion, it can be inferred that the critical strain energy release rate, G_c, can be measured by recording the load at which fracture occurs for a particular type of specimen, from a knowledge of the gradient dC/da at any pre-specified value of the crack length. The dC/da value is obtained from a calibration curve of the variation of the compliance with crack length for the same type of specimen and loading conditions as used for the fracture test. In principle there are three possible types of fracture modes, as shown in the diagram.

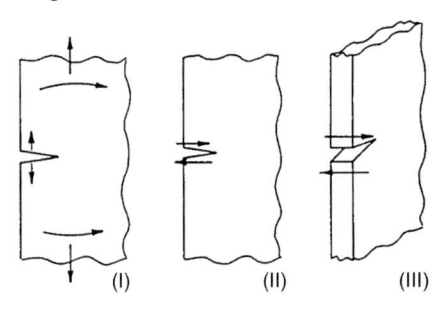

Fracture modes I, II and III for the evaluation of fracture toughness.

Mode I is the most widely used for the evaluation of the fracture toughness of materials. Mode II is sometimes used to measure the interlaminar fracture toughness of composites. Mode III is very rarely used.

57. Fracture Test

Destructive tests to measure the ultimate strength (stress at fracture) or the fracture toughness of a material. The strength is measured either on rectangular specimens for tests in three-point bending (flexural tests) or using dumbbell-shaped specimens for the case of tensile tests. (See Brittle strength.) Compression tests to measure strength are more widely used for high-performance composites than for conventional moulded polymer specimens. In all these cases the tests are carried out at low deformation rates using universal tensile testing machines. Fracture toughness is measured both at low deformation rates, using the same equipment, or at high speed, using pendulum or falling-weight (ball drop) testing equipment. (See Charpy impact strength.) In these tests one usually measures an empirical property, such as the ultimate tensile strength (stress at break) or the impact strength in terms of a specific energy to fracture (i.e. fracture energy recorded divided by cross-sectional area of specimen). With the introduction of fracture mechanics in the field of polymers in the 1970s, fracture tests have been adapted to measure fundamental parameters, such as K_c and G_c, instead of other empirical parameters. (See Fracture mechanics.) Tests at low speeds are carried out with the aid of compact rectangular specimens, as shown.

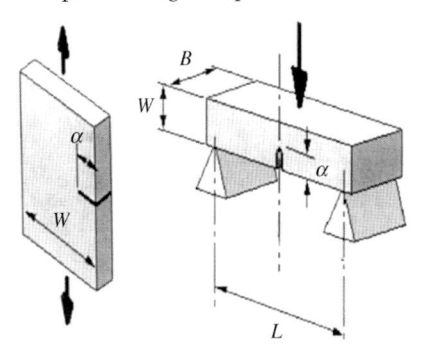

Single edge notch (SEN) specimens and loading modes in fracture toughness tests.

The Y values required to calculate K_c are obtained from the solution of a general polynomial

$$Y = C_0 - C_1(a/w) + C_2(a/w)^2 - C_3(a/w)^3 + C_4(a/w)^4.$$

The values of the constants C_0, C_1, C_2, C_3 and C_4 are obtained either experimentally or theoretically, assuming linear elastic behaviour. For the specimens shown, the values of the constants are given in the table.

Specimen type	C_0	C_1	C_2	C_3	C_4
Tension (compact)	1.99	−0.41	18.70	−38.48	53.85
Three-point bending ($L/W = 8$)	1.96	−2.75	3.66	−23.98	25.22
Three-point bending ($L/W = 4$)	1.93	−3.07	14.53	−25.11	25.80

The specific equations for K_c can be written directly in terms of the recorded fracture load P_f, knowing σ for the appropriate test.

- Tests in tension:

$$\sigma_0 = P_f / BW,$$

therefore

$$K_c = \frac{Y P_f}{BW} \sqrt{a}.$$

- Tests by three-point bending:

$$\sigma_0 = \frac{2}{3} P_f BW^2 / L,$$

therefore

$$K_c = \frac{3 Y P_f L}{2 BW^2} \sqrt{a}.$$

57.1 Pendulum Test

SEN rectangular specimens, shown on the right-hand side of the previous diagram, can also be used in Charpy and Izod tests to measure the fracture toughness of a material in terms of its G_c value. (See Charpy impact strength and Fracture mechanics.) With this approach, the energy recorded to fracture specimens (U) with different crack lengths is recorded and plotted against the product ϕBW, where ϕ is a geometrical calibration factor that depends only on the a/W ratio and on the span-to-thickness (S/W) ratio, as shown in the diagram.

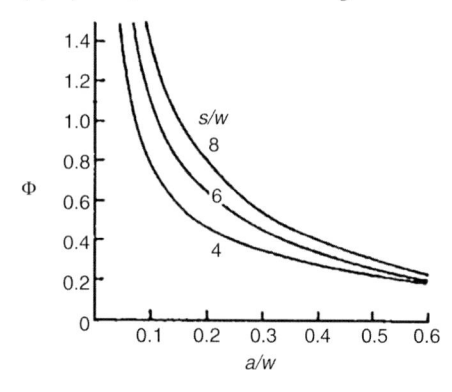

Variation of factor ϕ with a/W ratio at three S/W ratios in three-point bending tests. Source: Plati and William (1973).

The plot of the energy recorded to fracture specimens of different crack lengths against the product ϕBW gives a straight line. The slope of the straight line produced corresponds to the G_c value of the material, that is, $U = G_c \phi BW$. This equation is derived from the definition

$$G = \frac{P^2}{2B} \frac{dC}{da}$$

by letting

$$\phi = \left[\frac{C}{dC/d(a/W)} \right]^{-1},$$

where C is the compliance of the specimen at a specific a/W ratio. The values of ϕ can be obtained experimentally or can be calculated theoretically assuming linear elastic behaviour for the material tested.

A typical plot for the calculation of G_c from Charpy or Izod impact tests is shown in the diagram. The straight line obtained from the plot does not go through the origin, as the energy recorded by the instrument also includes the energy used to propel the specimen after fracture (i.e. the value of the energy at the intercept), which is constant and independent of crack length. (See Pendulum impact test.)

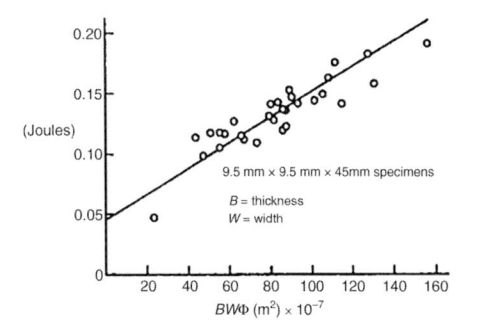

Plot obtained from Charpy impact tests on a PMMA cast sheet. Source: Plati and William (1973).

57.2 Falling-Weight (Ball Drop) Fracture Tests

In these tests, a sheet or disc is rested on a cylindrical support. A striker, or projectile, with a hemispherical tip is dropped from a given height on the sheet at the midpoint, as shown.

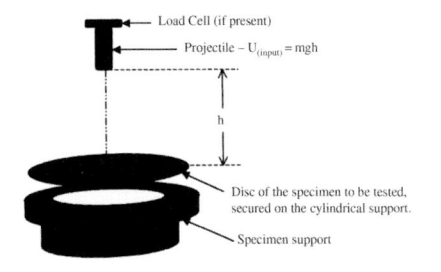

Sketch of the falling-weight (ball drop) impact test.

The evaluation of the impact strength of the material is carried out either by the 'staircase method' or on a 'probability of failure' basis at several levels of input energy, known as the 'probit' method. The staircase method consists of starting with a certain energy level ($U = mgh$) and examining whether or not this is sufficient to cause fracture of the sheet. With the 'probit' method, a number of specimens (usually 10) are subjected to impact at each of a series of input energies, and the fraction of specimens that have fractured at each level is recorded. These values are plotted against the input energy level, and the value that corresponds to 50% failure is used as the impact strength of the sheet. If a sufficiently large number of specimens is tested and the number of levels of input energies is realistic (at least five), the failure distribution curve should be of Gaussian type. Consequently, the plot of 'per cent failures' versus 'energy input' gives a sigmoidal curve. The median of the fracture energy (the so-called F_{50} value) gives a measure of the impact strength.

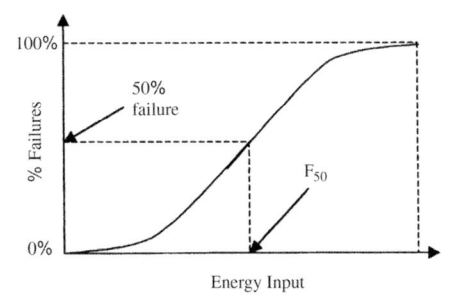

Probit plot for falling weight impact tests.

Nowadays most falling-weight apparatus are fitted with a load cell at the striker so that the force can be measured directly during the event and the energy can be computed from the area within the force–deformation curve recorded. The deformation is normally calculated from the velocity of the striker, assuming that it remains constant during the test. A typical force–deflection trace recorded in falling-weight impact tests is shown.

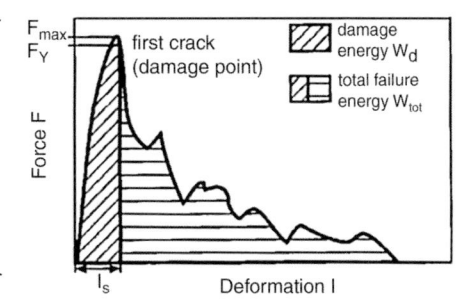

Typical force–deformation curve recorded in an instrumented falling-weight impact tester. Source: Ehrenstein (2001).

The trace identifies the peak load, which is normally assumed to correspond to crack initiation, and the rapid reduction in the load during the penetration of the indenter. The respective areas under the curve are also recorded and act as empirical parameters for the fracture toughness. The instrument can also be used for measuring the fundamental fracture toughness parameters, K_c and G_c, by using notched rectangular specimens tested in a cantilever mode, as in Izod tests, or in a three-point bending mode, as explained for Charpy tests. (See Fracture test and Fracture mechanics.)

58. Free Radical

Chemical species containing an 'odd' (unpaired) electron, formed by the symmetrical splitting of a covalent bond. Free radicals are very reactive, and for this reason they are used widely as initiators for polymerization. They are, however, also the cause of the rapid degradation reactions induced thermally or by UV light.

59. Free-Radical Polymerization

The reaction or process leading to the formation of high-molecular-weight polymers (thermoplastics) or networks (thermosets) from unsaturated monomers or oligomers via an 'addition' mechanism. Polymerization takes place in three stages.

Step 1. Initiation: The initiator (usually peroxides or azo-bis-isobutyronitrile, AIBN) decomposes to produce two free radicals, which add on to a double bond of the monomer(s) or oligomer molecules to start the polymerization, for example,

$ROOR \rightarrow 2RO^\bullet$ (peroxy radical) and then

$RO^\bullet + H_2C=CHX \rightarrow ROH_2C-CHX^\bullet$
(growing radical)

Step 2. Propagation:

$ROH_2C-CHX^\bullet + nH_2C=CHX \rightarrow RO(H_2C-CHX)_{n+1}^\bullet$ (high-molecular-weight chains)

Step 3. Termination: This represents the cessation of growth of the polymer chains through the extinction of free radicals, that is,

$RO(H_2C-CHX)_n^\bullet + RO(H_2C-CHX)_{n-1}^\bullet$
$\rightarrow RO(H_2C-CHX)_n(XHC-CH_2)_{n-x}OR$
(combination)

or

$RO(H_2C-CHX)_n^\bullet + {}^\bullet(XHC-CH_2)_{n-x}OR$
$\rightarrow RO(H_2C-CHX)_n-H_2C=CHX + (XHC-CH_2)_{n-x}OR$ (disproportionation)

However, termination often takes place through reactions of a growing polymer chain with other active species present, sometimes the solvent or additives used deliberately to control the size of polymer molecules. Other side reactions are also likely to take place alongside the propagation and termination reactions. In any case, a feature of free-radical polymerization is the presence of both original monomer(s) or oligomer at all times until the total conversion takes place. The use of mixtures of monomers results in the formation of random copolymers, whose composition depends on the relative reactivity of the monomers present and varies during the course of the polymerization reactions.

60. Free Volume

The volumetric fraction of spaces within the bulk of a polymer that is not occupied by molecular chains. These spaces are formed by the vibrational motions of segments of the polymer chains. The concept of free volume is used to explain the more rapid volumetric expansion taking place at temperatures above the T_g (glass transition temperature). The value of the free volume (f) increases linearly with temperature, T, according to the equation

$$f = f_0 + \alpha(T - T_0),$$

where α is the expansion coefficient and subscript 0 denotes a reference point. Above T_g, the value of α is much higher than that of the polymer in its glassy state.

61. Friction Coefficient

A parameter that denotes the resistance offered by the surface of a material to the sliding, or rubbing, action of a contacting object. The diagram shows the forces involved in defining the friction coefficient. From these, it is clear that the force P produces a compressive stress ($\sigma = P/A$) at the contact surface, while the drag force F_s produces a shear stress ($\tau = F_s/A$), where A is the contact area. The coefficient of friction (μ) is defined as $\mu = \tau/\sigma$.

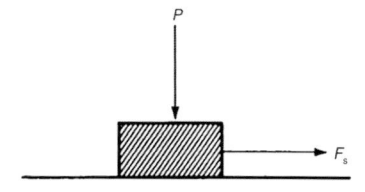

Forces involved in friction phenomena.

From the diagram, it can be observed that there are two coefficients of friction, a static coefficient (μ_s), related to the forces involved in starting the sliding action, and another, known as the dynamic coefficient of friction (μ_d), for situations where the forces considered are those involved in the movement of the sliding object. The latter is the parameter most widely used for comparison and design purposes. The μ_d values are usually slightly higher than the μ_s values. The coefficient of friction of a polymer is dependent not only on temperature but also on the sliding velocity and the magnitude of the load exerted on the object. The coefficient of friction measured at room temperature for the more common engineering polymers is around 0.2, with the exception of polytetrafluoroethylene (PTFE), for which the μ value is about 0.04–0.05.

62. Fringe Micelle Crystal

(See Crystalline polymer.)

63. Fumed Silica

Very fine amorphous silica particles with high surface area and high purity, obtained by precipitation from a sodium silicate solution through acidification. Used primarily as an antiblocking additive for films. (See Filler.)

64. Functional Filler

Fillers used in polymers to modify properties other than mechanical properties. Typical functional fillers are fire retardants, antitracking fillers and conductive carbon blacks.

65. Functional Polymer

Polymers exhibiting 'special' characteristics, such as intrinsic electronic conductivity and piezoelectric behaviour.

66. Functionality

A term that denotes the number of reactive groups or atoms present in reagents, such as monomers, oligomers, resins or hardeners. For instance, a monomer such as ethylene or styrene has a functionality equal to 2 insofar as the double bond can open to form two free radicals or two ions, which can react to produce linear polymers. A reagent such as triethylenetetramine,

$$NH_2-CH_2CH_2-NH-CH_2CH_2-NH-CH_2CH_2-NH_2,$$

on the other hand, has a functionality equal to 6, as there are six active hydrogen atoms in the amine functionality, capable of producing macromolecular networks with a high cross-linking density when used as a hardener for epoxy resins.

67. Functionalization

A chemical reaction that introduces functional groups (reactive groups) into a polymer, resin or oligomer.

68. Fundamental Property

Refers to a property defined and derived from fundamental principles. The value quoted for a fundamental property is independent of the test method used and of the geometry of the specimens used in the test.

69. Fungicide

An additive used to depress the growth of microorganisms. (See Antimicrobial agent.)

70. Furan Resin

Resins produced from condensation reactions of furfuryl alcohol in the presence of an acid catalyst to produce linear oligomers.

These oligomers can cross-link via condensation reactions in the presence of a strong acid to form a three-dimensional network.

71. Fusion Promoter

An additive used particularly in rigid PVC formulations in order to accelerate the fusion (melting) of particles in the screw channels of an extruder or injection moulding machine. Fusion promoters are also known as processing aids.

G

1. Gate

A small channel for the transfer of the melt from the runners to the cavities of a mould. Different types of gates are shown here schematically. (See Runner.)

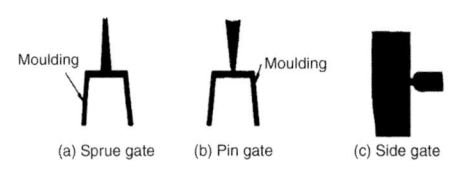

(a) Sprue gate (b) Pin gate (c) Side gate

Various types of gates used for the inlet of polymer melt into the cavity of an injection mould.

2. Gear Pump

A pump fitted between the barrel and the die of an extruder to increase the output and to prevent surging and irregular output. Gear pumps are used primarily for the extrusion of low-viscosity melts in the production of fibres and films. A schematic illustration is shown.

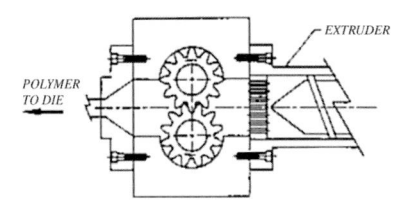

Positioning of a gear pump between screw tip and die. Source: Muccio (1994).

3. Gel

Describes the state of a polymer solution, a resin or an organic–inorganic hybrid that is intermediate between a solid and a liquid. Although the viscosity of a gel tends to infinity (the fluid loses its ability to flow), a gel is highly deformable. Gels consisting of mixtures of polymer and solvent, or polymer and plasticizer, are known as 'thermo-reversible' gels, which denotes their ability to go from the solid to the liquid state by changing the temperature. Gels derived from cross-linked polymers are called 'thermo-irreversible' gels owing to their inability to go back to a liquid state by increasing the temperature. The term 'gel' is also used to describe hard particles found in films, consisting of domains of cross-linked polymer, which are responsible for defects usually known as 'fish eyes'. (See Fish eye.)

4. Gel Coat

A term used to describe the coating of glass-reinforced plastic (GRP) hand-layup laminates produced by covering the inner surface of the mould prior to spraying the resin and the chopped glass rovings. Gel coats are usually based on resins that cure to a tough coating, and normally contain all the additives required to impart the required surface characteristics to the laminates, such as pigments, UV absorbers and antifouling agents.

5. Gel Permeation Chromatography (GPC)

A technique used to measure the molecular weight and molecular-weight distribution of polymers, which is also known as size exclusion chromatography (SEC). (See Molecular weight.)

6. Gel Time (or Gel Point)

Denotes the time taken for a thermosetting resin to assume the characteristics of a gel, through the formation of cross-links. The 'gel state' represents the transitional state between solid and liquid, which is identified by the point at which flow ceases as a result of the very rapid increase in viscosity. A sketch of a typical curve representing the change in viscosity with time for a liquid thermosetting system is shown.

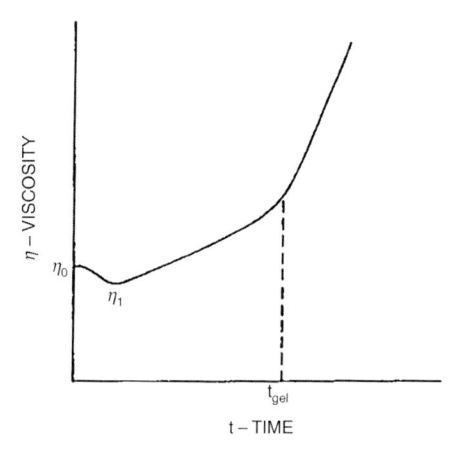

Change in viscosity with time for a typical thermosetting system. Source: Goodman (1998).

The initial decrease in viscosity results from an increase in temperature, arising from the exothermic nature of the reactions. The subsequent mild increase in viscosity takes place via the formation of isolated regions of highly branched or lightly cross-linked nuclei dispersed in the fluid, forming a 'sol'. As the number and dimensions of these domains increase, a stage will be reached at which they become immobilized and produce a 'gel', which is manifested as a very sudden and rapid increase in viscosity (gel point). The time required by the system for this to happen corresponds to the 'gel time'. A fundamental definition of 'gel time' can be derived from consideration of viscoelasticity principles under dynamic (cyclic) loading conditions. In these situations, it is possible to carry out the analysis in terms of either the complex shear modulus ($G^* = G' + iG''$, where G'' is the viscous component) or the complex viscosity ($\eta^* = \eta' - i\eta''$, where η'' is the imaginary viscosity associated with the melt elasticity component, that is, $\eta'' = G'/\omega$). The change that takes place in cross-linkable systems, from a liquid to a solid-like state, can be envisaged to have reached the transition point (gel point) for conditions in which $G' = G''$ or $\eta' = \eta''$ or tan $\delta = G''/G' = \eta''/\eta' = 1$. (See Viscoelastic behaviour, Dynamic mechanical thermal analysis and Complex compliance.)

7. Gelation

The event that leads to the formation of a gel from a liquid. This term is also used in PVC technology to denote the fusion of particles within the screw channels of an extruder or injection moulding machine, as well as the conditions achieved by a PVC paste in the first stage of a manufacturing process.

8. Gibbs Free Energy

Also known simply as free energy (G). It is a parameter of the second law of thermodynamics that defines the relationship between the internal energy (E) of a system and its entropy (S), that is $G = E - TS$, where T is the absolute temperature (kelvins). The concept is widely used in polymer science to determine whether certain phenomena are likely to occur. These include determinations of the solubility of a polymer in a solvent or the miscibility of a particular combination of two polymers. It is used as well to characterize the physical transitions of polymers resulting from changes in temperature.

9. Glass Bead

These are hollow beads, sometimes known as microspheres or Ballotini spheres, capable of providing apparent densities ranging from that of solid glass around 2.3 g/cm^3 down to 0.1 g/cm^3. Mostly produced from a chemically inert variety of silica–alumina glass.

10. Glass Fibre

Fibres used primarily for the production of fibre-reinforced polymers. Glass fibres are usually made from E glass, which has the following composition: silicon dioxide, 53%; calcium oxide, 21%; aluminium oxide, 15%; boron oxide, 9%; magnesium oxide, 0.3%; and other oxides, 1.7%. The structure is basically a lime–alumina–borosilicate glass consisting of a three-dimensional network of silica and other oxides. The letter E originates from the original designation for 'electrical grade' for its superior electrical properties compared to ordinary alkali glass (A glass). There is also considerable use of S glass fibres (the letter S stands for 'strong') for the production of high-performance composites owing to their higher modulus and higher strength than E glass fibres. Another variety, known as C glass fibres (C stands for 'chemically resistant'), is sometimes used for applications requiring stronger resistance to strong acids and alkalis. (See Composite.) Glass fibres have a monolithic amorphous structure as shown in the micrograph.

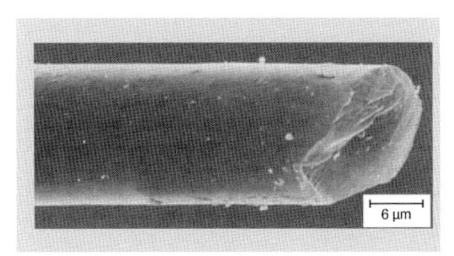

Typical monolithic structure of a glass fibre. Source: Ehrenstein (2001).

The commercial grades are coated with a thin layer (about 0.3–0.5 µm) of a 'size' to enable them to be handled during their production and the manufacture of composites. (See Size.) The diameter of glass fibres varies according to applications, from about 6 µm to about 20 µm.

11. Glass-Filled Polymer

Polymers reinforced with short glass fibres. (See Composite.)

12. Glass Flake

A platelet type of filler, usually made from chemically inert C glass, with a thickness around 3 µm and lateral dimensions of 0.5–3 mm. Glass flakes are widely used in coatings to provide barrier characteristics and chemical resistance to acids and alkalis.

13. Glass Transition Temperature

Also known as the glass–rubber transition (T_g), represents a reference temperature for the transition from the glassy state to the rubbery state. The T_g is thermodynamically classified as a second-order transition. (See Deformational behaviour.)

14. Glassy State

This is the state of polymers at temperatures below the glass transition temperature. In the glassy state, the polymer acquires its highest achievable Young's modulus, usually around 3 GPa. (See Deformational behaviour.)

15. Gloss

A property that denotes the shiny appearance of a surface, which is defined as the ratio of the intensity of light reflected within

16. Graft Copolymer

A block-type copolymer in which the second component emerges as a side chain from the backbone of the host polymer.

17. Graphene

The name for the structural sheet units of graphite lying in parallel planes at approximately 0.334 nm distance, separated by electrons in π–orbitals. The graphene planes are staggered as shown.

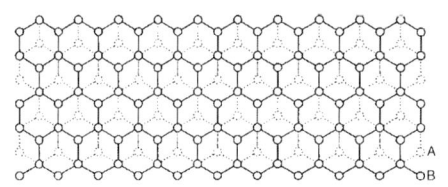

Stacking of graphene structural units in graphite. Source: Bell et al. (2006).

18. Graphite

A crystalline form of carbon characterized by a graphene layered structure found in carbon fibres. (See Carbon fibre and Composite.)

19. Griffith Equation

An equation that relates the strength (σ^*) of a brittle material to the size of internal cracks and to the intrinsic properties of the material, respectively, the surface energy (γ) and Young's modulus (E). For a thin infinite plate containing a central crack (a), the Griffith equation is usually written in the form

$$\sigma^* = \sqrt{\frac{2\gamma E}{\pi a}},$$

where a is the length of the central crack. (See Fracture mechanics.)

20. Gum Stock

The mixture of elastomer with all auxiliary ingredients and additives before curing.

21. Gutta Percha

A natural polyisoprene rubber. (See Natural rubber.)

H

1. Halogenated Fire Retardant

These are fire retardant additives containing either bromine or chlorine. (See Flame retardant.)

2. Hardener

The component of a resin mixture used to produce the network in cross-linked polymer products, normally known as thermosets. (See Epoxy resin.)

3. Hardness

An empirical mechanical property denoting the resistance of a polymer to the penetration of a sharp device made from a material with a much higher Young's modulus, such as a metal, glass or ceramic. In standard test methods, the hardness is numerically expressed in arbitrary units specifically derived from the method used.

The most widely used standard test methods for measuring the hardness of polymers are the Shore hardness (scales A and B) for soft polymers and the international rubber hardness (IRH) for elastomers, both of which use a needle penetration device.

For rigid polymers, the most widely used methods are the Rockwell hardness (scale R, L, M and E) and the Brinell hardness, both of which use a sphere penetration approach to assess the resistance to indentations. Widely used by research workers is the Vickers hardness, which is based on the penetration of a diamond-shaped indenter. The hardness value is expressed in terms of the load required to induce a specified amount of penetration of the 'indenter' from the surface into the bulk of a specimen.

For surface coatings, the hardness is often measured by the 'pencil' test, which identifies the minimum pencil hardness that can cause a scratch on the surface under specified conditions. More recently, nano-indentometers have been developed to measure the hardness in a manner similar to the Vickers method using a very fine indenter subjected to very small loads.

4. Haze

An optical property of polymer products, such as films, that describes the milky appearance resulting from the forward scattering of light from the surface at angles within the range of 2.5–90°. For this reason, haze is often referred to as 'multi-angle scatter', defined as the fraction of the total transmitted light. The scattering of light from the surface of films is usually attributed to surface irregularities resulting from the formation of spherulitic crystals. This is confirmed by the photograph, which shows an image of text seen through a polyethylene film exhibiting haze. The details of the print increase considerably in the central area of the film covered with a drop of cassia oil owing to the good match in refractive indices of the two media, which prevents the scattering of light from the surface of the film.

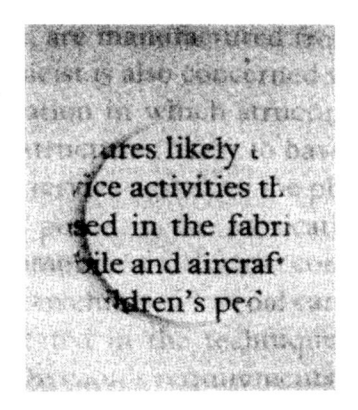

Evidence for the association of haze with surface light scattering. Source: Ross and Birley (1974).

Polymers in Industry from A–Z: A Concise Encyclopedia, First Edition. Leno Mascia.
© 2012 Wiley-VCH Verlag GmbH & Co. KGaA. Published 2012 by Wiley-VCH Verlag GmbH & Co. KGaA.

5. Head-to-Head

A term used in polymerization to denote the formation of a chemical bond between two free radicals attached to the side containing a bulky group (α position), such as in styrene or vinyl chloride, for example,

$$ClCH-CH_2-ClCH-CH_2$$
$$-CH_2CHCl-CH_2CHCl-$$

This type of reaction takes place primarily as chain-to-chain recombination or termination reactions.

6. Head-to-Tail

A term used in polymerization to denote the formation of a chemical bond between a free radical at the end of a growing polymer chain containing a bulky group, such as in styrene or vinyl chloride, and the unhindered end of a monomer molecule, so that the resulting polymer chain will have alternative heads and tails, for example,

$$-CH_2CHCl-CH_2CHCl-CH_2CHCl-$$

7. Heat Build-Up

A term used in the rubber industry to denote the conversion of mechanical energy into heat under cyclic loading conditions. (See Loss angle.)

8. Heat Distortion Temperature (HDT)

An empirical parameter that denotes the resistance of a polymer to deformations at high temperatures. It is measured by standard three-point bending methods on specimens immersed in an inert liquid and heated at a constant heating rate. The central deflection is plotted as a function of the temperature and the HDT value is registered as the temperature at which a certain specified deflection is reached. The values for the applied load and the target deflection are determined by the specific standard method, usually 1 mm.

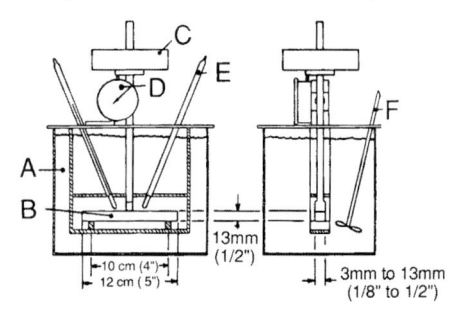

Typical HDT apparatus: A, heating liquid; B, test specimen; C, applied load; D, deflection monitoring dial gauge; E, thermometer; F, stirrer. Source: Unidentified original source.

9. Heat Setting

An operation carried out on filamentary products after they have been drawn in order to achieve dimensional stability. Heat setting is carried out by allowing the filaments to retract slightly by the application of heat at temperatures just below the melting point of the polymer in order to develop the maximum level of crystallinity through secondary crystallization and to relax any polymer orientation within the amorphous regions.

10. Heat-Shrinkable Product

Products usually in the form of flexible tubing, tapes or sealable wrappers that will shrink when heated above the glass transition temperature (T_g) for an amorphous polymer or above the melting point (T_m) for a crystalline polymer. This behaviour derives from the natural characteristic of the molecules in oriented polymers, which are driven by the second law of thermodynamics to assume a random coiled configuration to achieve the most stable high-entropy state. In practice, the polymer is usually slightly cross-linked to convert them into a rubber at the required shrinkage temperature, thereby protecting the product against damage and loss of dimensions.

11. Heterochain

A polymer chain where the constituents contain a mixture of carbon and other atoms, for example, in polyesters, polyethers, polyamides and polysulfones.

12. Heterogeneous Blend or Mixture

A blend or mixture of polymers that consists of two or more distinct phases. This situation arises when two or more polymers are not completely miscible.

13. Heterogeneous Nucleation

(See Crystallization.)

14. Heterogeneous Polymerization (or Heterophase Polymerization)

A generic term for polymerization taking place in a two-phase medium. This term includes the polymerization of liquid monomers in suspension or emulsion in water or supercritical CO_2.

15. High-Density Polyethylene (HDPE)

A polyethylene with density between 0.935 and 0.955, which is related to the degree of crystallinity, usually in the range 0.55–0.70. The higher degree of crystallinity in HDPE, relative to that in the corresponding medium- and low-density polyethylene grades, is responsible also for the higher rigidity and lower transparency. (See Ethylene polymer.)

16. High-Impact Polystyrene (HIPS)

(See Styrene polymer.)

17. High-Temperature Polymer

A polymer with a glass transition temperature $T_g > 150\,^{\circ}C$ or a melting temperature $T_m > 300\,^{\circ}C$ and exhibiting a high resistance to thermal oxidation to enable it to be used continuously at high temperatures. (See Polybenzimidazole, Polyimide, Polyketone, Polysulfone, Poly(phenylene sulfide) and Poly(amic acid).)

18. Hindered Phenol

Type of antioxidant based on phenol compounds in which the *ortho* and *para* positions are occupied by tertiary butyl groups to produce a hydrogen-free carbon atom in the aromatic ring. This is to stabilize the free radical formed on the phenolic oxygen after the hydrogen atom has been abstracted in the reaction with a free radical on the polymer chains formed through degradation reactions. The mechanism is illustrated.

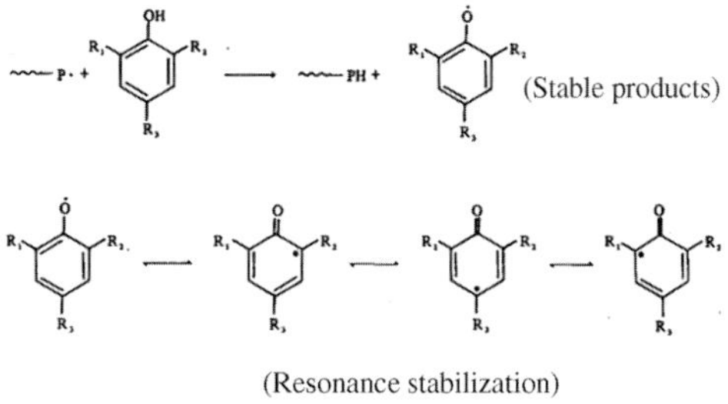

(Stable products)
(Resonance stabilization)

Mechanism for the free-radical quenching by phenolic antioxidants.

19. Homogeneous Blend

A blend of polymers exhibiting only one phase at microscopic level. This situation arises when two or more polymers are completely miscible.

20. Homopolymer

A polymer containing only one monomeric unit along the molecular chains.

21. Hooke's Law

A law that describes the relationship between the applied force and the resulting deformation of an elastic material. Accordingly, the deformation is directly proportional to the applied force, normally referring to forces acting in tension or compression. Hooke's law forms the basis on which Young's modulus is defined by converting the force into the related stress and the deformation into the related strain. (See Elastic behaviour.)

22. Hot-Melt Adhesive

(See Adhesive.)

23. Hot Runner Mould

An injection mould in which the runner region of the mould is kept hot by external heaters in order to prevent the contained polymer from solidifying along the actual moulded part. The ejected moulded part does not contain the usual sprue, thereby reducing the amount of polymer waste. (See Gate and Runner.)

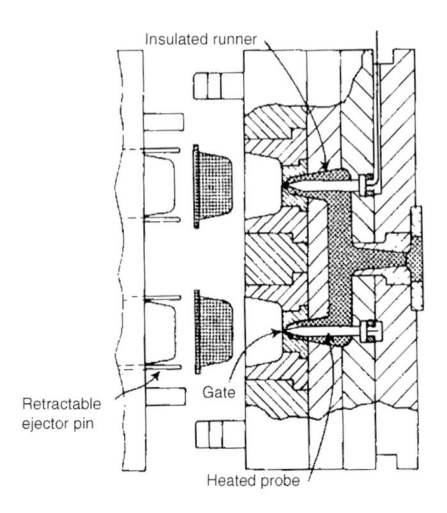

Example of hot runner mould used for injection moulding of thermoplastics. Source: McCrum *et al.* (1988).

24. Huggins Constant

A constant in the relationship between the intrinsic solution viscosity and the viscosity number, used in the determination of the molecular weight of polymers, that is,

$$(\eta_2 - \eta_1)/\eta_1 c = [\eta] + kc,$$

where $k = k'[\eta]^2$ and k' is known as the Huggins constant, with values between 0.3 and 0.4. (See Molecular weight.)

25. Hundred-Percent Modulus (100% Modulus)

The tensile modulus of a rubber or elastomeric material calculated by taking the value of the nominal stress at 100% extension in the force–extension curve.

26. Hydrogen Bond

The strong physical interaction between two groups from adjacent molecules or segments. These involve the attraction of a hydrogen atom chemically linked to an electronegative atom (usually N, O or halogen) from one molecule with an electronegative atom from another molecule. Hydrogen bonds can also occur within a polymer chain (intramolecular hydrogen bonds), as in the case of polyamides, which provide a good example to illustrate how hydrogen bonding can affect the properties and behaviour of polymers. In this case, a strong interaction occurs between the hydrogen atom of the amide group of one molecular segment and the carbonyl group of the amide group of another molecular segment. These interactions are responsible for the higher melting point of polyamides relative to that of polyesters with similar structure. One notes that the absence of such hydrogen atoms in the polyester chains can only provide polar attractions, which are weaker than hydrogen-bonding interactions.

27. Hydrolysis

The process by which chemical bonds within a polymer chain, or macromolecular network, are broken by the action of water. Hydrolysis is, therefore, the reverse of condensation. (See Condensation reaction.) When hydrolysis takes place within the polymer chains, as in the case of polyesters, polyamides, polyurethanes and polyimides, hydrolysis brings about a decrease in the molecular weight of the polymer, which causes a deterioration in mechanical properties.

28. Hydrolysis Stabilizer

Additives used to enhance the stability of a hydrolysable polymer, such as polyesters or polyurethanes, against degradation brought about by hydrolysis. The more widely used hydrolysis stabilizers are based on sterically hindered polycarbodiimides, which can react with water, according to the scheme:

$$RN=C=NR + H_2O \rightarrow RNH-CO-NHR$$

In the case of polyesters, the polycarbodiimide stabilizer can react also with any carboxylic acid groups present at the end of the chains, thereby reducing their catalytic activity on the rate of hydrolysis. The reaction of carbodiimides with carboxylic acid takes place according to the scheme:

$$RN=C=NR + R'COOH \rightarrow \underset{\underset{R'}{|}}{\underset{CO}{|}}{RN-CO-NHR}$$

29. Hydrophilic

A substance that attracts water, behaviour that is driven by the formation of hydrogen bonds between water molecules and highly polar groups in the substance.

30. Hydrophobic

A substance that does not attract or absorb water, behaviour arising from the presence of non-polar groups as in hydrocarbons and especially fluorocarbons.

31. Hydrostatic Pressure

Pressure applied evenly in all three directions.

32. Hydrostatic Stress

Tensile or compressive stresses evenly applied in all three directions.

33. Hydrothermal Stress

Internal stresses created in a product by changes in temperature.

34. Hydroxyl Equivalent (Number)

Defines the quantity of hydroxyl groups present in oligomers, resins or polymers, which corresponds to the weight in grams containing one gram equivalent (mole) of hydroxyl groups.

35. Hygroscopic

Capable of absorbing water, a characteristic related to the ability of water molecules to form hydrogen bonds with hydrophilic groups present in products.

36. Hyperbranched Polymer

Polymers whose chains branch out in three dimensions from a central multifunctional unit, continuing to form regularly structured branches as the chains grow in successive steps into highly ramified polymer molecules. Hyperbranched polymers are usually polyesters or polyamides, or mixed types, obtained from condensation reactions. Hyperbranched polymers are generally brittle and are used primarily as functionalized modifiers for thermosetting resin systems, producing regions with a well-defined structure and providing features that may give rise to enhanced properties and rheological behaviour. An example of the structure of a hyperbranched polymer is shown.

Hyperbranched aliphatic polyester produced from the polycondensation of bis(hydroxymethyl)propionic acid (bis-MPA) with 2-ethyl-2-(hydroxymethyl)-1,3-propanediol (TMP).

I

1. Impact Modifier

A general term for an auxiliary component of a polymer or resin formulation used to increase the toughness of products. In the case of thermoplastics, an impact modifier is incorporated into a brittle polymer matrix as pre-formed 'toughening' particles. The composition and morphology of the particles are designed in such a way as to promote strong interfacial bonds between the toughening particles and the surrounding polymer matrix, thus promoting an efficient stress transfer mechanism between the two phases. In this way the localized strain energy resulting from impacts can be redistributed through the bulk via the toughening particles, thereby preventing the formation and propagation of cracks. An example of this 'toughening' mechanism is the use of acrylic or ABS impact modifiers in rigid PVC formulations. There are, however, a number of theories that have been put forward for the toughening mechanism of polymers and, indeed, more than one mechanism can operate, depending on the nature of the polymer. For HIPS and ABS, for instance, it has been proposed that toughening takes place via the formation of crazes through the matrix between rubber particles, as a mechanism for absorbing the strain energy (see diagram).

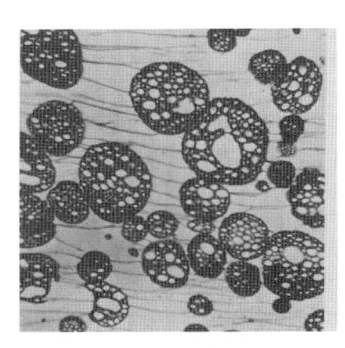

Crazes formed between dispersed particles of a high-impact polystyrene sample. Source: Unidentified original source.

It is possible, however, that the formation of crazes represents a 'second-stage' event, which follows the energy absorption mechanism by molecular relaxation within the 'interphase' regions, consisting of miscible domains of the two components. For thermosetting resins, the particles that bring about the toughening of the matrix are generated *in situ* through the addition of specially designed oligomers that react with the resin and hardener. The rubbery particles are nucleated before 'gelation' of the resin takes place and grow into larger particles as a result of the migration of reactive species from the surrounding resin mixture. The growth of precipitated particles ceases when the surrounding matrix 'gels' through the formation of a continuous network. The infusion of reactive species, such as the hardener, from the surrounding resin into the precipitated particles may nucleate the formation of other particles, which remain quite small owing to the constraints imposed by hardening of the matrix surrounding the larger particle agglomerates. (See Epoxy resin and Phase inversion.)

2. Impact Strength

Denotes the resistance of a material to impact loads, normally expressed in terms of the energy required to induce fracture of specific specimens. In this respect the term 'strength' is a misnomer insofar as strength normally denotes the value of the stress (force per unit area) required to fracture a specimen. Impact strength is usually measured with the use of pendulum equipment or by falling-weight methods. (See Impact test, Fracture test and Fracture mechanics.)

Polymers in Industry from A–Z: A Concise Encyclopedia, First Edition. Leno Mascia.
© 2012 Wiley-VCH Verlag GmbH & Co. KGaA. Published 2012 by Wiley-VCH Verlag GmbH & Co. KGaA.

3. Impact Test

Tests carried out by delivering loads at very high speed. (See Charpy impact strength, Izod impact test and Falling-weight impact test.)

4. Induction Time

A generic term used to denote the delay time for the start an event or process. Three specific examples of induction time experienced in polymer systems follow.

a) **Degradation and stabilization** The oxygen uptake of a polyolefin sample can be measured prior to reactions that lead to the formation of carbonyl groups in the chains, induced thermally or by the action of UV light. The efficiency of a stabilizer can, therefore, be assessed in terms of the increase in induction time that it brings about in a sample, as illustrated in the diagram.

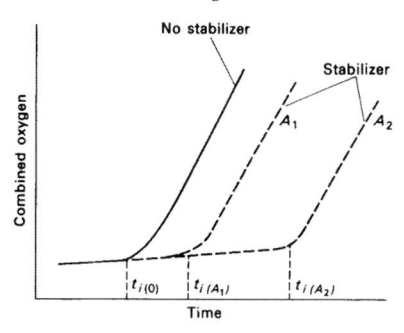

Evolution of combined oxygen with ageing time.

b) **Inhibition of polymerization** The inhibition of polymerization or curing reactions by free radicals can be achieved using quinone. Very small amounts of inhibitor are added to an unsaturated monomer or oligomer to react with the initial free radicals by exposure to light and/or oxygen in order to convert them into inactive radicals. (See Antioxidant.) In this respect, inhibitors must be distinguished from retarders, which decrease the rate of reactions rather than increasing the induction time for the reaction to take place.

c) **Liquid and gas absorption** Solvent can be taken up by a glassy polymer sample via a case II diffusion mechanism. The diagram shows the absorption of tetrahydrofuran (THF) in three samples of 'modified' epoxy resin systems. The top two curves refer to a homogeneous glassy structure, the second relating to the same resin containing pre-formed silica particles. In both cases absorption takes place without an induction time. On the other hand, in the third sample, a distinct induction time and a large reduction in total amount of absorbed THF are observed. There, the silica was formed *in situ* as three-dimensional domains, which represents a typical structure of organic–inorganic hybrids. (See Organic–inorganic hybrid.)

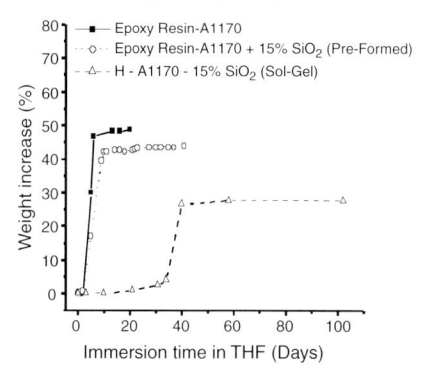

Solvent uptake of an epoxy–silica hybrid. Source: Prezzi (2003).

5. Infrared Spectroscopy (IR or FTIR)

An analytical technique, known also as vibrational spectroscopy, used to identify specific chemical groups. The technique detects the transition between energy levels in molecules resulting from vibration of interatomic bonds within a chemical group. At low temperatures, molecules exist in their ground vibrational state. Therefore, in order to bring them to a higher energy state, it is necessary for them to absorb energy from the surroundings. In vibrational spectroscopy, this is done by subjecting a sample to electromagnetic (EM) radiation of a range of frequencies (a procedure known as scanning), and monitoring the intensity of the transmitted radiation. Chemical groups will absorb energy at specific frequencies through resonance with the natural vibrations of the constituent atomic bonds, so that they can be identified through a calibration procedure, which involves scanning a compound of known composition. For a standard IR analysis, the wavelength of the EM radiation is in the range 2.5–25 µm, corresponding to a frequency with wavenumber 4000–40 cm^{-1}. (Note that the wavenumber in cm^{-1} is equal to 10^4/wavelength in µm.) The spectral position of the most common chemical groups in polymers that can be characterized by IR spectroscopy is indicated in the diagram.

In order to speed up the calculations required to process the acquired data, a mathematical technique, known as Fourier transforms, is often used in the software. The technique is referred to as Fourier transform infrared (FTIR) spectroscopy.

6. Inhibitor

An additive used to reduce the susceptibility of reactive species, such as monomers and resin–hardener mixtures, to undergo polymerization reactions during storage. The function of an inhibitor is to react with an active radical to give products with much lower reactivity. A classical inhibitor for free-radical polymerization is benzoquinone. Although hydroquinone is often used as a monomer 'stabilizer', it relies on the presence of oxygen to be transformed into an

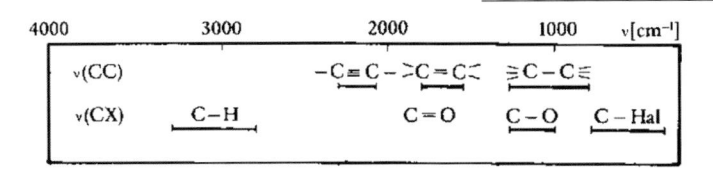

Spectral position, as absorption wavenumber (cm^{-1}), for (CC) and (CX) valency vibrations. Source: Kampf (1986).

Some spectrometers operate in the near-infrared (NIR) region 0.7–2.5 µm (4000–1400 cm^{-1}) and others in the far-infrared (FIR) region 50–800 µm (200–12 cm^{-1}). A typical IR spectrum obtained from a polymer sample is shown.

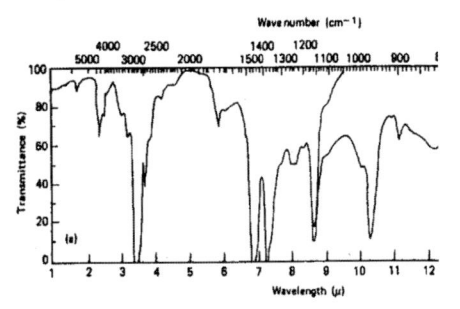

Typical infrared spectrum of a polymer.

inhibitor through oxidation to quinone. In effect, the inhibitor increases the induction time for the onset of propagation reactions.

7. Initiator

An additive that starts the reactions leading to polymerization or network formation in free-radical reaction systems. (See Free radical and Peroxide.)

8. Injection Blow Moulding

A moulding technique for the production of containers, consisting of two separate processes: respectively, the production of the 'pre-form' by injection moulding, and the subsequent 'blowing' operation into the final dimensions. This allows the blowing process to take place under highly controlled temperature conditions so that the optimum level of orientation can be introduced in the walls of the containers. A longitudinal pre-stretching operation is frequently used in order to obtain a well-balanced degree of biaxial orientation by the blowing operation. For this reason, the process is also known as 'stretch–blow moulding'. The most widely used polymer for injection blow moulding applications is poly(ethylene terephthalate) (PET). In this case the heating of the pre-forms for the blowing operation is carried out in-line by infrared radiation to temperatures in the region of 110–115 °C.

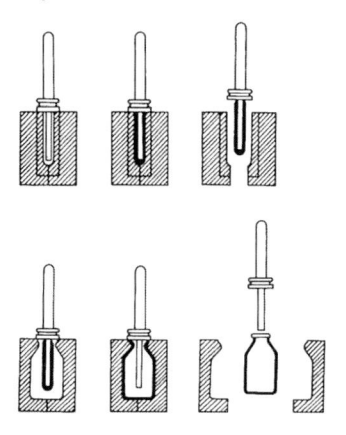

Sequential operations in injection blow moulding of bottles: (top) injection moulding of pre-forms; (bottom) stretch–blow operation. Source: Unidentified original source.

9. Injection Moulding

A technique used primarily for moulding thermoplastics. It is carried out by pumping a melt at high speed (producing shear rates around $10^4 \, s^{-1}$) into the cold cavities of a mould, where the polymer is allowed to cool to a suitable temperature before the moulded parts are ejected from the mould. The ejection temperature is below the glass transition temperature (T_g) for a glassy polymer and well below the melting point (T_m) for a crystalline polymer.

The injection process involves three consecutive operations carried out in a screw–barrel assembly similar to that used for extrusion. (See Moulding cycle.) First the polymer granules or powder are fed from a hopper into the feeding section of the screw. Then melting takes place while the screw rotates, delivering at the same time the required amount of melt to the front of the barrel via a non-return valve, which results in a pressure build-up in the melt at the nozzle. The screw then stops rotating and comes forward at high speed, as a plunger, to inject the melt into the cavities of the mould through the nozzle of the injection unit of the machine connected to the mould inlet, via a 'sprue', runners and gates. (See Runner and Gate.) A diagram of an injection moulding machine, showing also the mould with a hydraulic locking mechanism, is shown.

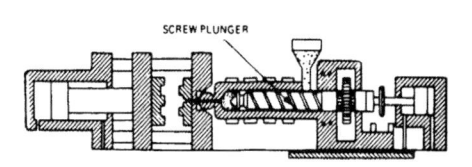

Simplified drawing of an injection moulding machine for thermoplastics, showing the plasticizing and injection unit (right), and the mould and hydraulic locking mechanism (left). Source: Unidentified original source.

In injection moulding of thermosetting plastics or rubber compounds, the mould is at a higher temperature than the barrel in order to produce fast 'curing reactions' after the cavity has been filled. The heat generated by friction in the injection unit and during the flow through the nozzle and the sprue–runner–gate sequence in the mould brings the temperature to the level required for curing the polymer in the cavities. Ejection takes place while the part is still hot owing to the cross-linked nature of the product, which prevents distortion of the moulded part during ejection. The screw, nozzle and mould channels carrying the melt to the mould cavities have to be designed in such a way as to prevent curing before the melt reaches the cavities. For this reason, obstructive features, such as a non-return valve at the front of the screw, have to be excluded. A ram-type injection unit is sometimes used for compounds that are too sensitive to the heating induced by the shearing action of the plasticizing screw.

There are several variants for the injection moulding of thermoplastics, two of which are described here.

a) **Co-injection moulding** This is used to obtain two-layer or three-layer moulded parts. This technique, also known as 'sandwich moulding', is designed in such a way as to enable the first shot entering the cavity to form the outer skin of a three-layer structure. After this event, the second unit injects the other polymer through the same 'sprue' of the mould, so that the incoming melt will flow through the central region, pushing the outer layers against the walls of the mould. The main steps are illustrated schematically.

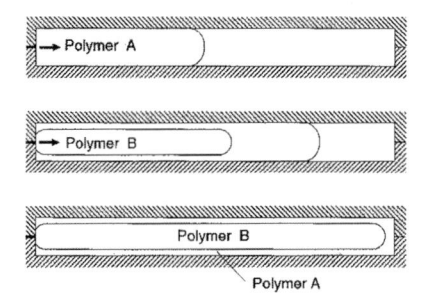

Principle of co-injection, whereby the second polymer (B) drives the first polymer (A) into the outer sections. Source: Osswald (1998).

One variant of the co-injection process is gas-assisted moulding, where the inner component is a gas (e.g. nitrogen), which 'blows up' the molten polymer against the cavity walls to produce hollow sections. Another version is the production of sandwich foamed mouldings, where the inner polymer contains a blowing agent to nucleate the formation of 'gas cells' and subsequent expansion into a foam.

b) **Structural foam moulding** This consists in the rapid injection of a melt containing the blowing agent into fairly thick section cavities of the mould, whereby the rapid cooling of the melt at the walls of the mould forms a solid outer skin. In this way, foaming is restricted primarily to the inner sections. (See Foam and Mould.)

10. Insulation

A component of a structure or an electrical circuit that prevents the escape of energy to the surroundings. The main type of insulation involving the use of polymers are: (a) thermal insulation, which prevents heat from escaping through the walls of a structure, and (b) electrical insulation, which prevents the leakage of current from a circuit. (See Foam, Thermal conductivity and Dielectric.)

11. Interaction

This term is often used to describe the attractions between certain groups or atoms present in polymer chains. Interactions can be intermolecular (between different molecules) or intramolecular types. The strength of the interactions varies from the weakest type, known as van der Waals, occurring in non-polar polymers, to the strongest type involving electrostatic forces between cations and anions, known as ionic interactions. Between the two there are dipole–dipole interactions and hydrogen bonds. The three main types of interaction are shown schematically. (See Hydrogen bond, Dipole and Ionomer.)

surface area. (See Dynamic mechanical thermal analysis and Law of mixtures.)

Under 'perfect' bonding conditions at the fibre–matrix interface of a composite, there is no contribution to viscoelastic losses from the fibres owing to their elastic nature.

Dipole-dipole *H-bond* *Ionic*

12. Intercalation

A term used to describe the insertion of chemical species into a crystal lattice containing empty lattice sites. (See Exfoliated nanocomposite and Nanofiller.)

13. Interfacial Bonding

A term that describes the attractive forces between the surfaces of two different materials in contact with each other. Typical examples of interfacial bonding in polymer-based products are in laminates, adhesives and composites. Although destructive mechanical tests are usually used to assess the interfacial bond strength, it is often difficult to discern from the results the relative contributions by the bulk of the adhering material and the actual interfacial bonding. The interlaminar shear strength test and lap shear test are described elsewhere. (See Composite test and Adhesive test.) An assessment of the interfacial bonding between fibres and matrix in composites can be obtained with the use of dynamic mechanical tests, in view of the large interfacial

Consequently, the loss component of the complex modulus of the composite (E_c'') is directly proportional to the volume fraction of the matrix (ϕ_f), that is,

$$E_c'' = (1-\phi_f)E_m''.$$

For 'poor' bonding situations, there will be additional losses resulting from the frictional (sliding) movements at the interface, which gives rise to an additional term in the loss modulus equation, that is,

$$E_c'' = (1-\phi_f)E_m'' + E_{m\text{-}f}'',$$

where $E_{m\text{-}f}''$ represents the related increase in loss modulus associated with energy losses at the interface. By comparing the dynamic mechanical spectrum obtained for the matrix component without fibres to that for the corresponding composites, it is possible to calculate the theoretical values for $E_{m\text{-}f}''$ over a wide temperature range of the spectrum. Any discrepancy between the measured and calculated E_c'' values can be attributed directly to interfacial losses. The schematic diagram indicates that very poor interfacial bonding can give rise to large values for

$E''_{m\text{-}f}$, particularly at temperatures around the T_g of the matrix. It follows that the strength of the interfacial bond can be quantified as the difference between the measured E''_c and that calculated from the law of mixtures for $E''_f = 0$. The larger the discrepancy, the lower the strength of the interfacial bond. Note that the geometric arrangement of the fibres is expected to have a minor effect on the measured E''_c value.

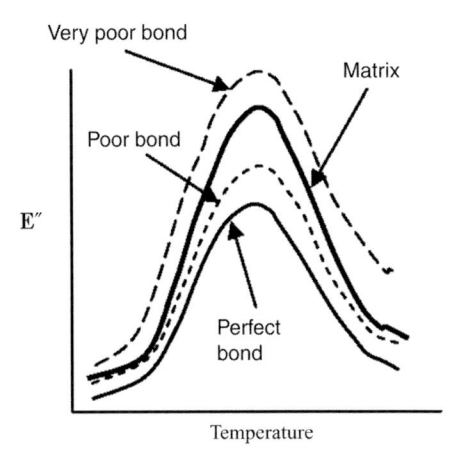

Effects of descriptive strength of interfacial bond on loss modulus of composites relative to the matrix. Source: Constructed from Mascia (1974).

14. Interfacial Polarization

This term refers to the polarization of dipoles present at the interface between filler and polymer or between the amorphous and the crystalline domains of a polymer, which results from an externally applied voltage. This is often the cause of unexpected higher losses than expected in polymers with low polarity.

15. Interlaminar Failure

Failures taking place by the breaking of the bond between layers of laminates. (See Delamination.)

16. Interlaminar Shear Strength (ILSS)

The stress required to break the bond between fibre layers in continuous-fibre composites. (See Composite test.)

17. Internal Energy

The total energy of a material, usually taken to be the sum of the thermal energy (enthalpy) and the product of external pressure and volume of the material, that is, $E = H + PV$. (See Gibbs free energy.)

18. Internal Lubricant

A long-chain alkyl compound with a hydrophilic terminal group that is miscible with the polymer at low concentrations and reduces the melt viscosity. (See Lubricant.)

19. Internal Mixer

The term indicates that mixing takes place inside a closed chamber for a specific length of time. An internal mixer is in effect a batch mixer. (See Compounding and Mixer.)

20. Internal Plasticization

Denotes the lowering of the glass transition temperature (T_g) through random copolymerization with a monomer containing flexible side groups, which results in an increase in the overall flexibility of the polymer chains and greater free volume. (See Plasticization and Plasticizer.)

21. Internal Stress

A generic term used to denote the stresses that are set internally within a product either during processing or through subsequent ageing. Internal stresses normally arise from the differential thermal expansion or contractions between adjacent layers strongly bound to each other. In injection moulding of thermoplastics, the low temperature of the mould causes a much more rapid cooling of the outer surface layer than the inner layers, owing to the low thermal conductivity of polymers. By the time the inner sections are cold enough for the moulded part to acquire sufficient rigidity to be ejected, the outer layers will be subjected to a much lower temperature drop than the central section. Accordingly, it is expected that the inner sections would undergo a much greater amount of shrinkage than the outer layers, were this not prevented by geometrical constraints. As a result, the outer layers will be subjected to compressive stresses and the inner sections will be in tension, as shown in the diagram.

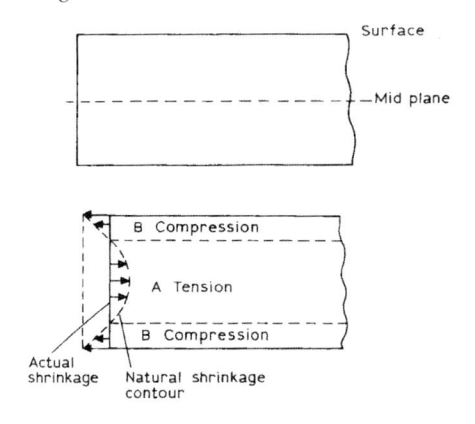

Illustration of development of internal stresses in injection-moulded articles.

This situation is particularly evident for the case of crystalline polymers, which undergo a much greater volumetric contraction during cooling than glassy polymers. In addition to the differences arising purely on the basis of natural contraction, the faster cooling rate in the outer layers results in a lower degree of crystallinity (hence, lower density) than in the central regions, owing to the direct relationship between degree of crystallinity acquired by the polymer and the cooling rate from the melt state. The existence of internal stresses can be verified by observations made on samples when the outer layers are removed from one of the surfaces by machining devices. This causes a previously flat moulding to bow outwards as a result of the tensile stresses in the central regions.

In any case, the magnitude of the internal stresses locked in as 'residual stress' is determined also by the extent of stress relaxation taking place while the moulded part cools to ambient temperature. In this respect, although the differential thermal contraction for glassy polymers is much less than for crystalline polymers, their high glass transition temperature reduces the extent of stress relaxation during cooling, which may ultimately result in a larger magnitude of 'frozen' internal stresses. The presence of metallic inserts in moulded products is another cause for the development of internal stresses in polymer products, owing to the large difference in thermal expansion coefficient between the two materials. The large shrinkage taking place within the bulk of a polymer during cooling, especially for the case of crystalline polymers, creates compressive stresses on the metal insert, which have to be counteracted by tensile stresses in the surrounding polymer area. This can also happen for the case of thermosetting polymers, as a result of the large shrinkage that occurs during curing. In general, these types of stresses are relatively small in thermoplastic products, because of the molecular relaxations that take place both during processing and in service.

The existence of internal stresses in moulded products can be examined under a polarizing microscope, owing to the anisotropy resulting from the small amount of molecular orientation of the polymer chains

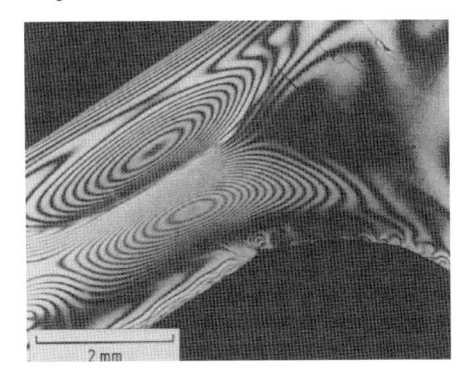

Residual stresses in a polycarbonate moulding viewed through crossed polarizers. Source: Birley et al. (1991).

22. Interpenetrating Polymer Network (IPN)

A type of network formed in a mixture of networking systems, which consist of two molecular interpenetrating domains of sub-micrometre dimensions. The nanosized dimensions of the domains is apparent from the transparent appearance of such a network mixture, while the heterogeneity of the networks is clearly identifiable in dynamic mechanical and dielectric spectrograms by the presence of two distinct 'loss' or tan δ peaks. The chemical interconnection of the two networks is identified by the narrowing of the distance between these peaks relative to those exhibited by the individual networks in isolation. IPNs can also be formed by single systems with different functionalities capable of producing networks by a dual mechanism.

23. Intramolecular

Features and events taking place within polymer chains. Examples of these are the intramolecular hydrogen bonds between the chains of polyamide (nylon) molecules within the lamellae of crystals, which are responsible for their high melting point.

24. Intrinsic Viscosity

A concept used for determining the molecular weight of polymers from measurements of the solution viscosity. (See Molecular weight.)

25. Intumescent Coating

A coating formulation that forms a carbonaceous (char) foam when heated up to high temperatures. This provides a heat barrier for the substrate, thereby reducing the temperature and, therefore, also the rate at which the formation of combustible volatiles are formed by pyrolysis. A typical intumescent formulation will contain: (i) the char-forming component, usually a polyhydric alcohol, a polyphenol or a polysaccharide; (ii) a blowing agent, usually dicyandiamide, melamine, urea or guanidine; and (iii) an acid, such as phosphoric acid or a Lewis acid. (See Flame retardant.)

26. Ion Exchange Capacity (IEC)

A measure of the total content of exchangeable ions in layered silicate minerals and in ion exchange resins, expressed as milliequivalents of ion per gram of sample (meq [ion]/g). (See Ion exchange resin and Nanoclay.)

27. Ion Exchange Resin

Cross-linked resins or polymers containing ionizable groups attached to the molecular chains. Accordingly they can be anionic or cationic. The anionic groups are usually SO_3^-, whereas the more frequently used cations are ammonium types. The solid resin, usually in the form of beads, can exchange ions with those from the surrounding solution, for example

$$\underset{\text{solid}}{M^-A^+} + \underset{\text{solution}}{B^+} \rightleftharpoons \underset{\text{solid}}{M^-B^+} + \underset{\text{solution}}{A^+}$$

These resins are frequently used to purify solutions by removing unwanted ions, for instance, in the deionization of water. The most common type of ion exchange resins are based on cross-linked polystyrene, produced from copolymerization of styrene and divinylbenzene. These are subsequently functionalized to produce the ion exchange capability. Typical structures are shown.

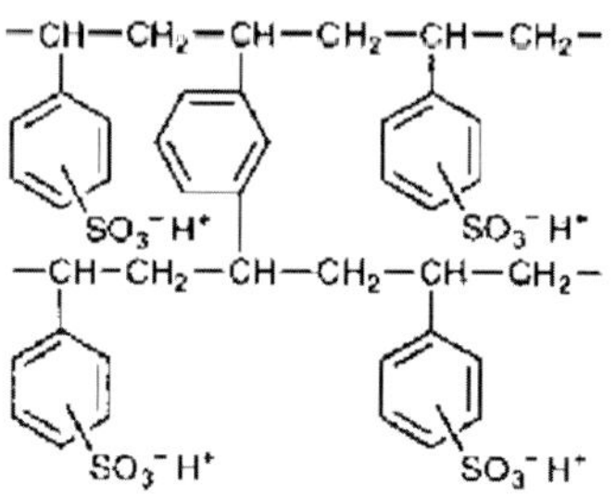

Cation exchange resin

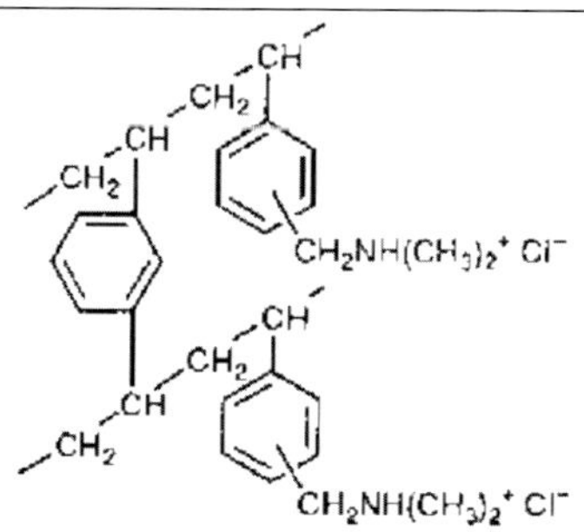

The resin on the left exchanges H^+ ions with a cations from the surrounding solution, while the resin on the right exchanges Cl^- ions with different anions, or OH^-, from the solution. The original ions in the resin will be regenerated through appropriate treatments, after they have been depleted of their original ions.

28. Ionic Conductivity

(See Membrane and Polyelectrolyte.)

29. Ionic Polymerization

A type of polymerization that takes place through the use of ionic initiators. (See Anionic polymerization and Cationic polymerization.)

30. Ionomer

A term used to describe those polymers containing 'clusters' of ionic species derived from the partial neutralization of acid groups dispersed as pendent groups along polymer chains. In addition to clusters, there are also 'multiplets' within the surrounding

Anion exchange resin

matrix, consisting of small regions of dispersed ionic species. The most widely known commercially available ionomers are those based on ethylene methacrylic acid (about 10–15 mol%) copolymers containing sodium or zinc carboxylate clusters. The chemical structure of the polymeric anions and the formation of clusters are shown.

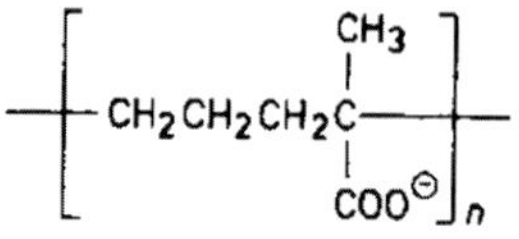

$$\left[-CH_2CH_2CH_2\underset{COO^\ominus}{\overset{CH_3}{\underset{|}{\overset{|}{C}}}}- \right]_n$$

Poly(ethylene methacrylate) anion.

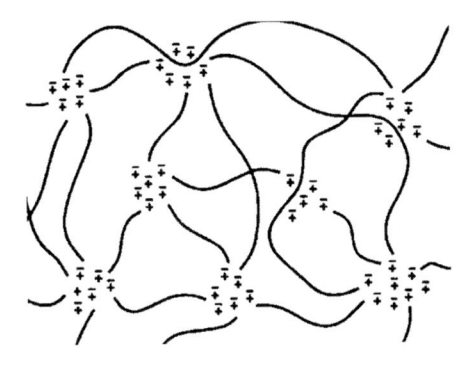

Ionic clusters of zinc or sodium carboxylate.

It must be noted, however, that not all the carboxylic acid groups are converted to carboxylate anions. The presence of ionomeric domains confers to the non-polar polymers (such as polyethylene) better resistance to non-polar liquid environments (such as oils) and provides a greater elastomeric character to ethylene–acrylate copolymers, with improved mechanical properties. Another case of an ethylene-based ionomer is represented by sulfonated EPDM elastomers. These contain only about 1 mol% sulfonate groups, which act primarily as physical cross-links and, therefore, make it unnecessary to carry out the usual vulcanization step. A quite different family of ionomers is that based on polytetrafluoroethylene containing ionic clusters of ionized sulfonic acid (H^+ and ^-O_3S ions), introduced along the chain in the form of a sulfonated perfluoroether copolymer. These are known commercially as Nafion and are widely used as membranes. (See Nafion.)

31. Isostrain Conditions

A term used to describe the deformation conditions of two or more contiguous phases under stress, where the magnitude and type of strain are the same in each phase. From this it results that the total stress is the sum of the stress acting on each phase. These conditions are used to stipulate the upper limits (bounds) for the prediction of the Young's modulus of laminates and unidirectional fibre composites. (See Law of mixtures.)

32. Isostress Conditions

A term used to describe the conditions of two contiguous phases, whereby the magnitude and type of stresses are the same. From this it results that, for members under mechanical stress, the total strain is the sum of the strains developed by each phase, so that

$$\varepsilon_c = \varphi_1 \varepsilon_1 + (1-\varphi_1)\varepsilon_2,$$

where φ_1 is the volume fraction of the high-modulus (reinforcing) phase. (See Law of mixtures.)

33. Isotactic Polymer

A term used to describe the stereoregular configuration of polymers containing side groups. (See Tacticity.)

34. Isothermal

Denotes events taking place under conditions in which the temperature is constant. The kinetics of chemical reactions or physical phenomena, such as crystallization, are usually carried out to obtain a fingerprint of the time evolution of the events monitored, as well as to determine the related rate constants and activation energy. (See Arrhenius equation.)

35. Isotropic

The state of a product or a specimen whose properties are equal in all three directions in space.

36. Izod Impact Test

Measurement of the impact strength of a material using rectangular specimens clamped as a cantilever and impacted at the free end by a mass delivered by a swinging pendulum. (See Impact strength.)

J

1. J integral

A fracture mechanics term used to denote the involvement of the loss of potential energy from a specimen in relation to material behaviour that does not satisfy the requirements of linear elastic fracture mechanics (LEFM), that is, a linear relationship between force and deformation up to the point at which fracture takes place. A lack of linearity is often found in situations where excessive yielding takes place in the crack region, so that a curve is generated, instead of a straight line, in the force–deformation relationship. The J integral is defined as the rate of loss of potential energy (U) from a specimen of thickness B with respect to the increase in crack length (a), that is,

$$J = \partial U/\partial A]_{\partial a} \to 0,$$

where A is the newly formed crack area ($A = Ba$). For conditions in which the crack propagates at constant extension (u), the J integral can be written as

$$J = \frac{u}{B}\frac{\partial P_Y}{\partial a},$$

where P_Y is the load at the yield point.

For tensile tests on specimens with suitable dimensions, the equation becomes $J = U/B(W-a)$, while for three-point bending tests, using the appropriate specimen geometry, the equation is $J = 2U/B(W-a)$, where U is the total energy involved in the fracture of the specimen (area under the force–deformation curve) and W is the specimen width. When applied to fracture conditions, the value of J is the fracture toughness parameter, equivalent to G_c for LEFM, representing the resistance of a ductile material to fracture propagation. For both cases, the units are J/m². (See Fracture mechanics).

2. Jeffamine

A tradename for a series of difunctional and trifunctional amine-terminated polyethers, used for curing of epoxy resins and for the production of polyureas in polyurethane systems.

3. Joint

An adhesive bonded structure or specimen, used to measure the adhesion strength.

K

1. K Value

An empirical parameter used in industry to represent the molecular weight of PVC.

2. Kaolin

A group of minerals used in particulate form as filler in polymer compositions and paper. Kaolin is essentially an aluminium silicate, consisting of alternating ionically interacting layers of alumina and silica.

3. Kapton

A tradename for polyimide films.

4. Kelvin–Voigt Model

A model for the viscoelastic behaviour of polymers. It consists of a spring and a dashpot acting in parallel, as shown.

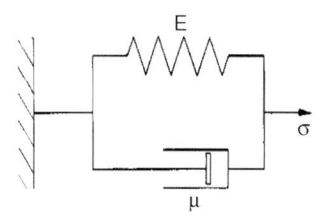

When a load (or stress) is applied to the model, the dashpot causes a retardation of the otherwise instantaneous response of the spring. As a result, the extension (or strain) will increase with time in a curved fashion until it reaches the equilibrium value. If the load (or stress) is removed after the creep period t_1, the deformation (or strain) will recover also in a curved fashion owing to the retardation of the retraction of the spring by the back-flow of the liquid in the dashpot, as shown.

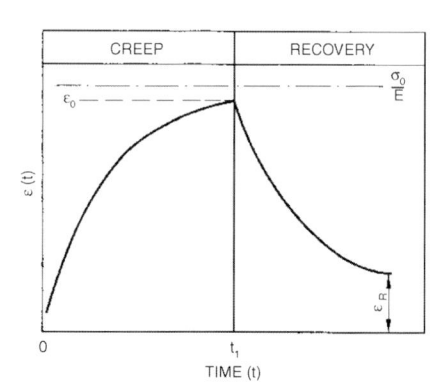

The Kelvin–Voigt model (often referred to as the Voigt model) describes, therefore, the behaviour of a material that exhibits a delayed or retarded elasticity similar to the behaviour displayed by polymers. The advantage of having a model of this nature is that it can be used to obtain an analytical (mathematical) expression to represent the relationship between stress, strain and time. In examining the response of the Kelvin–Voigt model, one notes that the stress (σ) is shared between the spring and dashpot, that is,

$$\sigma_{\text{applied}} = \sigma_E + \sigma_\eta. \qquad (1)$$

The strain for the whole model (ε), on the other hand, is equal to the strain in the spring, which is also equal to the strain in the dashpot, that is, $\varepsilon = \varepsilon_E = \varepsilon_\eta$. Substituting $\sigma_E = \varepsilon E$ and $\sigma_\eta = \eta\,d\varepsilon/dt$ into (1) gives the constitutive equation for the model as

$$\sigma = \varepsilon E + \eta\,d\varepsilon/dt, \qquad (2)$$

which can be solved to produce an equation for the strain as a function of time, that is,

$$\varepsilon = (\sigma/E)\left[1 - e^{-tE/\eta}\right]. \qquad (3)$$

This equation describes the increase in strain with time resulting from the applied stress, as shown in the previous diagram. Note that the term σ/E corresponds to the

Polymers in Industry from A–Z: A Concise Encyclopedia, First Edition. Leno Mascia.
© 2012 Wiley-VCH Verlag GmbH & Co. KGaA. Published 2012 by Wiley-VCH Verlag GmbH & Co. KGaA.

maximum strain that the model can reach, that is, at $t \to \infty$, and can be replaced by the symbol ε_V, where the subscript V stands for Voigt model. It is also important to note that the ratio η/E has the units of time, and is known as the 'retardation time', λ. Consequently, the increase of strain can be written as

$$\varepsilon = \varepsilon_V[1-e^{-t/\lambda}]. \qquad (4)$$

The constitutive equation can be solved also for conditions $\sigma = 0$ at $t = t_1$, that is, the 'recovery' period. The equation for the residual (remaining) strain becomes

$$\varepsilon = \varepsilon_1[1-e^{(t-t_1)/\lambda}], \qquad (5)$$

where ε_1 corresponds to the strain at time t_1, that is, at the end of the 'creep' period, and t is the total time, that is, from the start of the creep period. Obviously, one can divide both terms of (3)–(5) by the constant applied stress (σ) and obtain an expression that describes the linear viscoelastic behaviour in terms of the 'creep compliance', so that from (4) one obtains the expression

$$D(t) = D_\infty[1-e^{-t/\lambda}], \qquad (6)$$

where D_∞ is the creep compliance for $t \to \infty$, known also as the 'equilibrium compliance'. The physical meaning of the retardation time can be taken to represent the time required for the compliance to reach the value $D_\infty(1-1/e)$, which is approximately equal to two-thirds of its maximum value, D_∞.

5. **Kevlar**

A tradename for aromatic polyamide fibres.

6. **Kicker**

A term (jargon) used to denote an auxiliary additive that speeds up the decomposition of the main additive. The latter can be an initiator of a free-radical polymerization or curing process, and it can refer to the decomposition of a chemical blowing agent in the production of foams. The term 'kicker' is also used to describe the role of mercaptobenzothiazole (MBT) in speeding up the sulfurless vulcanization of a diene elastomer by tetramethylthiouram disulfide (TMTD) in order to produce monosulfidic cross-links.

7. **Kneading**

Describes the action of the blades on a melt or dough to induce mixing. (See Mixer and Compounding.)

L

1. Lamella

The crystal formed by chain folding during the crystallization of polymers. Individual polymer chains can participate in the formation of more than one lamella through interlayers of random coiled chains forming the amorphous domains. (See Crystalline polymer.)

2. Laminate

A product in the form of a sheet made up of layers of similar or dissimilar materials. Typical components of laminates are paper, wood and chopped strand glass-fibre mats.

3. Lamination

A process for the production of laminated products. An example of the production of laminated extruded sheets is shown.

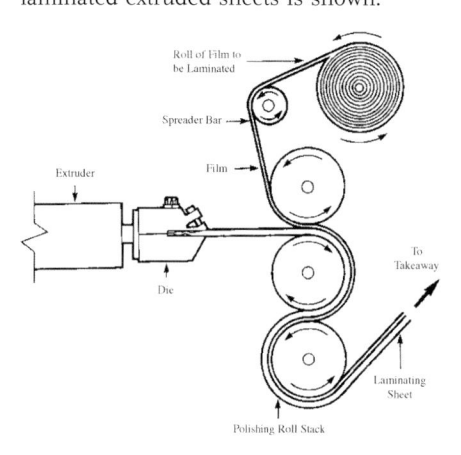

Example of film to sheet lamination. Source: Rosato (1998).

4. Land Length

A term used to denote the length of the parallel section of a die at the exit. (See Extrusion die.)

5. Lap Shear

A generic term for mechanical tests on adhesives use lap joints. (See Adhesive test.)

6. LARC

A tradename for a variety of products made at NASA-Langley via a technological process known as polymerization of monomer reactants (PMR).

7. Latex

A dispersion of elastomeric polymer particles in water, consisting of aggregates of nano-dimensioned primary particles. Particles have to be electrostatically charged to exert repulsive forces and prevent coagulation by interfacial diffusion during storage. This is usually achieved when the pH of the medium is greater than 7. Processing of a latex is normally carried out by spreading or dip coating techniques, where coagulation takes place by removing the water by evaporation. Once the particles are in continual contact, the thermodynamic drive (reduction in free energy) for molecular diffusion across the particles results in the formation of a continuous film or coating on an adherend substrate.

8. Law of Mixtures

A generic name for a type of law that stipulates that the physical properties of a mixture, or an array of ordered assemblies, can be estimated to be equal to the 'weighted' algebraic sum of the properties of the individual constituents. The weighting is expressed in terms of a fractional amount relative to the total, which can be taken as a volume fraction or weight fraction, depending on whether the particular property is related to geometric dimensions or to the mass of the material. The summation results from the principle that, when an external 'excitation' is applied to a mixture, the 'response' of the mixture as a whole entity is equal to the sum of the responses derived from the individual components. It is possible to distinguish two situations: one for mixtures with isotropic properties, and the other for anisotropic properties. In the latter case, the principle of the law of mixtures makes it possible to estimate the upper and lower limits (bounds) of the properties of the mixture with respect to the direction of the excitation relative to the orientation of the components of the mixture. This principle is now illustrated by examining the response of a laminar composite to: (i) mechanical forces, in order to derive expressions for the upper and lower bounds of the modulus of a composite; (ii) temperature gradient, to estimate the thermal conductivity; (iii) the flux of gases through a multilayered film or membrane, to determine its permeability; and (iv) electrical potential (voltage), as a means of deriving expressions for the volume resistivity of a layered dielectric. The law of mixtures can also be used to make estimates for mass-related properties.

8.1 Upper Limit for Modulus of Composite

If forces applied to the composite are transmitted along the plane of the laminae, the longitudinal extension is the same throughout the various laminar components and, therefore, the total forces acting on the composite are equal to the sum of the forces acting on the individual components. Taking into account the cross-sectional area of individual laminae, the forces can be converted to stresses, so that the stress acting on the composite becomes equal to the sum the stresses on each component, normalized by their volume fraction. This can be illustrated by reference to the diagram.

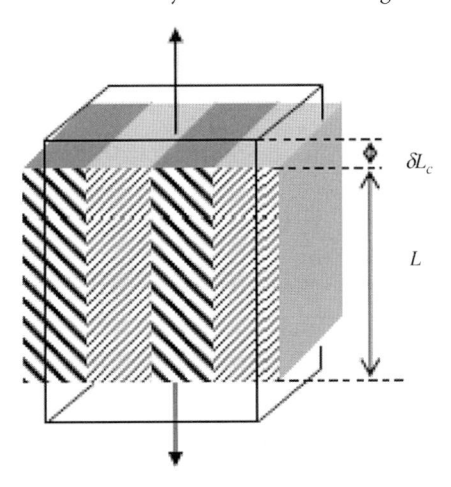

Illustration of isostrain conditions, arising from the equality of the extensions for each phase.

The diagram shows that a force applied in the direction of the laminae produces an extension δL_c that is equal in magnitude for all laminae (phases). Dividing δL_c by the length (L) gives the longitudinal strain, which is equal in both phases, that is, we have an 'isostrain' situation:

$$\delta L_c = \delta L_1 = \delta L_2,$$
$$\frac{\delta L_c}{L} = \frac{\delta L_1}{L} = \frac{\delta L_2}{L},$$
$$\varepsilon_c = \varepsilon_1 = \varepsilon_2.$$

In this case the total force acting on the 'composite' is the sum of the forces acting on the two phases. By dividing the forces by the cross-sectional area of the components, one derives an expression for the mechanical stress acting on the composite, corresponding to the weighted algebraic sum of the stresses acting on each phase, that is,

$$F_c = \sum_{i=1}^{i=n} F_1 + \sum_{i=1}^{i=n} F_2$$

and therefore

$$\sigma_c = \phi_1 \sigma_1 + (1-\phi_1)\sigma_2.$$

Dividing the stress terms by ε_c (where $\varepsilon_c = \varepsilon_1 = \varepsilon_2$) gives an expression for the upper limit of the Young's modulus of a composite, that is,

$$E_c = \phi_1 E_1 + (1-\phi_1) E_2,$$

which is the widely used law of mixtures for the prediction of the modulus of a composite.

8.2 Lower Limit for Modulus of Composite

If the applied forces act through the planes of the laminae, they will have the same magnitude in each lamina and, therefore, the total extension is the sum of the thickness increase of each individual lamina. Translating these conditions into related stresses and strains, and taking into account the volume fractions, one obtains the law of mixtures for the strain acting on the composite.

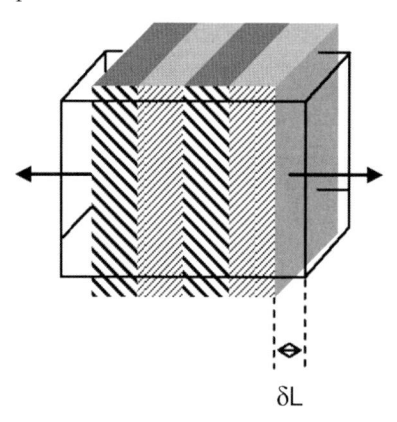

Illustration of isostress conditions, arising from the equality of the forces and interfacial areas for each phase.

The elucidation of this principle can be obtained by considering the diagram and the related deductions:

$$F_c = F_1 = F_2 \quad \text{and} \quad A_c = A_1 = A_2.$$

Since

$$\sigma = F/A,$$

then

$$\sigma_c = \sigma_1 = \sigma_2.$$

Hence the total deformation is the sum of the deformations occurring in each phase, that is,

$$dL_c = \sum dL_1 + \sum dL_2,$$

and the total strain is the weighted algebraic sum of the strain in each phase, that is,

$$\varepsilon_c = \phi_1 \varepsilon_1 + (1-\phi_1)\varepsilon_2,$$

where the subscript c refers to the composite, 1 and 2 to the two respective components, and ϕ is the volume fraction of the components, for example,

$$\phi_1 = \frac{\text{volume of phase 1}}{\text{total volume}}.$$

From the above it follows that the Young's modulus of the composite under isostress conditions is given by the equation

$$\frac{1}{E_c} = \frac{\phi_1}{E_1} + \frac{(1-\phi_1)}{E_2},$$

which corresponds to the lower limit of the modulus of a composite.

The comparison between the upper and lower limits of reinforcement is shown in the diagram for a two-component composite, where E_f = modulus of phase 1 and E_m = modulus of phase 2, with E_f/E_m ratios equal to 10 and 100, plotted in terms of the modulus enhancement factor, E_c/E_m, as a function of the volume fraction of phase 1, ϕ_f.

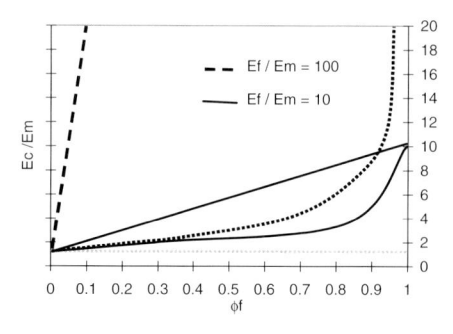

Comparison of upper limit and lower limit of modulus enhancement factor of composites.

From the diagram, it can be deduced that, for a composite in which the modulus ratio of the two components $E_2/E_1 = 10$, under isostrain conditions one can achieve a three-fold increase in modulus with a volume fraction of the high-modulus component of about 0.2. A volume fraction around 0.8, on the other hand, would be required to achieve the same level of reinforcement under isostress conditions. The diagram also indicates that under isostrain conditions the reinforcing efficiency would increase enormously when the modulus ratio of the two components increased, $E_2/E_1 = 100$. There is hardly any substantial increase in reinforcing efficiency if the same was done under isostress conditions.

8.3 Thermal Conductivity

- **Isoflux conditions** These are situations where heat is transferred through the planes of the laminated structure, where each layer receives the same quantity of heat from an external source. This is the situation where heat is transferred through the laminae of the composite. The resultant effect is that the overall temperature gradient is the sum of the gradients in the individual layers, that is,

$$\Delta T/\Delta X = (\delta t/\delta x)_1 + (\delta t/\delta x)_2,$$

so that the expression for the overall thermal conductivity, K_c, becomes

$$\frac{1}{K_c} = \frac{\phi_1}{K_1} + \frac{(1-\phi_1)}{K_2},$$

which corresponds to the 'lower limit'.
- **Isothermal gradient conditions** These situations arise when heat is transferred along the planes of the laminae of the composite, so that the total heat flux is the sum of the heat flux along each laminar component. Consequently, the equation for the thermal conductivity of the composite becomes

$$K_c = \phi_1 K_1 + (1-\phi_1) K_2,$$

which corresponds to the 'upper limit'.

8.4 Permeability

- **Isoflux conditions** Diffusion and heat transfer obey the same scientific laws and both are treated as transport phenomena. Consequently, all the considerations above for thermal conductivity apply also to the prediction of the permeability coefficient of laminates. Isoflux conditions imply that the permeation rate through the thickness of a laminated structure is the same as that occurring through each lamina. In other words, there is no loss of permeating species through the interface. This leads to the lower limit equation for the permeability (P_c) of a two-layer laminated structure as

$$\frac{1}{P_c} = \frac{\phi_1}{P_1} + \frac{(1-\phi_1)}{P_2}.$$

For a multilayer structure it is simply a question of writing the related weighted algebraic sum of the inverse of the permeability of each component, that is,

$$\frac{1}{P_c} = \frac{\phi_1}{P_1} + \frac{\phi_2}{P_2} + \frac{\phi_3}{P_3} + \cdots + \frac{\phi_n}{P_n}.$$

- **Isoconcentration gradient** Note that the other extreme condition corresponding to isoconcentration gradient conditions is unlikely to be encountered in practice in view of the thin sections of the constituent laminae.

8.5 Volume Resistivity of Composite Dielectric

- **Isostress conditions** In a laminar composite, when a voltage is applied along a direction parallel to the planes of the laminae, the resulting current is the sum of the currents transmitted by each phase, so that the total volume resistivity will correspond to that derived for the lower limit of the law of mixtures, that is,

$$\frac{1}{\rho_c} = \frac{\phi_1}{\rho_1} + \frac{(1-\phi_1)}{\rho_2}.$$

- **Isocurrent condition** If the voltage is applied across the laminae of the composite, the current density is equal in each lamina. Hence the total voltage gradient is the sum of the voltage gradients that exist through each lamina. This results in an expression for the volume resistivity of the composite corresponding to the upper limit, that is,

$$\rho_c = \phi_f \rho_f + (1-\phi_f)\rho_m.$$

8.6 Mass-Related Properties

An example of a mass-related property is density. In this case, deviations from additivity predictions based on weight fractions can only arise if the presence of voids has not been taken into account, or if there are interactions between the two components that lead to a change in density of one or both of them, causing changes in degree of crystallinity (thermoplastics) or an increase in cross-linking density (thermosets). The equations derived from the law of mixtures for the density of composites (d) differ according to whether volume fractions (φ) or weight fractions (ω) are considered. The equations are

$$d_c = \varphi_f d_f + (1-\varphi_f) d_m$$

and

$$\frac{1}{d_c} = \frac{\omega_f}{d_f} + \frac{(1-\omega_f)}{d_m},$$

where the subscripts c, f and m stand for composite, filler and matrix.

There are cases where the law of mixtures has been applied on an empirical basis, that is, without any theoretical reasons to justify its applicability. One such case is the prediction of the glass transition temperature T_{gb} of a homogeneous blend, known as the Fox equation, written as

$$\frac{1}{T_{gb}} = \frac{\omega_1}{T_{g1}} + \frac{\omega_2}{T_{g2}},$$

where ω_1 and ω_2 are the respective weight fractions of the two components of the blend. Note, however, that deviations from the above equations are often quantified with the addition of an interaction term, so that the above equation would be rewritten as

$$\frac{1}{T_{gb}} = \frac{\omega_1}{T_{g1}} + \frac{\omega_2}{T_{g2}} + k\omega_1\omega_2,$$

where k is an empirical interaction parameter for the particular system considered.

9. Lay-Flat Film

This term refers to tubular films insofar as they are flattened at the nip of the take-off rolls. (See Blown film and Blow up ratio.)

10. Lead Stabilizer

(See PVC.)

11. Leakage Flow

The flow that takes place within the clearance between the flights of the screw and the barrel of an extruder. This occurs in the opposite direction to the flow in the channels (i.e. back-flow). (See Extrusion theory.)

12. Life Cycle Analysis (or Assessment)

A methodology that seeks to identify the environmental impact of a product by considering the environmental effect at every stage of its life cycle. This includes the impact of extracting the raw materials, transforming them into new products, using them and then disposal and/or recycling.

13. Life Prediction

The estimation of the longevity or serviceability of a product from accelerated tests carried out at higher temperatures. The methodology is based on the principle that the deterioration of the properties of a material results from degradation reactions that can be related directly to the ambient temperature by the Arrhenius equation. (See Arrhenius equation.) In this respect the time to failure t_f (i.e. the longevity) is inversely related to the rate of the reactions that cause the deterioration in properties. Therefore, the Arrhenius equation can be written as $t_f = B \exp(\Delta E/RT)$, where B and ΔE are parameters related to the structure of the material and its interaction with the environment, R is the universal gas constant and T is the absolute temperature. Therefore, if the time, t_f, required to produce a specified reduction in a selected property (usually a 50% reduction in ductility) is measured, after the specimens have been exposed to an environment at different temperatures, an Arrhenius plot can be made to obtain the extrapolated t_T value for the expected life at ambient temperature T. Alternatively, the lifetime (t_T) can be specified (usually between 5000 and 20 000 hours) and the extrapolation is made to obtain the temperature T_t, representing the maximum temperature at which the article can be used in the particular environment considered, as shown in the diagram.

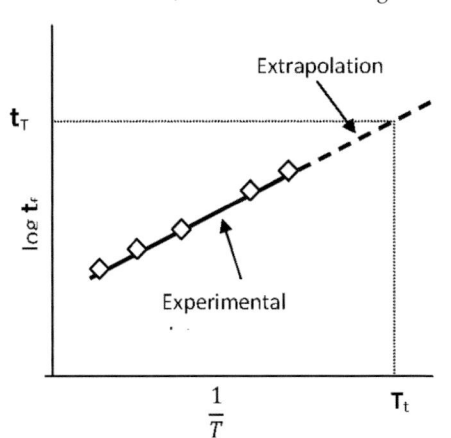

Example of Arrhenius-type plot for the life prediction of polymers in a particular environment.

This method works particularly well with cross-liked polymers, as they can be exposed to very high temperatures without distortions or destruction of the shape of the specimens used in the tests. The method is even more accurate for vulcanized rubber, as the extrapolation is carried out from data obtained within the same deformational state of the polymer, which is characterized by a constant activation energy. This may not be the case for glassy cross-linked polymers, where the extrapolation may cross two deformational states, that is, rubbery and glassy, that have different activation energies. Hence the extrapolation used for these situations may produce much less reliable predictions. A 'rule of thumb' is sometimes used to estimate the life prediction from a few accelerated tests at higher temperature. This assumes that the life expectancy of a product increases two-fold for every 10 °C decrease in temperature.

14. Light Microscopy

A microscopy technique that uses visible light, occasionally UV light, as the incident radiation to view an object through magnifying 'lenses'. Various methods of examination are available:

a) direct illumination;

b) dark-field observations, whereby only those rays diffracted in the object contribute to the formation of the image;

c) phase contrast microscopy, by which observation of objects is made with the use of a plate in the image side to produce a $\pm 90°$ phase shift in the zero-order diffraction maximum;

d) fluorescence microscopy using UV light to detect objects, or features within the object, capable of emitting fluorescent radiation; and

e) polarization microscopy, a technique that is widely used for morphological examinations and quantitative estimation of the amount of orientation in polymer products.

In these examinations, two polarizing filters are used, that is, a polarizer and an analyser at right angles to each other. The polarizer is placed between the light source and the condenser, and the analyser is between the objective and the eyepiece. The sample is placed in such a way that the major refractive indices are at 45° and 135° to the direction of the polarizer. The polarized incident light is split into two waves of equal amplitude in the direction parallel to the directions of the major refractive indices, traversing the sample with different velocities and, after emerging from the sample, combine to form an elliptically polarized wave. However, only the portion of polarized light that travels parallel to the analyser is transmitted. The difference T in optical path lengths nd in the sample (where n is the refractive index and d is the thickness) is a measure of the anisotropy, that is,

$$T = (n_1 - n_2)d.$$

When $T = a\lambda$ (where $a = 0, \pm 1, \pm 2, \ldots$, and λ is the wavelength) the emerging wave is polarized in the direction of the polarizer and cancelled. For isotropic samples $T = 0$ and the image appears dark. This gives rise to the classical 'isogyral cross' ('Maltese cross') images of spherulites present in crystalline polymers. (See Spherulite, Orientation and Birefringence.)

15. Light Scattering

A term that denotes the reflection of light in multiple directions from the surface of objects, such as particles, whose refractive index is greater than that of the surrounding medium, provided that the dimensions of the scattering centres are larger than the wavelength of the light. Light scattering techniques have been used for measuring the weight-average molecular weight of polymers in solution and to examine the morphology of polymers in the solid state. Neither technique is widely used these days, as there are more rapid and more accurate methods available.

16. Light-Sensitive Polymer

A polymer capable of absorbing light (visible or UV), which results in the production of cross-linked structures. A typical example is poly(vinyl cinnamate), where the unsaturation in the side groups produces the light-absorbing characteristics of the polymer through the formation of conjugated double bonds with the benzene ring and, at the same time, provides sites for the production of cross-links. The structure of derivatives of cinnamic acid is shown. (See UV stabilizer.)

Derivatives of cinnamic acid, where $Y = COOH$; and X and Z are other substituents.

Cross-linking of polymers by light can be brought about also with the use of light-absorbing additives, such as difunctional azides, which decompose (releasing nitrogen) into very reactive free radicals that will react with labile hydrogen atoms in the polymer chains, as indicated schematically.

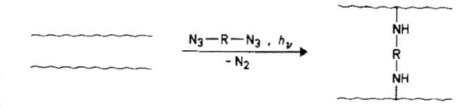

Cross-linking of polymers with azides through light absorption.

Light-sensitive polymers are used in photofabrication (e.g. photoresist) and for the formation of printing plates and microcircuits. (See Photoresist polymer.)

17. Light Transmission Factor

An index, T, that defines the intensity of light transmitted through a medium, that is, $T = 1 - F_{abs} - F_{sc}$, where F_{abs} and F_{sc}, respectively, are the fractions of the intensity of light absorbed and scattered by the medium. (See Optical properties.)

18. Limiting Oxygen Index (LOI)

A parameter that describes the fire resistance of a polymer and is defined as the concentration of oxygen (%) in an oxygen–nitrogen mixture capable of sustaining the burning of a polymer specimen mounted in a draft-free chamber. The higher the LOI value of a polymer, the greater is its fire resistance. A typical apparatus is shown.

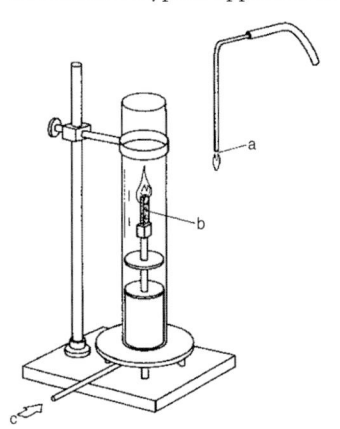

Apparatus for measuring the Limited Oxygen Index, Method ASTM 2863. Source: Courtesy of ASTM International (formerly American Society for Testing and Materials).

19. Linear Behaviour

Denotes a relationship between two variables that remains the same irrespective of the value or magnitude of the variable considered.

20. Linear Elastic

Denotes the behaviour of a material for which the stress is proportional to the resulting strain at all times and irrespective of the type and level of stress applied. This implies that there is no loss of strain energy under cyclic loading conditions. (See Elastic behaviour.)

21. Linear Polymer

A polymer whose molecular chains are not linked and do not contain long-chain branches.

22. Linear Viscoelastic

A viscoelastic behaviour by which the time dependence of the modulus or compliance of a polymer does not change irrespective of the level of stress or strain that is imposed on a structural member This implies that there is a linear relationship between stress and strain. (See Viscoelasticity, Viscoelastic behaviour and Nonlinear viscoelastic behaviour.)

23. Liquid-Crystal Polymer (LCP)

A polymer containing regular rod-like rigid units, known as mesogenic units, which can assemble in ordered arrays when the polymer is in its melt state or in solution.

The mesogenic units may be either located within molecular chains or attached as pendent groups to flexible polymer chains, as shown in the diagram.

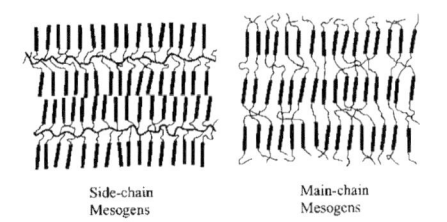

Side-chain Mesogens   Main-chain Mesogens

Schematic representation of the organization of mesogenic units in liquid-crystal polymers.

The liquid-crystal domains, known as the mesophase, are not true crystals insofar as the order of the constituent units is only short-range. When the mesophase is dispersed in the polymer melt, they are also known as thermotropic LCPs. Examples of these LCPs are polyesters and poly(ester amide)s whose mesophase is obtained with the use of acetoxybenzoic acid (ABA), acetoxynaphthoic acid (ANA) and acetoxyacetoanilide (AAA). The synthesis and structure of a typical polyester-based LCP available commercially is shown.

Reaction scheme and structure of a typical polyester LCP available commercially.

When the mesophase is dispersed in a polymer solution, they are called lyotropic LCPs. An example of a lyotropic LCP is poly(p-phenylene terephthalamide)s dissolved in sulfuric acid, used for the production of reinforcing fibres, known as 'Aramids'. (See Composite.)

24. Liquid Rubber

Low-molecular-weight polymers or liquids, containing NH_2 or $COOH$ end groups, used to increases the toughness of thermosetting resins, particularly epoxy resins. (See Amine-terminated butadiene–acrylonitrile and Epoxy resin.)

25. Load–Deflection Curve

A graph obtained when performing flexural (three-point bending) tests presented in the form of plots of the recorded load (force) against the central deflection of the specimen up to the point of fracture. These provide a fingerprint of the mechanical characteristics of the material, by calculating the outer-skin values of the stress from the applied load, and the strain from the deflection at the point where the load has been applied. For three-point bending tests, stress is given by $\sigma = 3PL/2WB^2$ and strain by $\varepsilon = 6B\Delta/L^2$, where P is the load, Δ is the central deflection, L is the span, and B and W are respectively the thickness and width of the rectangular specimen. From the load–extension curve it is possible to calculate: (i) the Young's modulus from the slope of the tangent to the curve starting from the origin (Young's modulus = stress/strain); (ii) the yield strength (if failure is not brittle) taken as the stress value at the peak or at the point there is a rapid change in slope; and (c) the yield strain as the value of the strain corresponding to the yield stress. Note, however, that the equations that are normally used for the calculation of the outer-skin central stress and strain, on both the tension and compression sides of the specimen, have been derived on the basis that the deflection is very small. This may bring about considerable errors in the estimate of the yield strength and yield strain of polymers from flexural tests. For this reason the data obtained in these tests are quoted as flexural values, and are generally higher than the equivalent data obtained from tensile tests.

26. Load–Deformation Curve

A graph obtained when performing compression tests presented in the form of plots of the recorded load (force) against the reduction in thickness of the specimen. Compression tests are rarely used for measuring fundamental properties of polymers, such as Young's modulus or yield strength. They are more usually used for the evaluation of specific characteristics of materials under compression loads, such as, for instance, the fibre–matrix debonding in composites or the compression set of rubbers.

27. Load–Extension Curve

A graph obtained when performing tensile tests presented in the form of plots of the recorded load (force) against the extension of the specimen up to the point of fracture. These provide a fingerprint of the mechanical characteristics of the material, as shown in the diagram.

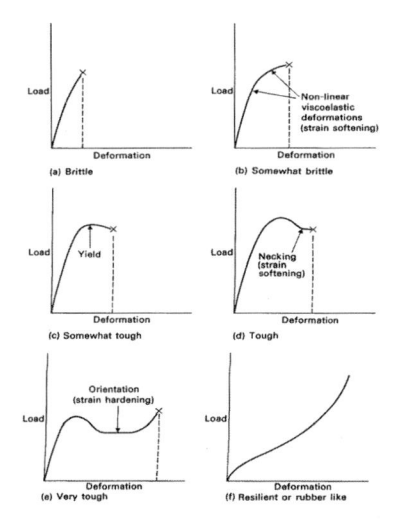

Schematic illustration of the behaviour of different polymers in a tensile test. Source: Mascia (1974).

By converting the required values for the load into stress (load/cross-sectional area) and extension into strain (extension/initial gauge length of the specimen), it is possible to calculate: (i) the Young's modulus from the slope of the tangent to the curve starting from the origin; (ii) the yield strength (if failure is not brittle) taken as the stress value at the peak or at the point there is a rapid change in slope; (iii) the yield strain as the value of the strain corresponding to the yield stress; (iv) the extension at break as the strain at break, expressed as a percentage of the total extension divided by the gauge length; (v) the tensile strength at break, known also as the ultimate tensile strength; and (vi) the relative toughness by taking the area under the curve up to the point of fracture.

28. Locking Mechanism

A mechanical device designed to open and close the mould of an injection moulding machine. (See Injection moulding.)

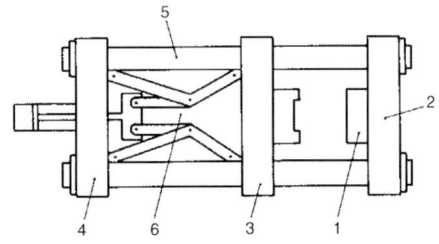

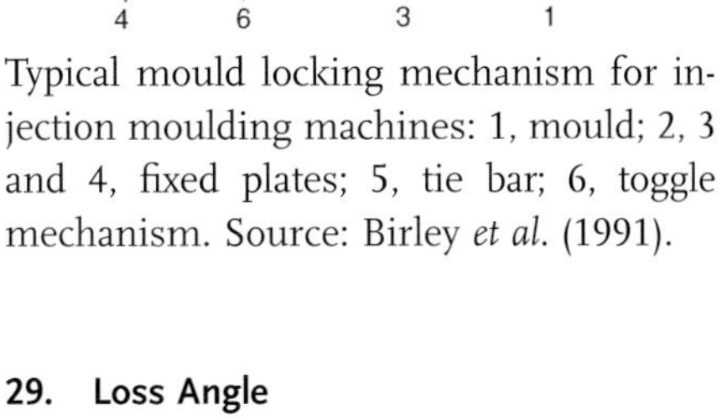

Typical mould locking mechanism for injection moulding machines: 1, mould; 2, 3 and 4, fixed plates; 5, tie bar; 6, toggle mechanism. Source: Birley et al. (1991).

29. Loss Angle

The angle, δ, formed between the stress and the strain in dynamic mechanical tests arising from the viscoelastic behaviour of polymers. The value of the loss angle lies between 0, corresponding to that for an ideal solid (elastic material), and $\pi/2$, which is equivalent to the value for an ideal liquid (Newtonian). For structural polymers, the loss angle at low temperatures is much closer to 0 than to $\pi/2$. (See Dynamic mechanical thermal analysis and Viscoelasticity.)

Representation of elastic behaviour in cyclic stress situations involving tension and compression: the stress is always in phase with the strain.

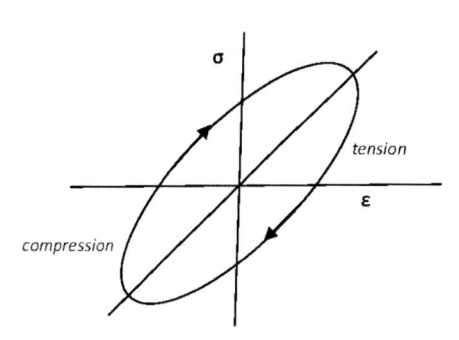

Representation of viscoelastic behaviour in cyclic stress situations involving tension and compression: stress and strain are always out of phase.

The same concept applies to the angle formed between the voltage gradient (electrical stress) acting on a dielectric and the resulting current density in an alternating current field. The theoretical values also lie between 0 and $\pi/2$, corresponding respectively to the values of an ideal dielectric (resistive material) and that of a capacitor. The electrical loss angle values for non-polar polymers, such as polyethylene, are much closer to zero than the mechanical loss angle of any rigid polymer. For both situations, that is, mechanical and electrical stresses, the loss angle values are usually quoted as loss tangent (tan δ). (See Dielectric thermal analysis and Insulation.)

30. Loss Factor

Another term for the loss tangent or tan δ. (See Loss angle.)

31. Loss Modulus

The imaginary component of the complex modulus of polymers. (See Dynamic mechanical thermal analysis and Viscoelasticity.)

32. Loss Tangent

(See Loss angle.)

33. Low-Density Polyethylene (LDPE)

Polyethylene with density in the range 0.910–0.925 kg/m^3. (See Ethylene polymer.)

34. Low-Profile Additive (LP Additive)

A polymer that is usually mixed with an unsaturated polyester resin in the production of bulk moulding compounds and sheet moulding compounds. The role of the LP is to prevent the formation of slight undulations on the surface resulting from an uneven distribution of the resin/fibre ratio in the uncured pre-impregnated chopped strand glass mats. An LP functions by migrating towards the surface during curing as a result of its reduced solubility in the cross-linked resin, which allows it to fill the gaps created at the interface with the walls of the mould. A variety of polymers have been used as LP additives, including poly(vinyl acetate), poly(vinyl chloride) and acrylic polymers. These additives have been found to have beneficial effects also with respect to the control of mould shrinkage.

35. Lower Bound

Corresponds to the lower limit value of the properties of composites predictable by the law of mixtures. (See Law of mixtures.)

36. Lower Critical Solution Temperature (LCST)

The temperature corresponding to the minimum in the curves of temperature versus volume fraction representing the border between the miscible region and the two-phase region. (See Miscibility.)

37. Lubricant

An additive used in polymer formulations to enhance the processing characteristics of polymers by reducing the friction between the melt and the walls of the processing equipment. There are two types of lubricants:

a) **External lubricants** are additives that exhibit a limited miscibility at high temperatures with the polymer, so that the molecules can diffuse to reach the surface of the metal equipment, forming a multi-molecular layer or micelle attached to the surface of the processing equipment.

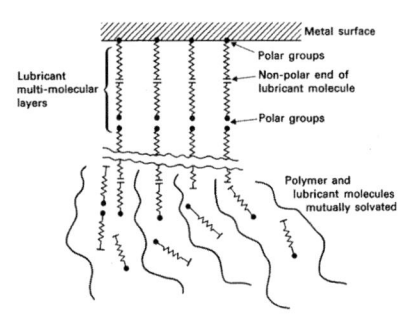

Migration of fatty-acid-type lubricants to the surface of the equipment and the formation of a micelle and a weak boundary layer. Source: Mascia (1974).

The intermolecular forces within the micelle formed are very weak and can be easily overcome by the action of the shear stresses at the wall of the processing equipment, thereby providing a lubrication mechanism for the flow of the high-viscosity polymer melt. External lubricants are usually high-molecular-weight fatty acids, amines, amides or metal soaps, containing 12–18 carbon atoms in the aliphatic chain. In more recent years, hydrocarbon waxes, consisting of low-molecular-weight copolymers of ethylene with highly polar comonomers, have also been used for this purpose. These do not form micelles but produce a low-viscosity interlayer between the melt and the metal surface.

b) **Internal lubricants** have a certain level of miscibility with the polymer and will bring about an appreciable reduction in melt viscosity. An internal lubricant, however, is usually used in amounts that exceed the solubility limit in the polymer, so that the soluble portion reduces the melt viscosity (from which the term 'internal lubricant' is derived) and the rest migrates to the surface of the processing equipment to produce lubrication. An example of the use of internal lubricant is the incorporation of stearic acid in rigid PVC formulations. Sometimes two different lubricants, an internal lubricant and an external lubricant (long-chain carboxylate salt, which is almost totally insoluble in the polymer), are used in polymers with a high melt viscosity that are susceptible to thermal degradation, such as PVC and heavily filled thermosetting moulding compounds.

38. Lubrication Approximation

An approximation made in the solution of fundamental flow equations through channels of variable cross-sectional area along the flow path. The approximation requires that there will be no change in the axial velocity on the basis that the rate of change in momentum is very small. (See Momentum equation.)

39. Lüder Lines

Also known as shear bands, these consist of dislocated shear planes observed on specimens tested under plane-strain compression conditions. They are often used to support the theory that the onset of yielding failures takes place through shear deformations along the plane where the shear stresses are highest. (See Yield failure.)

M

1. M_{100} and M_{300}

Terms denoting the modulus of rubbery materials at 100% and 300% extension. (See Rubber elasticity.)

2. Machining

A manufacturing operation by which an article is produced by mechanically removing material, using a cutting device, known as a tool. Thermoplastics have to be in their glassy state to be able to tolerate the high local heat generated by the interfacial stresses and to ensure that the resulting temperature rise at the cutting surface does not bring the polymer into the rubbery state. For the same reason, crystalline polymers must have a fairly high melting point to be able to be machined. The cutting speed, rake angle of the tool and thickness of the chip have to be carefully controlled to prevent excessive temperature rises, which would bring the polymer into the rubbery state. Machining operations include drilling and blanking.

3. Macromolecule

A generic term for large molecules with regularly spaced monomeric units, arranged either linearly (polymers), in highly branched fashion or in the form of an infinite network.

4. Magnesium Hydroxide

An inorganic filler, with particle size within the range 0.5–5 μm, used primarily as a non-toxic fire retardant additive at levels in the region of 40–65 wt%. (See Flame retardant.) It is also used, however, as a cross-linking agent for chloroprene rubber, chlorosulfonated elastomers and some fluoroelastomers. (See Curing and Vulcanization.)

5. Magnetic Filler

A filler used to impart magnetic properties to polymers. The magnetic moment of the compound produced is directly proportional to the volume fraction. The most commonly used magnetic filler is magnetite (Fe_3O_4). Normally, a loading of 25–45 vol% (corresponding to 60–80 wt% Fe_3O_4) is required to obtain polymeric magnets. Other magnetic fillers include barium ferrite, Alnico (an alloy of aluminium, nickel and cobalt, with some iron and copper), samarium cobalt and rare-earth iron borides.

6. Mandrel

The core part of a tubing die or pipe die. For the case of cross-head blow moulding dies, the mandrel and die assembly can be subjected to programmed cyclic vertical movements to adjust the die gap as a means of increasing the flow rate. In this way it is possible to compensate for the reduction in the wall thickness, caused by sagging due to the weight of the parison, thereby ensuring that the produced container has a uniform wall thickness along its length. The mechanism of thickness adjustment with movable mandrel is shown.

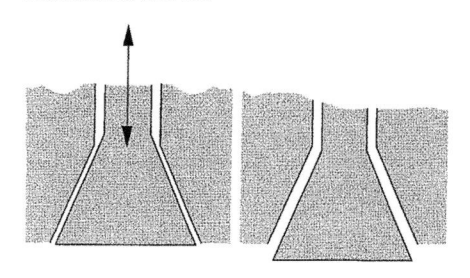

Schematic diagram of a movable mandrel.

Mandrels can also be made to rotate during extrusion of tubular products as a means of destroying the weld lines formed by the spider legs and to achieve a uniform pressure gradient in the flow direction along the entire circumference. Rotating mandrels can also be used to produce special products, such as nets or perforated tubings, as shown in the diagram. (See Extrusion die and Blow moulding.)

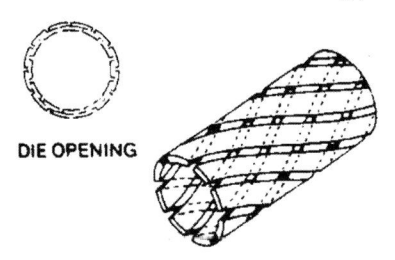

Rotating mandrel die makes perforated tubing

Example of extruded products made with the use of rotating mandrels. Source: Rosato (1998).

7. Mark–Houwink Equation

An equation for the relationship between solution viscosity, $[\eta]$, and the molecular weight of a polymer, corresponding to the so-called 'viscosity-average molecular weight', M_v, that is,

$$[\eta] = KM_v^a,$$

where K and a are characteristic constants for a specific type of polymer that have to be determined experimentally. (See Molecular weight.)

8. Mass Polymerization

Known also as bulk polymerization, this is the polymerization of a monomer in the absence of any solvents or other fluid media, such as water or supercritical CO_2. This type of polymerization is used primarily for systems that can be polymerized by an addition polymerization mechanism, that is, without the elimination of volatile species. Although mass polymerization is commonly used for the production of castings in moulds, coatings, adhesives or matrices for composites, it is often used also for continuous polymerization in reactors for the production of moulding powders. One of the difficulties of bulk polymerization is the heat of reaction, which can produce large temperature rises. For this reason, this method of polymerization is restricted to very thin sections, as indicated by the examples given earlier. (See Polymerization, Curing and Cross-linking.)

9. Master Batch

A polymer compound, usually in granular form, containing large quantities of additives. A master batch is pre-mixed in small amounts with a virgin polymer (i.e. one free of additives) at the processing stage, as a means of incorporating additives into polymer products without having to undergo an expensive compounding operation. One master batch can often be used for a variety of polymers.

10. Master Curve

A curve for the variation of a deformational parameter (e.g. modulus or compliance) with time of duration of an applied load, derived by the extrapolation of experimental results obtained at higher temperatures. (See Time–temperature superposition.)

11. Mastication

An operation carried out on raw natural rubber as a means of reducing the molecular weight. (See Mechanochemical degradation.)

12. Matrix

The polymer phase of a composite. (See Composite.)

13. Maxwell Model

A mechanical analogue, consisting of a spring and a dashpot connected in series, used to model the stress relaxation behaviour of polymers when held under constant strain, as shown.

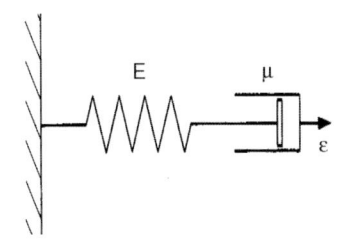

If a strain (ε) is suddenly imposed onto the Maxwell model and is held constant in time, it will be observed that at time zero (t_0) the spring will stretch instantaneously while the dashpot will remain in its original position. As time progresses the spring will retract because the force acting on it will cause the liquid in the dashpot to flow. This means that the force (and corresponding stress σ_E) on the spring decreases with time. The stress σ_η acting on the dashpot is always equal to that acting on the spring, so that $\sigma = \sigma_E = \sigma_\eta$. This stress will gradually decay until the spring has retracted completely and the stress becomes completely relaxed (i.e. becomes equal to zero). This 'stress relaxation' situation is illustrated in the diagram.

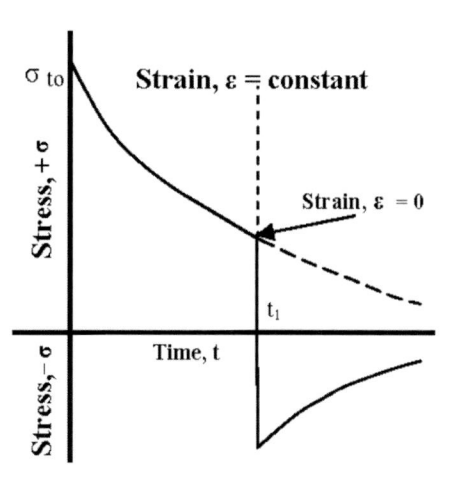

Illustration of the relaxation of stresses according to the Maxwell model.

With this model it is possible to obtain a constitutive equation for the stress relaxation process, noting that, at any time t, the total imposed strain ε is the sum of the strain on the spring and that on the dashpot,

$$\varepsilon = \varepsilon_E + \varepsilon_\eta, \qquad (1)$$

that is, the strain is shared between the two components. Note that the corresponding rates at which the strain increases with time ($d\varepsilon/dt$) are also additive. The constitutive equation is derived by making the appropriate substitutions in (1), that is,

$$(d\varepsilon/dt)_E = (1/E)\, d\sigma/dt \quad \text{and} \quad (d\varepsilon/dt)_\eta = \sigma/\eta,$$

where E is the modulus of the elastic solid represented by the spring and η is the viscosity of the liquid in the dashpot. The solution of the resulting constitutive equation for the condition $\varepsilon = $ constant becomes

$$\sigma = \sigma_0\, e^{(-tE/\eta)}, \qquad (2)$$

where σ_0 is the stress at time zero, while the ratio η/E can be replaced by a constant λ, known as the 'relaxation time', as it has the units of time. Dividing both sides of the equation by ε, which is constant (i.e. independent of time), and rewriting (2) in terms of the modulus as a function of time, known as the 'relaxation modulus', one obtains

$$E = E_0\, e^{-t/\lambda}. \qquad (3)$$

From the previous graph it can be inferred that if, at time t_1, the strain is to be instantaneously forced to become equal to zero, this would require an equal stress in the opposite direction (i.e. a compressive stress). The applied compressive stress, however, would decay to zero at infinite time. This is because the relaxation rate is determined only by the relaxation time λ, which is constant and independent of the directional nature of the stress. It is understood that these are general models and, therefore, can be used for either tension, compression or shear deformations, so that the derived equations would be identical. For shear deformation situations, for instance, one would use the symbol G for the modulus.

14. Mechanical Properties

Properties that characterize the response of a material to mechanical stresses, expressed in terms of stress at failure (strength), gradient of the plot of stress versus strain (modulus) and parameters related to the energy required to cause fracture (toughness). (See Fracture mechanics.)

15. Mechanical Spectroscopy

A technique for determining the deformational characteristics of polymers when subjected to cyclic stresses. The data are presented as plots of the complex modulus and loss modulus, or tan δ, against temperature. (See Dynamic mechanical thermal analysis.)

16. Mechanochemical Degradation

Degradation reactions that take place by continually shearing a polymer melt. For polymers that degrade by chain scission via free-radical reactions, such as polyisoprene rubber and polypropylene, mechanochemical degradation is often used as a technique for deliberately decreasing the molecular weight of a polymer. In the case of polypropylene, the technique is used to produce the so-called 'controlled rheology grades', mostly used for the production of fibres. An early example of mechanochemical degradation is the 'milling' of natural rubber to reduce the molecular weight in order to make it processable in the melt state. This operation is generally known as 'mastication', which is a process carried out at low temperatures in order to ensure that degradation is induced by the high shear stresses developed during shearing, which can increase further through 'stress-induced' crystallization effects. The process is further assisted by the infusion of oxygen from the atmosphere. Mechanochemical degradation has also been used to enhance the compatibility of polymers in the production of blends, owing to the possibility of producing a certain amount of graft copolymers through chain transfer reactions. These will subsequently act as compatibilizers for the unaffected polymer chains.

17. Melamine Formaldehyde

Abbreviated to MF, a thermosetting resin obtained from the reaction of melamine with formaldehyde. (See Amino resin.)

18. Melt

The state of a polymer at temperatures above that of the rubbery plateau. It represents the liquid sate of the polymer, as it denotes the ability of the polymer to flow, which is manifested through a net movement of molecules from one site to another. Flow takes place via rotations of chain segments involving uncoiling and reptation movements. (See Deformational behaviour and Non-Newtonian behaviour.)

19. Melt Elasticity

A term used to describe a characteristic of polymer melts that causes partial recovery of the imposed deformation when the stresses are removed. A manifestation of this behaviour is the swelling of an extrudate at the exit of a die, where the wall shear stress suddenly vanishes due to the elimination of the pressure gradient that existed during the flow though the die. The die swell ratio (i.e. for circular dies, diameter of extrudate/diameter of die) is frequently quoted as an empirical parameter characterizing the elasticity of polymer melts. (See Die swell ratio and Normal stress difference.) The phenomenon originates from the long-chain nature of polymer molecules, which have to undergo a certain degree of alignment during flow.

The consequence of the recovery characteristics of polymer melts can be illustrated by the swelling of an extrudate emerging from a die. At the die exit, the pressure on the melt becomes equal to the atmospheric pressure and, therefore, the pressure gradient (and also the associated shear stress at the wall) becomes equal to zero. The sudden removal of the 'flow stresses' causes the partially aligned molecules to retract back into a random configuration, resulting in a lateral swelling of the extrudate. This indicates that, during the flow of the melt through the die, other stresses are developed in the direction perpendicular to the flow direction, which can be directly associated with swelling.

In drag flow situations, as in the case of the flow in rotational rheometers, the effect is evidenced directly from measurements of the pressure at walls. This pressure is not experienced by Newtonian liquids. A parameter, known as the normal stress coefficient, ψ_1, is defined to describe the melt elasticity characteristics of the flow of polymers, that is, $\psi_1 = N_1/\dot{\gamma}^2$, where $\dot{\gamma}$ is the shear rate and N_1 is the difference between the normal stresses acting along the flow direction and that perpendicular to the flow direction.

In dynamic flow situations the shear stress (τ) lags behind the shear rate ($\dot{\gamma}$) by a phase angle (δ), so that if $\dot{\gamma} = \dot{\gamma}_0 \sin(\omega t)$ then $\tau = \tau_0 \sin(\omega t - \delta)$, where ω is the angular velocity and ωt is the angle of deformation at time t. This results in an equation for the viscosity in complex notation in the form of $\eta^* = \eta' - i\eta''$, where η' is the real viscosity component, i is the complex number ($\sqrt{-1}$) and η'' is the imaginary component of the viscosity (a measure of the melt elasticity characteristics), corresponding to G/ω, where G the elastic shear modulus.

20. Melt Flow Index (MFI)

Also known as the melt index or melt flow rate (MFR), is an empirical parameter that describes the ease of flow of polymer melts. It is defined as the grams of polymer extruded in 10 minutes through a capillary die of specified dimensions, resulting from an externally applied load. The temperature at which the measurements are made and the magnitude of the applied load varies according to the nature of the polymer. For the same polymer, the value of the MFI is often measured at two different loads. The conditions to be used are stipulated by appropriate standard test methods. For the case of polyethylene, irrespective of its density, the measurements are made at 190 °C and the two loads normally used are 2.16 and 5.0 kg. A typical set-up for the apparatus used for measuring the MFI of polymers is shown schematically.

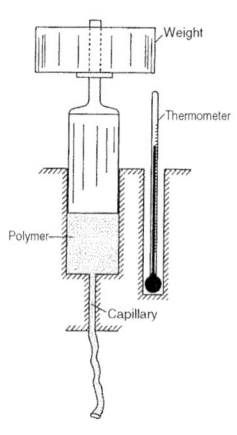

Schematic diagram of the apparatus for measuring the melt flow index of a polymer. Source: Osswald (1998).

21. Melt Fracture

A phenomenon that manifests itself in the form of undulations or gross deformities in extrudates, particularly those obtained from orifices with a small cross-sectional area. These are associated with die-entry instabilities, which create pressure fluctuations as a result of a slip–stick effect on the wall of the die. If the die length is not sufficiently large, the pressure fluctuations will not dampen out by the time the melt reaches the die exit and will create gross distortions in the emerging extrudate. These can take different geometric shapes, as shown in the photographs.

It is widely accepted that melt fracture occurs when the shear stress at the wall of the capillary in the entry region reaches a critical value, which depends on the viscosity and, therefore, on the molecular weight of the polymer. The higher the viscosity (hence the lower the temperature and the higher the molecular weight), the lower the value of the critical shear stress for melt fracture to occur, $(\tau_w)_{cr}$. While the value of the critical shear stress for melt fracture decreases with increasing temperature, owing to the reduction in melt viscosity, the product of wall shear stress for melt fracture and weight-average molecular weight (M_w) remains constant, that is, $(\tau_w)_{cr} M_w = $ constant. This implies that it is possible to alleviate the melt fracture problem either by increasing the processing temperature or by using a polymer with a lower molecular weight.

22. Melt State Polymerization

Refers to the increase in molecular weight of condensation polymers that can be brought about through extension reactions during processing.

23. Melt Strength

A term that denotes the capability of a polymer melt to resist fracturing at the exit

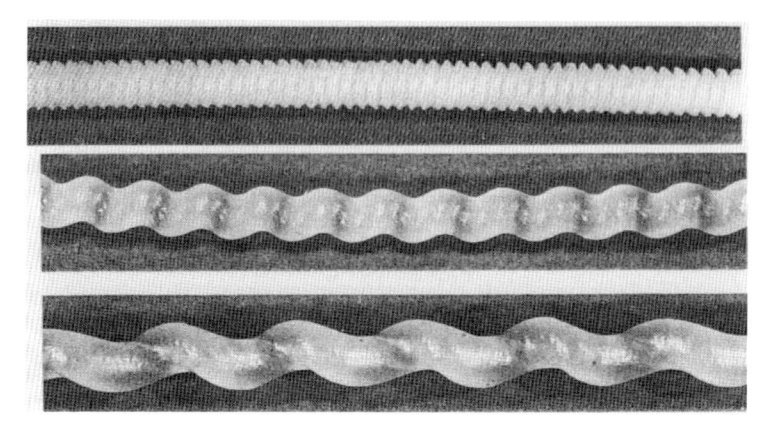

Examples of extrudate distortions resulting from melt fracture at the die entry of a capillary die.

of the die when it is drawn by the take-off equipment. Although it does not have the same meaning as melt fracture, the two characteristics are related. (See Drawdown resonance.) A typical test for evaluating the drawdown characteristics, as well as elongational viscosity and melt strength of polymer melts, is illustrated.

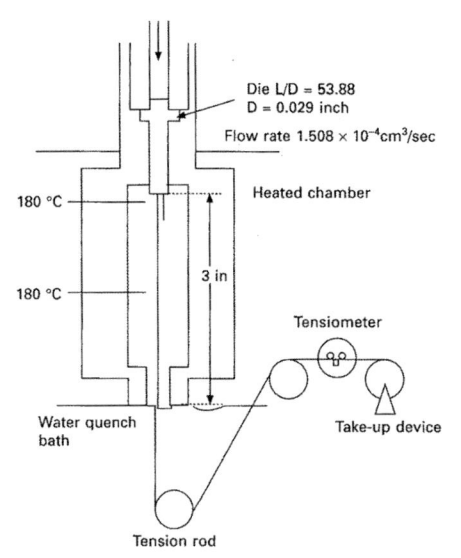

Schematic diagram of the apparatus used to measure the elongational viscosity, melt strength and drawdown properties of polymer melts. Source: Minoshima et al. (1980).

The drawdown characteristics are assessed by recording the maximum draw ratio that can be achieved, determined from the ratio of peripheral velocity of the take-off roll to the exit velocity of the melt emerging from the extrusion die. In these experiments, it is also possible to measure the forces for drawing the filament at different drawing speeds, with the aid of a 'tensiometer'. The critical drawdown ratio can be estimated from the ratio of the velocity of the filament at drawing rolls to that at the die exit (manifested as a regular waving pattern along the length), while the melt strength is calculated by recording the force at which the filament breaks away from the die. From a plot of the stress (calculated from the recorded force) against drawing rate of the filament between die and rolls, it is also possible to obtain an estimate of the elongational viscosity of the melt. (See Elongational viscosity and Melt fracture.)

24. Melting

Denotes the change from solid-like behaviour to a state in which flow can take place. In the case of crystalline polymers, melting takes place through the breaking up of crystals. In general melting results from the increase in heat content, which increases the internal energy associated the with rotational movements of molecular chains. In the case of amorphous polymers, this takes place via a gradual increase (rather than a sudden jump) in enthalpy.

25. Melting Point (T_m)

The temperature at which the crystals of a solid are disrupted to become a liquid. This is also known as a first-order transition insofar as it is a transition that occurs via a sudden increase in enthalpy (heat content), as shown.

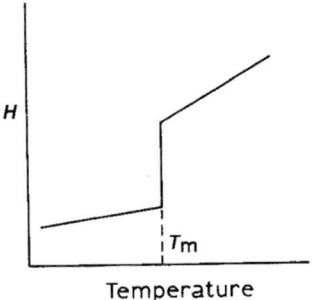

Change in enthalpy with temperature for an ideal crystalline solid.

Polymers differ from ideal crystalline solids insofar as melting takes place over a wide range of temperatures, so that the enthalpy starts to increase more rapidly when melting of the crystals begins and slows down again with a discontinuity in the trace at the point when the melting of the crystals is complete. The temperature at which the latter takes place is taken as the melting point, as shown.

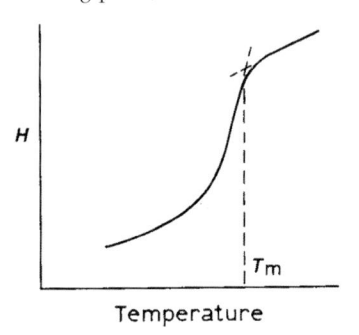

Change in enthalpy with temperature for a crystalline polymer.

Amorphous polymers do not have a thermodynamically definable melting point, hence the transition from the rubbery state to the melt state should be regarded as the 'flow temperature'.

26. Membrane

Films with selective characteristics with respect to permeation of liquids or gases. (See Nafion.)

27. Memory Polymer

A polymer that, after stretching and cooling to ambient temperature, will regain its original dimensions when reheated. (See Heat-shrinkable product, Deformational behaviour and Cross-linked thermoplastic.)

28. Mercaptan

Chemical compound containing the group SH. (See Vulcanization.)

29. Mercerization

A process named after the inventor, John Mercer, involving the treatment of cellulose with alkalis (mostly sodium hydroxide) to allow swelling of the fibres through penetration of water between the polymer chains.

30. Metal Deactivator

An additive used to reduce the catalytic degradation effect of metal ions, present in polymer products in the form of impurities, which may also be contained in other components of the formulation, such as fillers or pigments. Metal deactivators act as 'complexing' agents (known also as 'chelating agents') so that the ions become immobilized and will not reach the radicals present on polymer chains, which may have been generated by oxidative degradation reactions. The most widely used complexing agents in polymers are organic phosphines or phosphites and more highly nitrogenated organic compounds, such as melamine, bis-salicylidene diamines and oxamides. (See Chelating agent.)

31. Metal Powder

Used as filler for the production of conductive compositions. Many metals have to be excluded or suitably coated to prevent the migration of metal ions into the polymer matrix, which could adversely affect the thermal oxidation stability. A wide range of different of particle sizes and geometries can be obtained through various manufacturing routes, including condensation of metal vapours, atomization of molten metal and electrolytic deposition from metal compounds.

32. Metallization

The process by which a thin metal layer is deposited on the surface of a polymer product. There are several methods used for the metallization of polymers, the main ones of which follow.

a) **Vacuum deposition** is carried by evaporating a metal, usually aluminium, heated by tungsten filaments, and depositing the atomized metal layers under vacuum on the cold surface of a polymer substrate. Often the metal layer is in the form of a sandwich between two layers of a lacquer, a base coat to provide a smooth surface, and a top coat for the protection of the metal coating. The top coat is necessary as the metal particles do not form a continuous film and may be abraded very easily from the surface of the substrate.

b) **Electroplating** is a rigorous process that involves several steps, respectively surface preparation, electroless deposition of copper or nickel, followed by the electrolytic deposition of the final metal layer. This is to make the surface conductive for the subsequent electrolytic deposition. A good bond with the substrate is achieved by etching the surface with strong oxidizing acids, which create pits and crevices for the electroless copper coating. Chemical bonds may also be formed through salt formation with the carboxylic acid groups, but these tend to be unstable and produce blistering in service. (See Blistering.) Polymers containing etchable inclusions, such as alloyed diene elastomer particles, as in ABS, usually provide the best sites for the creation of the required crevices. Crystalline polymers rely on the higher etchability of the amorphous regions, relative to the crystal lamellae, to create the features required for the mechanical keying of the metal coating. The subsequent electroless deposition is carried out by copper or nickel salt solutions with a reducing agent, such as formaldehyde, in a caustic soda solution at pH around 11–13. Other additives may be used to control the reactions, such as hydrazine, or auxiliaries, such as phosphites, to increase the ductility of the metallic coatings. The electroplating process *per se* may include several steps depending on the performance requirements of the plated product. A chromium plate deposition for automotive exterior trim can be made up of several layers with different thicknesses in the following order (thickness values in µm): electroless copper 0.75, copper or nickel strike 2.5, copper plate 20, semi-bright/bright nickel 21, nickel for microporous chromium 2.5, and chromium 0.25 (information from Margolis, 1986). The multilayered structure of the coatings illustrates also the complexity of the entire procedure involved in electroplating operations, which often include several intermediate and final treatments.

c) **Hot stamping** is carried out via the transfers of metal coatings, usually in the form of labels or images, from a film carrier by application of localized heat and pressure from a tool or 'die' (made in metal or silicone rubber) onto a cold substrate. The carrier film (or foil) is made from a high-melting-point polymer, such as poly(ethylene terephthalate).

33. Metallocene Catalyst

Evolved from Ziegler–Natta catalyst systems used for the production of narrow molecular-weight polyolefins by a gas-phase polymerization technique. Metallocenes are complexes of transition metals, mostly zirconium or titanium, and two cyclopentadienyl (Cp, C_5H_5) ligands coordinated in a sandwich structure. The compounds used as catalysts are derivatives containing an intramolecular bridge between the Cp rings. By varying the details of the substituent groups around the Cp rings, and others connected to the metal ion, it is possible to control the type and sequence of the stereoregularity in the polymer chains, which provides the chemist with the ability to 'tailor' the molecular structure to different property requirements, particularly transparency and toughness. An elucidation of the mechanism for building up isotactic polypropylene chains from the 'active' sites of the catalyst is shown.

Mechanism for the polymerization of olefins by metallocene catalysis. Source: Adapted from Kaminsky et al., as cited in Ugbolue (2009).

34. Methylnadic Anhydride (MNA)

A curing agent for epoxy resins, having the chemical structure shown. (See Epoxy resin.)

35. Mica

A platelet reinforcing filler with excellent dielectric properties, derived mainly from 'muscovite', essentially an aluminium silicate mineral with density of 2.8 g/cm^3. Although it is possible to produce sheets a few centimetres wide for special electrical applications, the filler variety is available in the form of flakes about 1–3 μm thick and 10–500 μm wide. A micrograph of a mica filler sieved through a 200 mesh net is shown.

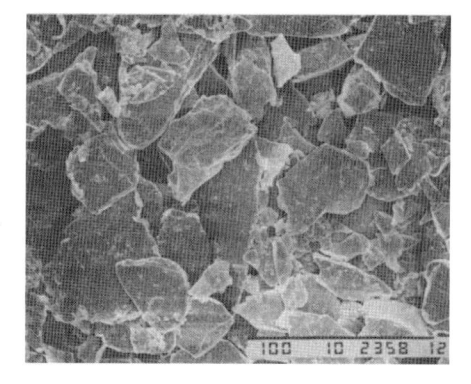

Micrograph of a muscovite mica filler. Source: Wypych (1993).

Because of the fairly high aspect ratio, mica fillers display a fairly high reinforcing efficiency without the orientation features provided by glass fibres. For this reason mica is often used in combination with glass fibres in mouldable thermoplastic composites, as a means of reducing the warping that results from differential shrinkage associated with the monoaxial orientation of the fibres.

36. Microcavitation

A phenomenon describing the creation of a multitude of small voids during ductile deformations of heterogeneous materials, such as HIPS, ABS, or particulate composites with poor interfacial bonding.

37. Microemulsion

A dispersion of solid particles or liquid droplets in a liquid medium, distinguished from an ordinary emulsion by their thermodynamic stability and their nanostructure. The latter feature is responsible for their optical transparency.

38. Microhardness

Hardness measured with small indenters and to low levels of penetration.

39. Micrometre (Micron)

A linear dimension corresponding to 10^{-6} m. The usual symbol for micrometres is μm (sometimes μ alone is used for micron).

40. Microscopy

A technique that takes magnified views of geometric features of specimens. The images can be taken by reflection of incident radiation (reflection microscopy) or by transmission of incident radiation (transmission microscopy). (See Light microscopy, Transmission electron microscopy and Scanning electron microscopy.) The range of heterogeneities that can be examined by the various microscopic techniques is indicated in the diagram.

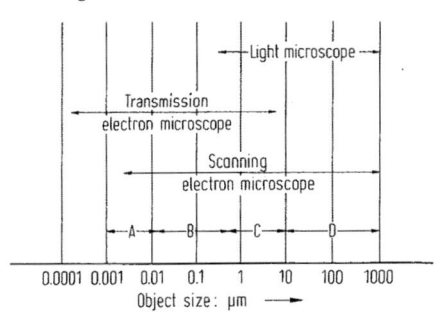

Working range of various microscopy techniques: A, polymer coils as amorphous or crystalline domains; B, nanocomposites; C, composites and pigmented polymers; D, foams and laminates. Source: Kampf (1986).

41. Microvoid

Small voids present in polymer products. Microvoids are frequently found in composites as a result of air entrapment during the impregnation of the fibres with resin. Microvoids are also formed in crystalline polymers and polymer blends subjected to high stresses, owing to the development of hydrostatic tensile stresses at the interface between crystals and amorphous domains, or between the glassy and rubbery domains of blends such as HIPS. The formation of microvoids is usually manifested as whitening regions in the stressed areas.

42. Microwaving

Using a very energetic form of radiation at very high frequencies (2–15 GHz), primarily for the continuous vulcanization of rubber, also known as ultra-high-frequency (UHF) vulcanization. It follows the same principles as dielectric heating, which relies on the energy losses to generate heat. Specific additives are often used to increase the dielectric losses to the required levels. (See Dielectric, Polarization and Loss angle.)

43. Mineral Filler

A filler from mineral sources. (See Filler.)

44. Mineral Pigment

A pigment from mineral sources. (See Colorant.)

45. Miscibility

A term used to describe the ability of a combination of two polymers, or an additive and a polymer, to form homogeneous mixtures, that is, the components become intimately dispersed at the molecular level. Miscibility is achieved under conditions in which the related free energy of mixing (ΔG^m) is negative. According to the second law of thermodynamics,

$$\Delta G^m = \Delta H^m - T\Delta S^m,$$

where ΔH^m is the heat of mixing, T is the absolute temperature and ΔS^m corresponds to the change in entropy. For mixtures of a polymer and an additive, the heat of mixing can be calculated theoretically from knowledge of the respective solubility parameters of the components of the mixture, that is,

$$\Delta H^m = \varphi_1 \varphi_2 V_1 (\delta_1 - \delta_2),$$

where φ_1 and φ_2 are the respective molar volume fractions, V_1 is the volume occupied by the polymer, while δ_1 and δ_2 are the solubility parameters. For the case of mixtures of two polymers in which the volume fractions of the two components are not vastly different, it is more appropriate to use the Flory–Huggins equation to calculate the heat of mixing, that is,

$$\Delta H^m = kT\chi'_{12} n_2 \varphi_1,$$

where n_2 is the number of molecules of polymer 2, k is the Boltzmann constant and χ'_{12} is a characteristic interaction parameter. The term $kT\chi'_{12}$ represents the difference between the enthalpy of one molecule of polymer 1 when surrounded by molecules of polymer 2, and that of one molecule of polymer 2 when surrounded by molecules of polymer 1.

From the above considerations it follows that, since mixing results in an increase in entropy, the ability of the two components to form a homogeneous mixture is favoured if the heat of mixing is negative, that is, when heat is evolved during mixing. Even so, however, the Flory–Huggins equation fails to predict the possibility of phase separation taking place with changes in temperature. The related theories have later taken into account the change in free energy with temperature and have stipulated that phase separation can take place at a 'critical solution temperature', T_c. There are two curves that represent the boundary between a miscible (single-phase) system and a two-phase mixture, known respectively as binodal and spinodal phase separation. The idealized diagram of critical solution temperature as a function of the volume fraction of the components shows that there can be an upper critical solution temperature (UCST) and a lower critical solution temperature (LCST). Accordingly, phase separation takes place when the second derivative of the free energy with respect to volume fraction is equal to zero, that is,

$$\partial^2 (\Delta G^m)/\partial \varphi_2 = 0,$$

where φ_2 is the volume fraction of the minor component. A schematic diagram of the phase separation conditions is shown.

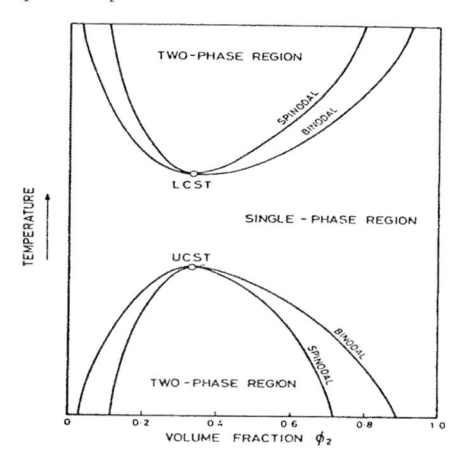

Schematic diagram showing the boundaries between miscibility and phase separation as upper (UCST) and lower (LCST) critical solution temperatures. Source: Unidentified original source.

46. Mixer

Equipment and devices used for mixing components of a polymer formulation. These can be 'static mixers', for systems where the required breaking up and movement of the components takes place through a series of complex channels. More usually, mixing is carried out in 'rotational mixers', where the stretching and movement of the components takes place between the stationary walls and moving walls of channels, that is, the surfaces of rotors. The majority of mixing devices include both complex flow paths and rotation features.

An example of a static mixer is the 'Kenic' mixer, which consists of a tube containing helical elements of alternating reverse pitch at 90° to each other, as illustrated.

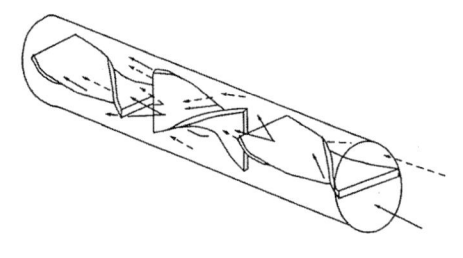

Flow pattern in a Kenic static mixer. Source: Baird and Collias (1998).

In a Kenic static mixer the elements of the fluid entering the channel are stretched and folded through helical rotations so that elements in the centre are distributed to the wall, while the fluid at the wall is moved to the centre along the flow path. In this way a stream entering the first element is split into two streams and each of them is further divided as the fluid moves along the channel, as illustrated in the diagram.

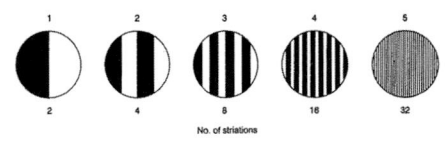

Consecutive stretching, splitting and refolding into 32 layers in a static mixer. Source: Baird and Collias (1998).

Rotational mixers can be dived into laminar flow mixers, operating at relatively low speeds, which are used for high-viscosity fluids, such as pastes and melts, and turbulence mixers, used mainly with liquid dispersions and powders. The extent, or degree, of mixing in laminar flow mixers is determined by the weighted-average total strain (WATS) that a fluid element undergoes during mixing, and the residence time distribution (RTD). These two parameters are related to the average velocity of the fluid. For a simple shearing situation experienced by a melt in a drag flow between parallel plates, the total strain, γ, can be calculated from the velocity of the plate, V_z, the gap, H, between the stationary plate and the moving plate, and the duration of the flow, t (i.e. the residence time), as $\gamma = (V_z/H)t$. For the case where the flow pattern is complex, each term is taken as an average value.

46.1 Melt Mixer

The first melt mixing device to be used in industry was the *two-roll mill*, originally used for the mastication of natural rubber and for the incorporation of vulcanizing agents. It has been widely used also for plastics, particularly for laboratory work. The rolls rotate in the opposite direction and at different speeds at a ratio between 1.5 and 3.0 in order to create a velocity gradient (shear rate) across the gap. The rolls are heated to the required temperature, allowing a few degrees temperature difference between the two surfaces to ensure that the melt will preferentially adhere to one roll and form a continuous band, known as the 'hide' or 'crepe'. A rolling bank is formed between the two rolls by the differential velocity, where new material is continually fed from the hide. To ensure that mixing takes place also across the width of the rolls, the hide has to be cut frequently at an angle, so that the rolling bank breaks up and re-forms with fresh material. The mechanics of the operation is shown in the diagrams.

c - rolling bank,
f - hide or crepe

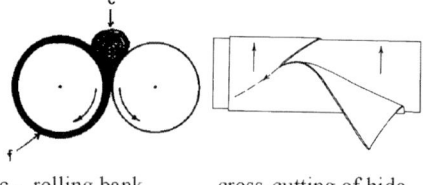

cross-cutting of hide

Principle of operation of a two-roll mill. Source: Matthews (1982).

For compounding operations of rubbers and many soft thermoplastics and blends with elastomers, the most widely used process is the *internal mixer*. In essence the internal mixer operates on the same principle as the two-roll mill with a built-in mechanism for continually cutting, folding and transferring the melt contained in the chamber from one rotor to another. This is assured by feeding a volume of melt that is less than the actual volume of the chamber (the volumetric ratio is known as the fill factor). The role of the ram is primarily to feed the polymer and keep a seal on the chamber while mixing takes place. For the mixing of rubber compounds, at the end of the mixing cycle the melt is transferred onto a two-roll mill, from which it is removed in the form of a sheet or as strips.

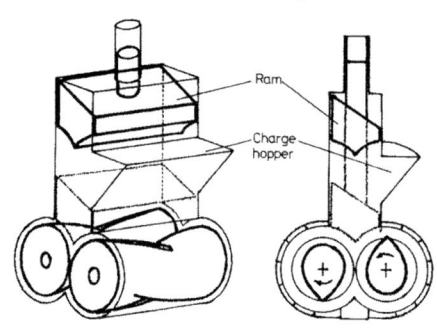

Principle of mixing in an internal mixer. Source: Matthews (1982).

In the case of thermoplastics, the melt is discharged into an extruder where it is pelletized as it exits the multi-orifice die (die-face cutting) and immediately drenched in water for cooling. Alternatively the laces formed by the die are cooled and then cut (lace cutting). For thermoplastics the most widely used mixers for compounding are twin-screw extruders and the reciprocating interrupted-flight single-screw extruder. (See Compounding and Twin-screw extruder.)

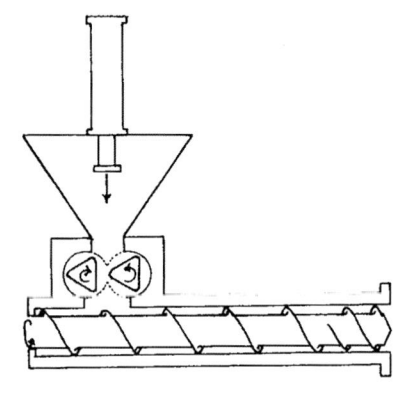

Illustration of the discharge of a mix from an internal mixer into an extruder. Source: Matthews (1982).

Some co-rotating twin-screw extruders may contain sections containing intermeshing self-wiping kneading blocks or other forms of mixing elements to enhance the mixing efficiency by the creation of multidirectional flow patterns. Photographs of these special sections of the twin screw are shown.

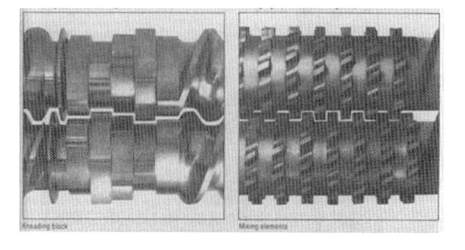

Kneading blocks (left) and mixing elements (right) of intermeshing co-rotating twin screws. Source: Baird and Collias (1998).

An illustration of the operation principle of the reciprocating screw extruder with internal pins (usually referred to as a Ko kneader) is shown. The screw has the normal rotational movement with an additional reciprocating action extending over the length coinciding with the incidence of the pin–slot arrangement. These produce both intensive and extensive mixing due to the complex flow pattern as well as the shearing of the melt in the intermesh of the barrel pin and flight slot of the screw. Since the reciprocating action of the screw produces a pulsating drag flow pattern, the melt is transferred at the end of the mixing channels to the metering zone of a conventional single-screw extruder to smooth out the pulsations before the melt reaches the exit orifice die for granulation.

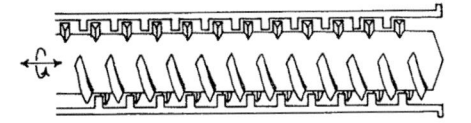

Reciprocating action of a screw with interrupted flights and barrel pins of a Ko kneader. Source: Matthews (1982).

46.2 Paste Mixer

There are basically two types, the 'parallel shaft' mixer, known also as the Z-blade blender, and the 'planetary mixer'. Schematic diagrams are shown.

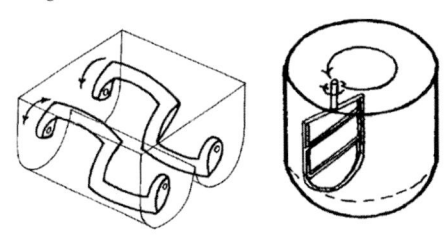

Paste mixers: (left) Z-blade blender; (right) paddle planetary mixer. Source: Matthews (1982).

Note that, in the planetary mixer, the impeller shaft not only rotates in the normal way but also moves in a circular path around the vertical central line of the mixing vessel. In many cases the mixing vessel also rotates and, for this reason, these mixers are sometime known also as 'pony' mixers.

46.3 Low-Speed Powder Mixer

A common type is the 'ribbon mixer', which operates at low speeds using spiral blades, as illustrated.

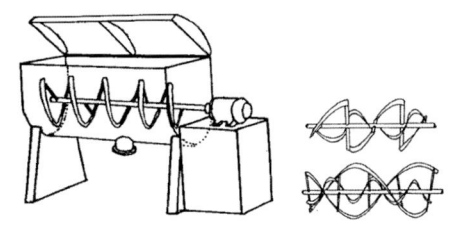

Low-speed powder mixer: (left) general layout of a ribbon blender; (right) other blade designs. Source: Matthews (1982).

46.4 Fluid Dispersion Mixer

A widely used mixer for dispersion of solid particles, such as suspensions in low-viscosity fluids, is the three-roll mill. The breaking up of agglomerates takes place in the very small gap between the rolls, which operate at fairly high speeds. While the first two rolls rotate in opposite directions, the last roll can operate in either direction. The dispersion adheres to the rollers by surface tension effects rather than dripping through the gap. More than one pass may be necessary to obtain the desired degree of particle break-up and dispersion.

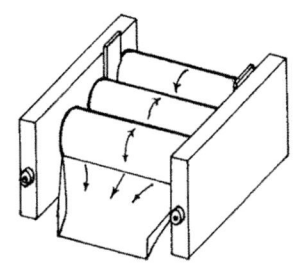

Typical set-up of a three-roll mill. Source: Matthews (1982).

There are also mixers that operate at very high speed, referred to as 'colloid mills', which operate as continuous mixers. Intensive shear is achieved by the high rotational speed of the rotors, which are placed in close proximity to a stator to provide high shear rates. One type of colloid mill comprises a hollow cylinder that rotates at high speed within another concentric cylinder, forming only a small gap between the surfaces of the two. The fluid enters at one end and is forced to exit at the other end through centrifugal forces and rotor design (left-hand diagram). Another type is the 'disc colloid mill', which is based on the same principle but employs two circular discs instead of cylinders (right-hand diagram). The fluid suspension is fed through the centre and leaves tangentially at the bottom end by high centrifugal forces, which can produce peripheral speeds of the order of 60–70 m/s.

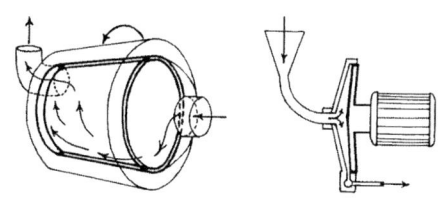

Cylindrical (left) and disc-type (right) colloid mills. Source: Matthews (1982).

46.5 High-Speed Powder Mixer

These produce mixing through the formation of vortices created by impellers rotating at very high speed. The main difference between the different types of high-speed powder mixers is the design of the impeller and drive system.

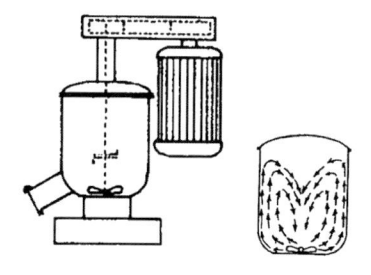

Example of drive and impeller arrangement for intensive powder mixers (left) and the flow pattern of powder (right). Source: Matthews (1982).

The frictional forces created by repeated impacts causes a considerable increase in temperature, which can be very useful in producing a better dispersion of solid additives with a low melting point. This feature is exploited in the production of PVC dry blends as a means of speeding up the infusion of plasticizer into the micropores of the powder particles. This assists the formation of a solid skin on the outer surface of the particles, thereby enhancing the free-flowing characteristics of the powder. A thermocouple is used to record the temperature continuously, which serves as a means of monitoring the mixing cycle and for stopping the rotors to discharge the powder into a larger cooling chamber operating at lower agitation speed.

47. Mixing

An operation or process that reduces the non-uniformities, concentration gradients or size of dispersed components, which leads to the randomization of the system, driven by the increase in entropy that the system is seeking to achieve. The components can be considered in terms of individual molecules, as in the case of the formation of solutions (liquid–liquid and solid–liquid), or supramolecular entities that can break up to limiting dimensions, as in the case of immiscible liquids, a dispersion of solid particles in a liquid, or particles intermixed with other solid particles. Accordingly, the movement of components to different parts of the flow field can take place by molecular diffusion or redistribution of phases. (See Miscibility.) The latter can take place through 'intensive mixing' (requiring external forces for the breaking up of the constituent components) or by 'extensive mixing' (requiring only the transportation of components, such as particles, along the imposed flow field). Extensive mixing includes operations that are often referred to as distributive, simple mixing or blending, whereas intensive mixing includes terms such as compounding and dispersive mixing. Another term frequently used in mixing operations is 'kneading', where randomization of components takes place through continually elongating and folding layers over one another. (See Mixer.)

48. Modulus

A property of materials that denotes their resistance to the deformations resulting from the application of external forces. There are three parameters required to fully define this behaviour, respectively: Young's modulus (E), shear modulus (G) and bulk modulus (K).

48.1 Young's Modulus

Defined as the gradient of the plot of stress versus strain for conditions below those leading to yielding or fracture, that is, $E = $ stress/strain (σ/ε). Stress is the force acting on a unit cross-sectional area ($\sigma = F/A$) and strain is the extension per unit length ($\varepsilon = \delta L/L$), where δL denotes an extension that is much smaller than the length. In the case of polymers, the gradient is taken either from the tangent of the straight line taken at the origin (tangent modulus), or as the ratio of the stress at a pre-specified level of strain (usually 1–2%) divided by the strain (secant modulus), as shown in the diagram.

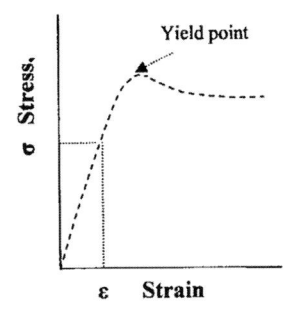

Stress versus strain plot for materials failing by yield.

For most materials, the Young's modulus measured in tension is equal to that measured in compression. For an ideal material, that is, one with an elastic behaviour and isotropic properties (equal in all directions), the Young's modulus is constant and can be described by a single coefficient. Steel and glass behave very closely to an ideal material, hence one can find in the literature or handbooks the values of their Young's modulus, that is, 200 GPa for steel and 70 GPa for glass. The Young's modulus of wood depends on the type of wood but also on whether the Young's modulus is measured along the grain (longitudinal direction) or across the grain (transverse direction); hence one usually finds a range of values quoted, typically 2–40 GPa.

48.2 Shear Modulus

Defined as the ratio of shear stress, τ, to shear strain, γ, that is $G = \tau/\gamma$. An illustration of shear deformations occurring by in-plane displacements and through torsions is shown. In either case the force acts in the direction of the plane and shear stress is defined as the force divided by the area.

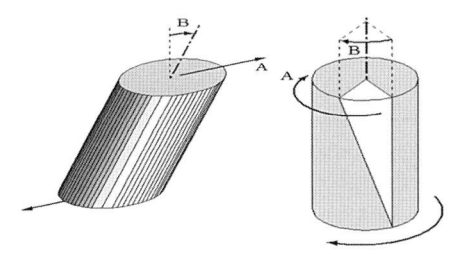

Shear deformation by in-plane shear (left) and through torsion (right).

Shear strain is determined by the extent of distortion, expressed in terms of the displacement of the plane relative to the distance from the fixed plane. This is also equal to the tangent of the angle of the distortion, $\delta X/Y$, or the extent of circumferential torsion, $\delta C/C$. These normalized distortions correspond to the tangent of the angle, B, formed by the distortion and the torsion, respectively.

48.3 Bulk Modulus

When a component is subjected to three perpendicular tensile or compressive stresses, one can consider the total change in volume to calculate the volumetric strain using the concept of 'bulk modulus'. The definitions of the terms involved are as follows.

- volumetric strain

$$\varepsilon_v = \varepsilon_1 + \varepsilon_2 + \varepsilon_3$$

- mean stress (hydrostatic tension or pressure)

$$\sigma_m = \sigma_1 + \sigma_2 + \sigma_3$$

- bulk modulus

$$K = \sigma_m/\varepsilon_v$$

The bulk modulus can be related to Poisson's ratio v, shear modulus G, and Young's modulus E, through the following expressions:

$$E = 3K(1-2v) \quad \text{and} \quad K = 2G(1+v)/3(1-2v).$$

Poisson's ratio is the ratio of the lateral strain to the longitudinal strain, resulting from a uniaxial tension or compression stress σ_1, that is, $v = \varepsilon_1/\varepsilon_2 = \varepsilon_1/\varepsilon_3$. Note that the value of Poisson's ratio is related to the volumetric expansion (tension) or contraction (compression) that results from the application of the stress. The maximum value (denoting no change in volume) is 0.5, which corresponds to the value obtained for rubbers. For crystalline polymers, such as polypropylene or HDPE, the value is in the region of 0.40–0.42, while for glassy polymers, such as polystyrene and polycarbonate, the value is around 0.33–0.55.

49. Modulus Enhancement Factor

Denotes the increase in modulus of a polymer resulting from reinforcement with fibres or fillers. It is defined as the ratio of the modulus of the composite (E_c) to the modulus of the matrix (E_m). The graph shows the effects of the strength of the fibre–matrix interfacial bond on the variation in E_c/E_m ratio as a function of the duration of the applied load for the case of a thermoplastic matrix composite. Comparison is made with the ideal behaviour expected from a composite exhibiting elastic behaviour and without fibre–matrix interactions.

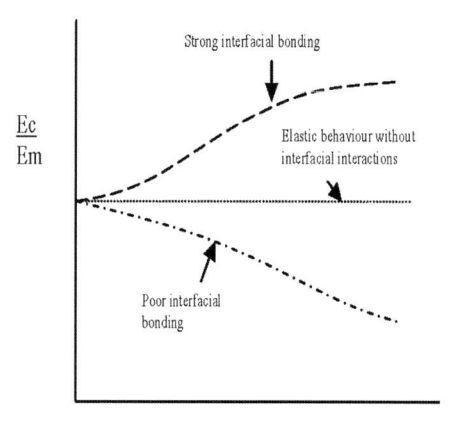

Variation of modulus enhancement factor with the duration of the applied load for thermoplastic composites. Source: Derived from author's unpublished work and Mascia (1974).

50. Molar Mass

(See Molecular weight and Degree of polymerization.)

51. Molecular Weight

Abbreviated to MW, corresponds to the mass of a molecules expressed in grams. This is the same as molar mass. The molar mass of ethylene, C_2H_4, is 28 (i.e. $2 \times 12 + 4 \times 1$). The molar mass, MW, of a moulding grade of polyethylene can be as high as 280 000, which corresponds to 10 000 ethylene units in the chain. (This is known as the degree of polymerization.) The MW of an ultra-high-molecular-weight grade, on the other hand, can be 10 times higher. For polymers, the size of the molecules is not uniform but statistically distributed over a wide range. The distribution can be symmetrical or skewed at either side of the median value. Normally an average value is reported for MW and, preferably, also the method used for the measurement. The different average molecular weights are defined according to the way the average is calculated.

The 'number-average' MW is the average calculated on the basis of the number of molecules N_i having molecular weight M_i. Calling α_i the fraction of molecules with molecular weight M_i, the equation for calculating the number-average MW is

$$\bar{M}_n = \sum \alpha_i M_i = \sum \left(\frac{N_i}{\sum N_i}\right) M_i = \frac{\sum N_i M_i}{\sum N_i}.$$

If the weight fraction of such molecules, φ_i, is used for the computation of the MW, then the expression for the 'weight-average' MW becomes

$$\bar{M}_w = \sum \varphi_i M_i = \sum \left(\frac{W_i}{\sum W_i}\right) M_i = \frac{\sum W_i M_i}{\sum W_i},$$

and, therefore,

$$\bar{M}_w = \frac{\sum N_i M_i^2}{\sum N_i M_i}.$$

If all the molecules are equal in size, the two averages are the same, and the polymer is said to be monodisperse, that is, $M_w/M_n = 1$. With broadening of the distribution, which is usually the case, this ratio becomes larger than 1 and the system is said to be polydisperse. The ratio M_w/M_n, therefore, quantifies the degree of polydispersity. The diagram depicts the MW curve for a system that exhibits a distribution skewed towards low MWs and normalized with respect to M_n, in which are reported the various average MW values. In this diagram there appears also an M_z average, which represents the degree of skewness of the curve. For a symmetric distribution this value would be the same as M_w.

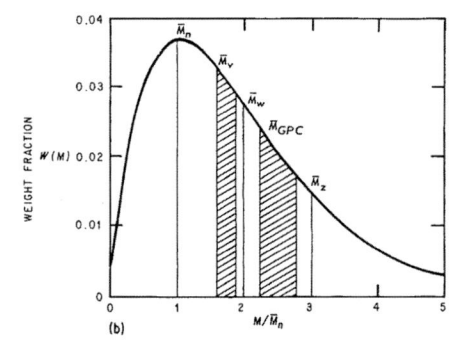

Typical molecular-weight distribution of a polymer, showing various average molecular weights on the curve, respectively M_n, M_v, M_w, M_{GPC} and M_z. Source: Ver Strate (1978).

Note that, for the particular polymer represented in the diagram, the measured values M_v (from solution viscosity measurements) and M_{GPC} (from gel permeation chromatography) are both higher than M_n, and M_{GPC} is even higher than M_z. For the latter, the molecular-weight values are expressed in terms of polystyrene equivalents and not in absolute terms.

51.1 Measurement of Average Molecular Weight

There are different ways of measuring average molecular weight. The more widely used are 'solution viscosity' to measure the 'viscosity-average molecular weight', M_v, and 'gel permeation chromatography' for determination of the entire molecular-weight distribution. There are also methods specifically designed for particular systems, such as, for instance, end-group analysis for polymers or resins containing functional end groups that can be measured by titration methods or by spectroscopic techniques. Typical systems that make use of end-group analysis are epoxy resins, unsaturated polyesters, CTBN and ATBN, polyamides and polyols. The molecular weight is calculated from basic principles, for example,

[mole equivalent of functional groups]
= weight of sample/molecular weight.

This method has some practical limitations for high-molecular-weight polymers owing to the very low concentration of functional groups in the samples that can be used for chemical analysis. There are also several empirical methods that are used to measure parameters related to molecular weight, but these are used primarily for quality control purpose. These include 'K value' and 'viscosity number' for PVC, MFI for polyolefins and styrene polymers, and 'intrinsic viscosity' for polyamides and PET. In most cases, relationships are available to convert these parameters to nominal number-average molecular weight.

51.2 Solution Viscosity Method

The method is based on the Mark–Houwink relationship between the intrinsic viscosity and molecular weight, $[\eta] = KM^\alpha$, where k and α are constants that depend only on the nature of the polymer. The viscosity of dilute polymer solutions at different concentrations, and that of the solvent, are measured with specifically designed viscometers, such as the Ostwald and Ubbelohde viscometers, by simply recording the time for the solution to flow between two marks in the capillary, as shown.

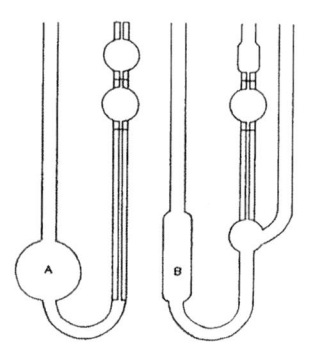

Oswald (left) and Ubbelohde (right) viscometers. Source: Unidentified original source.

The relationship between the ratio of the viscosity of the solution, η_2, to the viscosity of the solvent, η_1, and the polymer concentration, c, is given by the expression

$$\eta_2/\eta_1 = 1 + [\eta]c + kc^2 + \text{higher-order terms},$$

where $[\eta]$ and k are constants, and the higher-order terms are very small relative to the previous first- and second-order terms. The equation can, therefore, be rearranged to give

$$(\eta_2 - \eta_1)/\eta_1 c = [\eta] + kc,$$

so that a plot can be made of $(\eta_2 - \eta_1)/\eta_1 c$ against concentration, c. This will give a straight line, whereby the intercept corresponds to the value of $[\eta]$, known as the 'limiting viscosity number' or 'intrinsic viscosity', while the slope gives the value of k. These parameters depend on molecular weight, which is known for most polymers from the Mark–Houwink equation. The viscosity-average molecular weight so obtained is related to the number-average value by the expression

$$\bar{M}_v = \left(\frac{\sum N_i M_i^{1+\alpha}}{\sum N_i M_i}\right)^{1/\alpha}.$$

For the special case where $\alpha = 1$, the viscosity-average MW would correspond to the weight-average value. In practice, the value of α is between 0.5 and 0.8 and, therefore, the viscosity-average MW is lower than the weight-average value, but higher than the number-average value.

51.3 Gel Permeation Chromatography Method

The method relies on the ability of a column (packed with a porous cross-linked polymer gel swollen with solvent) to retard the rate of flow of a dissolved polymer to an extent that depends on the size of the polymer molecules (i.e. their molecular weight). Pores consist of regions of the network with a low degree of cross-linking, therefore, containing larger amounts of solvent. Smaller molecules experience a greater reduction in flow rate and, therefore, take longer to elute through such a column than larger (higher-molecular-weight) species. This effect results from the ability of smaller molecules to take up positions in the smaller pores of the gel and, therefore, they are retained longer in a packed column than the larger molecules, that take up positions in the larger pores. Detectors are employed to record the retention time of species of different molecular weights within the column and produce elution curves, representing the concentration of polymer in the eluted solvent. The detectors normally used include refractive index measurements, low-angle light scattering, UV and IR radiation. A typical elution curve obtained by measuring the increase in refractive index over that of the solvent, which is proportional to the concentration of the polymer eluted, is shown.

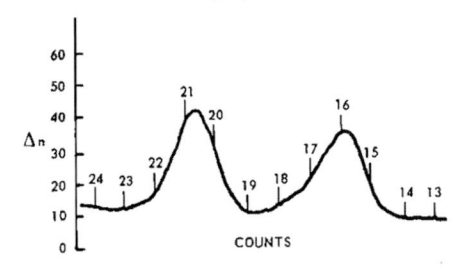

Example of an elution curve (GPC chromatogram). Source: Ver Strate (1978).

The conversion of the elution curves into a molecular-weight distribution curve (i.e. concentration, or weight fraction, versus molecular weight) is achieved by measuring the elution time of standard solutions containing fractionated monodisperse polymers of known molecular weights. A column has to be calibrated with a series of well-characterized polymer solutions, usually solutions of monodisperse polystyrene. For this reason, the molecular weight of the polymer examined is referred to as 'polystyrene equivalent molecular weight'.

52. Molybdenum Oxide

Corresponds to Mo_2O_3, widely used as a smoke suppressant in fire retardant polymer formulations. The Mo_2O_3 acts as a catalyst for the oxidation reactions occurring at high temperature, so that less carbonaceous matter (smoke and soot) is produced during combustion.

53. Molybdenum Disulfide

MoS_2 used as a solid lubricant in engineering polymer grades, such as PTFE, to enhance the wear resistance.

54. Momentum

A mathematical concept related to moving bodies or the flow of gases or fluids. It is the product of mass (m) times velocity (V), that is, $M = mV$.

55. Momentum Equation

A fundamental equation on which all flow calculations are based. It corresponds to Newton's second law, which states that the rate of change of momentum of a moving body is equal to the sum of all forces (F) causing the motion, that is,

$$\frac{d(mV)}{dt} = \sum F,$$

where m is the mass and V is the velocity. (See Flow analysis.)

56. Monodisperse Polymer

A polymer with dispersity index equal to 1, indicating that all polymer chains have the same molecular weight.

57. Monofilament

Filamentary products containing only one filament.

58. Monomer

Chemical compound used for the production of polymers by polymerization reactions. For instance, styrene is the monomer required to produce polystyrene.

59. Monsanto Rheometer

An instrument used to evaluate the curing characteristics of a rubber gum. (See Curemeter.)

60. Montmorillonite

A nanoclay consisting of hydrated sodium calcium aluminium magnesium silicate hydroxide, $(Na,Ca)(Al,Mg)_6(Si_4O_{10})_3(OH)_6 \cdot nH_2O$, used primarily for the production of polymer nanocomposites. The interesting characteristic of montmorillonite for nanocomposites is the layered structure of alternating silica tetrahedra and alumina octahedra, as shown.

This structure allows the penetration of a cationic surfactant (a quaternary ammonium compound) between the layers through the exchange of surface cations (known as intercalation), which brings about the separation of the layers (known as exfoliation) into platelets with a thickness of a few nanometres, when they are incorporated into a polymer. (See Exfoliated nanocomposite and Intercalation.)

61. Mooney Equation

An equation that relates the increase in viscosity of a polymer resulting from the incorporation of a particulate filler, that is,

$$\ln(\eta_c/\eta_m) = K_c V_f/(1-V_f/\varphi_{max}),$$

where η_c and η_m are the respective viscosities of composite and matrix, K_c is a geometric constant for the filler particles, V_f is the volume fraction of the filler, and φ_{max} is the packing factor (defined as the ratio of the true volume to the apparent volume occupied by the filler).

62. Mooney–Rivlin Equation

(See Rubber elasticity.)

63. Mooney Viscometer

An apparatus used to evaluate the curing characteristics of a rubber. (See Curemeter.)

64. Morphology

A term used to describe the characteristic heterogeneous nature of a polymer, such as that arising from the presence of crystalline domains, fillers or other fine components not molecularly miscible with the host polymer.

65. Mould

The component of a moulding equipment or manufacturing line that provides the shape of the product. It consists of two 'halves', respectively, the 'fixed half' attached to the stationary part (often referred to as platen) and the 'moving half' attached to the moving part. The impressions formed between the two 'halves' are known as cavities. (See Compression moulding, Injection moulding and Transfer moulding.)

66. Mould Design

Two important aspects to consider in mould design are the 'flow analysis' for cavity filling and the heat transfer for 'cooling' the moulded part. There are two aspects that need special consideration in cavity filling, the formation of a solid skin as the melt front moves forwards from the gate and the radial flow path. (See Cavity filling.) In terms of pressure requirements, the main consideration is the packing pressure, while the solid skin formation is particularly important for the moulding of fibre-reinforced polymers, as it has considerable effect on the anisotropy resulting from the orientation of the fibres in the outer layers. (See PVT diagram.)

67. Mould Shrinkage

The amount of shrinkage (linear %), relative to the dimensions of the mould cavities, that a polymer undergoes in a high-pressure moulding operation. (See PVT diagram.)

68. Moulding

A shaping process that produces articles through deformations induced on a material feedstock within the confines of the cavities of a mould through the application of pressure. The most commonly used moulding processes for the processing of polymers are compression moulding, transfer moulding, injection moulding and blow moulding. (See Compression moulding, Transfer moulding, Injection moulding and Blow moulding.)

69. Moulding Cycle

The sequences and times required for the various steps of a moulding operation. The sequential events that take place in a typical injection moulding operation are illustrated.

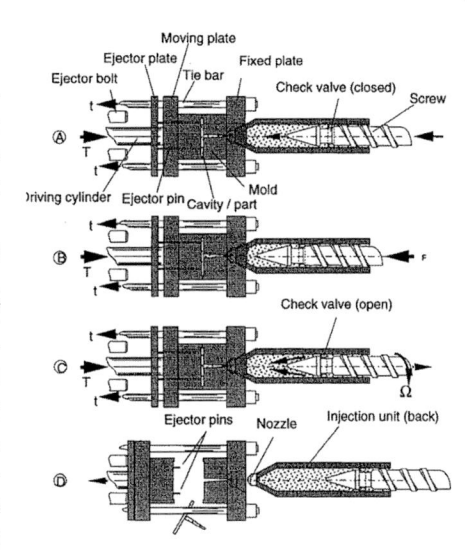

Various stages of an injection moulding cycle. Source: Osswald (1998).

The moulding cycle starts with the injection of the melt from the front end of the barrel into the cold mould by a rapid forward motion of the screw (events A and B in the diagram). The screw then rotates to plasticate the cold polymer granules residing in the feed zone, and to convey the polymer melt in the metering zone to the front of the barrel. In doing so it also moves axially back into the previous position (event C). The mould opens and the moulded parts, together with runners and sprue, are ejected from the cavity (event D).

The cycle time is the total time required to carry out the sequence of events shown in the diagram. The time for the mould to open and close in order to eject the parts is known as the 'dead' time and is usually the shortest fraction of the total time. The longest part of the cycle is the 'cooling time', which is defined as the time between injecting the melt and opening the mould. This is due to the intrinsically low thermal conductivity of polymers. The only effective expedient available to shorten the cycle time is to operate with the lowest possible mould temperature, which is limited by the freezing of the melt while flowing into the cavities of the mould, preventing the cavity from being filled completely.

In relation to the heat transfer analysis, for determining the cooling time, although thermal diffusivity is a predominant factor, there is not a very large variation in the values exhibited by polymers, deriving from their chemical structure. More important in this respect are the temperature at which the moulded part can be ejected (T_D) and the temperature at which the polymer ceases to flow, that is, the onset temperature for the rubbery state (T_R), when the viscosity becomes extremely high. (See Deformational behaviour.) The latter determines the maximum flow path length achievable and the minimum pressure required to obtain a satisfactory weld of two meeting melt fronts. Neither temperature (T_D and T_R) can be defined from fundamental principles, nor can they be measured experimentally with any degree of accuracy to produce data that can be used universally. An equation that has been widely used for estimating the minimum cooling time for a moulded part, and that would require a knowledge of the value of T_D, is

$$t_{cooling} = (h^2/\pi\alpha)\ln[(8/\pi^2)(T_M - T_W)/(T_D - T_W)],$$

where h is the thickness, α is the thermal diffusivity, T_M is the melt temperature, T_W is the mould temperature and T_D is the allowable ejection temperature. The value of T_D is particularly difficult to specify, as it may depend also on the ejector system used for removing the moulded parts from the cavities.

70. Moulding Defect

Visual defects that appear in a moulded part, particularly in the injection moulding of thermoplastics. The more common type of defects are (i) weld lines, (ii) sink marks, (iii) internal voids, (iv) distortions, and (v) warping. (See Weld line.) Sink marks and internal voids are experienced particularly in relatively thick sections of a moulded part produced from a crystalline polymer or thermosetting 'compound'. These defects arise primarily as a result of the differential cooling rate between the outer skin layers and the middle section of a moulded part, which results in a differential density through the thickness due to different degrees of crystallinity or cross-linking density developed. In both cases the faster cooling in the outer layers gives a lower density than the slow cooling rate in the middle. The diagram shows that the type of defects, that is, whether in the form of sink marks (inward suction of the outer surface layers) or internal voids, depends on the modulus of the polymer, which determines the rigidity of the moulded part

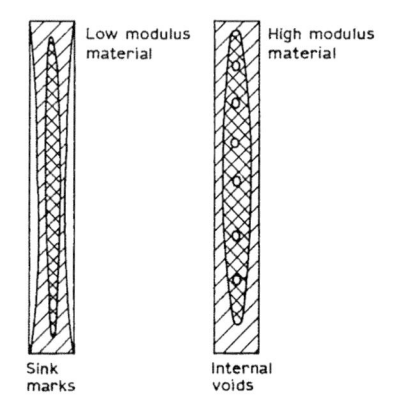

Illustration of the consequences of differential density through the thickness of a moulded part. Source: Mascia (1989).

71. Mullins Effect

A term used to describe the strain softening behaviour of filled rubbers attributed to the deterioration of the adhesion of the polymer chains from the surface of filler particles.

72. Mylar

A tradename for biaxially oriented films of poly(ethylene terephthalate).

N

1. Nafion

A tradename for a particular ionomer, used primarily for the production of the proton exchange membrane (PEM) for fuel cells, consisting of polytetrafluoroethylene backbone chains with regularly spaced perfluoroether side chains terminated by sulfonic groups. The chemical structure can be represented by the following formula.

$$-(CF_2CF_2)_x-(CF_2CF_2)_y-\underset{\underset{CF_3}{|}}{(OCF_2CF)_z}-OCF_2CF_2-SO_3H$$

The term Nafion has become almost a household name, like nylon, used even in scientific publications. There are, however, several other perfluoroether-based membranes available commercially. The important parameters that characterize the structure of Nafion are (i) the equivalent weight (EW), corresponding to the number of grams of dry Nafion per mole of sulfonic acid groups contained in the structure, and (ii) the ion exchange capacity (IEC), which represents the number of milliequivalents of H^+ ions present per gram of polymer. This can, therefore, be related to average EW as $IEC = EW/1000$.

The morphology of hydrated Nafion consists of ionic clusters, about 4 nm in diameter, containing the sulfonated perfluoroether side chains, which are organized as inverted micelles arranged in a lattice and interconnected by narrow channels, about 1 nm in diameter. Both the diameter of the clusters and that of the interconnecting channels increase with the absorption of water, so that the movement of protons or hydronium ions (H_3O^+) from one cell to another can readily take place under the influence of an externally applied voltage. The amount of water absorbed, and the associated swelling of the membrane, are controlled by the surrounding crystalline hydrophobic polytetrafluoroethylene domains. The actual detailed morphological structure and the proton conductivity are controlled by the amount of water absorbed. At low water contents, not all the $-SO_3H$ groups are dissociated, and the interaction between water molecules via hydrogen bonds is somewhat low, so that the protons or hydronium ions will not be able to move very fast.

2. Nanoclay

A term used for clay fillers that can be exfoliated to produce platelets with thickness in the region of 3–20 nm.

3. Nanocomposite

A composite containing reinforcing agents with at least one dimension in the region of 3–100 nm.

4. Nanofiller

A term used to describe a filler that can be exfoliated to produce platelets with thickness in the region of 3–20 nm.

5. Nanometre

A linear dimension corresponding to 10^{-9} m. The symbol for nanometre is nm.

6. Natural Rubber

Rubber obtained from the sap of many different types of trees in the form of a water-based latex. The most common type is from the tree known as *Hevea brasiliensis*, which is what is usually referred to as 'natural rubber'. The chemical structure corresponds to the *cis*–1,4–polyisoprene configuration. The other naturally occurring latex, known as 'gutta percha', corresponds to the *trans*–1,4–polyisoprene configuration, a harder rubber originally used for cable insulation, which has been replaced by polyethylene. The structures of the two types are shown.

natural rubber
(*cis*-1,4-polyisoprene)

gutta percha
(*cis*-1,4-polyisoprene)

Natural rubber has a degree of polymerization in the region of 5000 and a broad molecular-weight distribution. The naturally occurring latex has a solids content in the range 25–45%, which is precipitated into a coagulum with the addition of acetic acid and rolled into sheets, known as crepe. (See Rubber and Elastomer.)

7. Network

A term used to describe the cross-linked structure of a thermoset polymer or that of a vulcanized elastomer.

8. Newtonian Behaviour

Characteristic of liquids whose shear flow behaviour can be described by Newton's law, which stipulates that the shear stress (τ) acting on a lamina of fluid during flow is directly proportional to the shear rate ($\dot{\gamma}$) corresponding to the velocity gradient (dV/dt) in the plane perpendicular to the flow direction, that is, $\tau = \eta\, dV/dt$, where η is the shear viscosity, also known as the Newtonian viscosity, which is a parameter that defines the resistance of a fluid to flow and depends only on temperature. Any deviation from this characteristic gives rise to a non-Newtonian behaviour. In the case of polymer melts, the viscosity decreases with increasing shear rate, giving a characteristic that is often referred to as pseudoplastic behaviour. (See Rheology.)

9. Nomex

A tradename for a paper made from an aromatic polyamide, represented by the chemical structure shown. These polymers have a very high melting point and a high resistance to thermal oxidative degradation, coupled with intrinsic fire retardant properties. (See Aramid.)

The aromatic polyamide, *m*-phenylene isophthalamide.

10. Non-Destructive Test

A test that does not result in any changes in chemical structure, or in the destruction, of the specimen used for the examination.

11. Nonlinear Dielectric Polymer

Consists of a polymer composition that exerts dielectric characteristics at low voltages but becomes conductive when the voltage increases substantially above the line voltage. These systems are useful, therefore, for the production of devices for the protection of circuits against overloads. A good example is their use in the construction of cable joints and terminations in order to provide a mechanism for the leakage of spikes of current resulting from lightning strikes, which would otherwise create failures in the insulation through tracking. The high nonlinearity of the resistivity in relation to the applied voltage can be brought about through the incorporation of intrinsically nonlinear fillers, such as silicon carbide, iron oxide or zinc oxide. Another way to produce a nonlinear dielectric is via the incorporation of carbon black at levels just below the critical concentration to achieve percolation conditions.

A crystalline polymer is preferable for this purpose as it would enable the filler to be located predominantly within the amorphous regions, thereby reaching the required threshold conditions at lower levels of addition than would be required with an amorphous polymer.

At low voltages, the current cannot be transmitted across the conductive carbon black particles, as these are separated by a non-conductive polymer interphase. On increasing the voltage, the conditions are eventually reached whereby electrons can be transferred through the separating dielectric gap by a 'tunnelling' effect, thereby producing conductive paths. Cross-linking the polymer can bring about an increase in the stability of the nonlinear characteristics by preventing changes in the morphological structure. These could result from overloads in the circuit, which would raise the temperature of the dielectric.

12. Nonlinear Viscoelastic Behaviour

Denotes a deformational behaviour by which the relationship between stress and strain, at any given temperature, depends not only on the duration of the excitation or the history of the deformation but also on the level of stress and strain. The difference between linear and nonlinear viscoelastic relationships between stress and strain is shown in the diagram. The isochronous relaxation modulus (stress/strain at time $t=$ constant) decreases with increasing level of stress used to produce the strain considered. Similarly, the isochronous compliance (strain/stress at time $t=$ constant) increases with increasing level of strain considered. The time indicated on the various curves increases in the order $t_1 < t_2 < t_3 < t_4$.

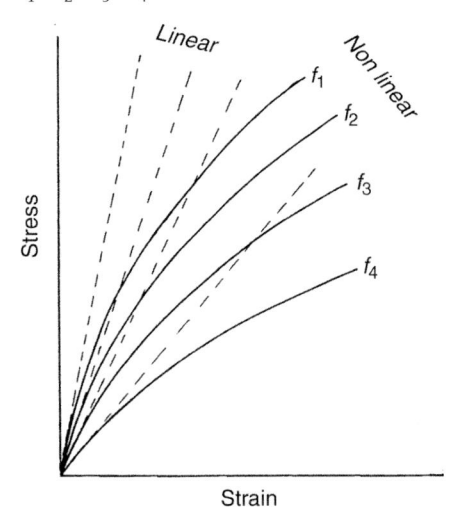

Linear and nonlinear stress–strain relationship under isochronous conditions.

The implication of the nonlinear viscoelastic behaviour of polymers is that it is not sufficient to specify a single value for the compliance or modulus for a given specified temperature and duration of the load acting on the material. It is required to specify also the level of stress or strain. In other words, the modulus will be stated as $E(T, t, \varepsilon)$ and the compliance as $D(T, t, \sigma)$, where the appropriate values of T, t, ε and σ are specified. It is worth noting that the deviation of D or E from the linear behaviour becomes increasingly more pronounced at higher temperatures and at longer times. This means that the compliance or modulus curves split up into divergent curves as time increases.

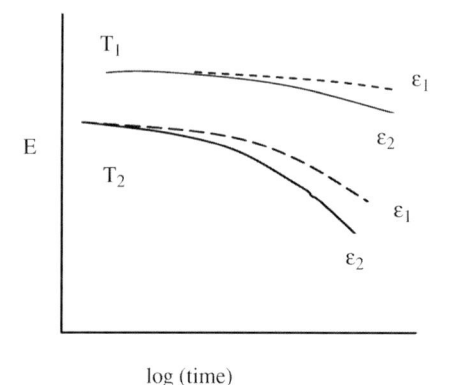

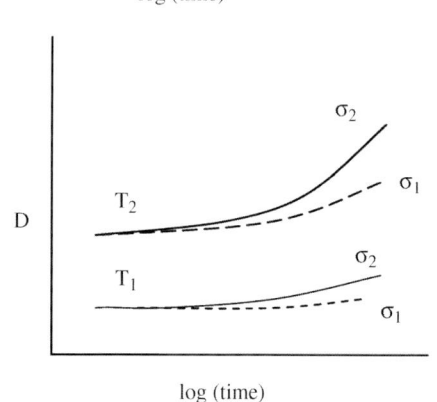

Nonlinear variation of relaxation modulus E (top) and creep compliance D (bottom) with time at two different temperatures ($T_2 > T_1$), two strains ($\varepsilon_2 > \varepsilon_1$) and two stress levels ($\sigma_2 > \sigma_1$).

It has to be borne in mind that, since the nonlinearity becomes a prominent feature only at fairly high strain levels, it becomes an important consideration only under conditions where the modulus is relatively low. It is unlikely, therefore, that substantial nonlinearity will be experienced for the case of engineering polymers (e.g. polyamides, polycarbonates, acetals and PEEK). It is even less likely in the case of structural adhesives and composites, where the strain levels reached are quite low.

In any case the level of strain in engineering products is deliberately kept at low levels as a safety factor, as a means of preventing the occurrence of fracture failures arising from defects, such as crazing and microvoids. (See Crazing.)

13. Non-Migratory Plasticizer

A type of plasticizer that does not diffuse easily into adjacent or surrounding materials.

14. Non-Newtonian Behaviour

A behaviour of fluids by which the relationship between shear stress and shear rate, defining viscosity, is not linear. This is to say that the ratio of shear stress to the shear rate, that is $\eta = \tau/\gamma$, at any given temperature, is not constant but is a function of the shear stress acting on the fluid. Polymer melts are typical fluids exhibiting non-Newtonian behaviour. The most widely used models to describe the relationship between shear stress and shear rate for polymer melts are the power-law equation and the Carreau model. Note that the flow behaviour of a liquid affects the velocity profile through the channels. In the absence of slip or lubrication, the velocity at the wall is zero and rises to a maximum in the centre. For a Newtonian liquid, the velocity profile is parabolic, giving rise to a linear increase in velocity gradient from zero at the centre to a maximum at the wall. For liquids such as polymer melts exhibiting a power-law or Carreau behaviour, the velocity profile becomes flatter towards the centre, resulting in a curved increase in velocity gradient, rather than a linear one, from the centre to the maximum at the wall.

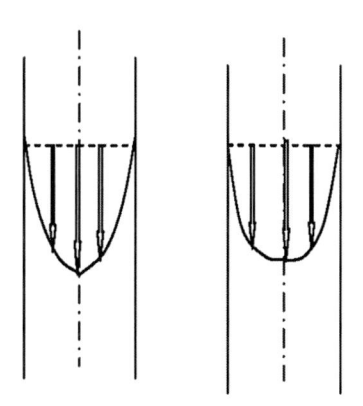

Velocity profile of liquids flowing through cylindrical and rectangular (slit) channels: (left) Newtonian behaviour; (right) power-law and Carreau behaviour.

15. Non-Newtonian Liquid

A liquid that does not exhibit Newtonian behaviour.

16. Non-Polar Polymer

A polymer whose chemical structure does not contain polar groups, or where the polar groups are arranged in such a way as not to produce permanent dipoles along the molecular chains. Typical non-polar polymers are the polyolefins (PP, PE, TPO and EPDM) and PTFE. Note that in the case of PTFE there are four symmetric C–F groups in each monomeric unit, each of which produces individually a strong dipole arising from the large difference in electron density between the two atoms forming the C–F bonds. However, the dipoles so formed are symmetrical and act in the opposite direction, so that they cancel each other and, therefore, produce a zero dipole moment.

17. Normal Stress Difference

The difference between normal stresses, exhibited by polymer melts and solutions, acting perpendicularly to the drag flow direction. This phenomenon is usually illustrated by observations of the upward flow along a rotating spindle immersed in a viscous polymer solution, as depicted in the diagram.

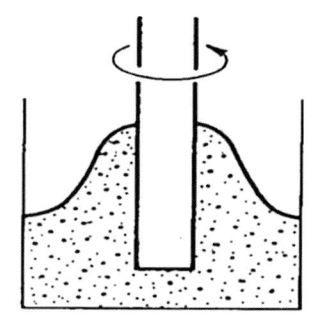

Upward rising of a polymer solution around a rotating spindle, attributed to stresses acting in the perpendicular direction to the rotational flow.

Since this phenomenon does not occur when the solvent alone (a Newtonian fluid) is subjected to the same rotational flow, it is clear that the solvent does not exhibit the normal stress characteristics of the polymer solution (a viscoelastic fluid). The behaviour is attributed to the melt elasticity of polymer melts and solutions. While the rotational flow is caused by the imposed shear stress (τ) exerted by the torque on the spindle, the upward motion results from the resulting difference in normal stresses ($\sigma_\Theta - \sigma_L = -N_1$, where N_1 is known as the 'first normal stress difference'). The parameter that is used to determine the melt elasticity characteristics of polymer melts and solutions in drag flow situations is known as the 'first normal stress coefficient' ψ_1, which is related to the shear rate $\dot{\gamma}$ (associated with the shear stress τ) by the expression $\psi_1 = N_1/\dot{\gamma}^2$.

18. Norrish I and Norrish II

These are mechanisms for the chain scission of polymer chains caused by the presence of carbonyl groups as a result of the absorption of UV light. The carbonyl groups may already be present in the chain as constituent units of a copolymer derived from an olefinic monomer and carbon monoxide, or they may be formed through oxidation reactions along the polymer chains of a hydrocarbon or other ethenoid polymers. The two mechanisms are shown.

Norrish I

$$-CH_2-\overset{O}{\overset{\|}{C}}-CH_2- \xrightarrow{h\nu} -CH_2-\overset{O}{\overset{\|}{C}}{}^{\bullet} + {}^{\bullet}CH_2-$$

$$-CH_2-\overset{O}{\overset{\|}{C}}{}^{\bullet} \longrightarrow -{}^{\bullet}CH_2 + CO$$

Norrish II

$$-CH_2-CH_2-\overset{O}{\overset{\|}{C}}-CH_2-CH_2-CH_2- \xrightarrow{h\nu} -CH_2-CH_2-\underset{CH_2-CH_2}{\overset{O\cdots H}{\overset{|}{C}}}C-$$

$$\longrightarrow -CH_2-CH_2-\overset{O}{\underset{H}{\overset{\nwarrow}{C}}} + -CH_2{=}CH{-}CH_2{-}$$

It is noted that, whereas the Norrish I mechanism produces carbon monoxide as a by-product, the Norrish II mechanism produces double bonds along the polymer chains.

19. Notch

An artificial crack machined in a specimen used to evaluate the fracture resistance characteristics of materials. Notches can be slit types or V types. The latter are more widely used to measure the impact strength of materials with pendulum methods, such as Charpy and Izod.

20. Notch Sensitivity

Denotes the reduction in fracture energy observed when a notch is introduced into a specimen. Although the phenomenon is not amenable to interpretations in terms of fracture mechanics principles, it is often used by design engineers as a criterion for the selection of materials for the manufacture of articles that are likely to be subjected to impact loads in service.

21. Novolac

A phenol formaldehyde resin that does not contain CH_2OH groups and, therefore, requires the addition of a hardener capable of generating formaldehyde, typically hexamethylene tetramine, to produce cross-links for the network in moulded products. (See Phenolic.)

22. Nozzle

A device fitted in front of an injection moulding machine or spraying equipment.

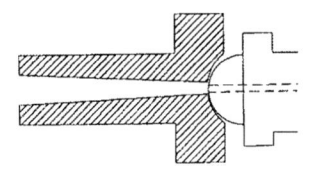

Nozzle position of an injection moulding unit relative to the sprue of the mould. Source: Unidentified original source.

23. Nuclear Magnetic Resonance (NMR)

A spectroscopic analytical technique involving the monitoring of absorbed or emitted electromagnetic (EM) radiation of a compound, in the radio-frequency range between 900 MHz and 2 kHz, by stimulated transitions between energy levels in the system, which are influenced by the environment of the nuclei. Nuclear magnetic resonance (NMR) arises from the interaction of the applied EM radiation with nuclear spins when the energy levels are split by the external magnetic field. For this to happen it is essential that the nuclei of the atoms possess a nuclear spin, which arises when there is an odd number of protons or neutrons. For the vast majority of polymers, the hydrogen atoms in the molecular chains have these characteristics and, for this reason, the technique is often referred to ^{1}H NMR, where the 1 denotes the number of protons interacting.

24. Nucleating Agent

An insoluble additive used to nucleate the formation of crystals in polymers cooled from the melt state. This is also referred to as heterogeneous nucleation insofar as the formation of crystals starts at heterogeneous sites, that is, at the surface of the particles of nucleating agent. Nucleating agents are mainly inorganic in nature, typically silica, talc, clay, metal oxides and pigments in general. Organic nucleating agents are solids with a high melting point, particularly salts such as sodium, potassium or aluminium benzoate.

A nucleating agent accelerates the formation of crystal nuclei when the polymer is cooled from the melt, which is manifested in an increase in the temperature for the onset of the crystallization process, as indicated in the thermogram.

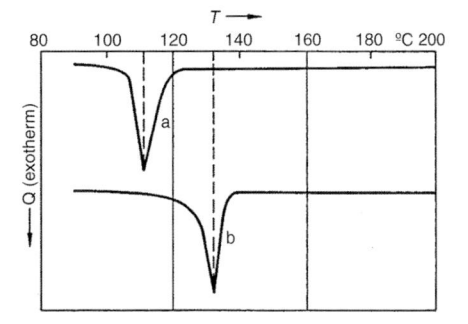

DSC thermogram for samples of polypropylene cooled from 200 °C at a rate of 8 K/min: (a) sample without nucleating agent; (b) sample containing 1% sodium p-tert-butyl benzoate. Source: Jansen (1990).

The larger number of nuclei that are formed in the presence of nucleating agents brings about a reduction in the size of the sperulites, as shown in the micrographs.

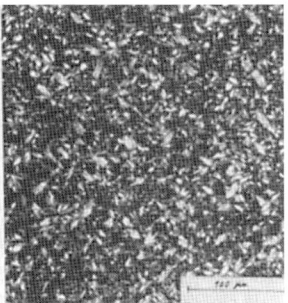

Effect of a nucleating agent on the size of spherulites in polypropylene samples: (left) without nucleating agent; (right) with nucleating agent. Source: Jansen (1990).

The smaller spherulites bring about an enhancement in optical clarity, owing to the reduced amount of internal light scattering, and an increase in fracture toughness, which arises from the great ability of the crystals to dissipate strain energy through sliding of crystals before the formation of cracks. It is important to note, however, that the presence of a nucleating agent has only a marginal effect, if any, on the actual rate of crystallization, as shown in the thermograms for the isothermal crystallization of samples of poly(ethylene terephthalate) (PET) at 100 °C on supercooled samples in the glassy state. (Note that the T_g of PET is around 75 °C.)

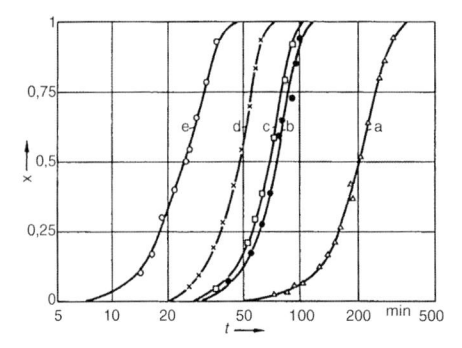

Cold crystallization isotherms for samples of PET containing 0.5% of different types of nucleating agents: (a) sample without nucleating agent; (b) TiO_2; (c) SiO_2; (d) kaolin; (e) talc. Source: Jansen (1990).

The isotherms are plots of the degree of crystallinity (i.e. fraction of crystalline domains) as a function of time. The diagram shows that the rate of increase in degree of crystallinity is practically the same in all cases, while the samples with nucleating agents start to crystallize much sooner. The data shown highlight the very strong nucleating power of talc relative to other inorganic mineral particles.

25. Nucleation

A term used to describe the formation of the 'nuclei', in the shape of nanoscopic polymeric domains, relative to the crystallization of polymers from the melt or solution, as well as the phase separation of two miscible polymers on cooling from the melt state through evaporation of the solvent from a solution mixture. This term is also used for the initial step in the formation of cells in the production of foams.

26. Nucleation and Growth

A mechanism for the formation of a particulate morphology in polymer blends resulting from phase separation of one of the components, frequently in the form of networks, to produce the 'nuclei', followed by their growth into particles. This mechanism is widely found in the toughening of a thermosetting resin from miscible mixtures with appropriate oligomers. During curing there is a precipitation of cross-linked species from this latter component before gelation of the surrounding resin matrix takes place, which brings about the formation (nucleation and growth) of toughening particles. (See Impact modifier.) The term 'nucleation and growth' is also used for crystallization phenomena and for the formation of cells in the production of foams.

27. Number-Average Molecular Weight

Also known as number-average molar mass, corresponds to the molecular weight of a polymer where the average is based on the number of molecules taken to calculate the average value. (See Molecular weight.)

28. Nylon

A term for aliphatic polyamides derived from an early tradename for the homonymous synthetic fibres.

The general structure of aliphatic polyamides produced from dicarboxylic diacids and diamines, known as polyamide x,y, PA x,y or nylon x,y, where x and y are the number of CH_2 groups in each monomer unit, is represented by the formula:

$$-[NH]-(CH_2)_x-NH-\underset{\underset{O}{\|}}{C}-(CH_2)_y-\underset{\underset{O}{\|}}{C}]_n-$$

The structure of polyamides produced from α,ω-amino acids or from the corresponding cyclic amide, known as PA x or nylon x, where x is the number of CH_2 groups in the monomer unit, is represented by the formula:

$$-[\underset{\underset{O}{\|}}{C}-(CH_2)_x-NH]_n-$$

The linear structure of aliphatic polyamides allows these polymers to form crystalline domains through close packing, as in the case of linear polyethylene, while the presence of amide groups in the molecular chains produces strong intermolecular forces, via the formation of hydrogen bonds, which are responsible for the very high melting point in comparison to linear polyethylene. The number of hydrogen bonds that can be produced and, therefore, the value of the melting point depends on the distance between and the symmetry of the amide groups along the chains. This effect is illustrated in the table, for the two types of polyamides.

Nylon x,y+2	Melting point (°C)	Nylon x+1	Melting point (°C)
Nylon 4,6	278	Nylon 6*	225
6,6*	265	7	235
8,6	235	8	195
7,7	205	9	210
8,8	215	10	178
9,9	177	11*	190
10,10	206	12*	175

*See text.

An even number of CH_2 groups in the chains can produce one hydrogen bond for every repeating unit, while an odd number of CH_2 groups will only produce a maximum of one hydrogen bond for every two repeating units. The latter will, therefore, result in polymers with a lower melting point. The two most important aliphatic polyamides available commercially, used mostly for the production of fibres or as engineering thermoplastics, are nylon 6,6 and nylon 6, nylon 11 and nylon 12. The glass transition temperature (T_g) values of the more common polyamides are approximately 60 °C for PA 6, 70 °C for PA 6,6, and 50 °C for PA 11 and PA 12. More recently a nylon 4,6 has also been introduced. Nylons are also available in the form of blends with ethylene–methacrylic acid ionomers and other ethylene–acrylate copolymers, as well as nitrile rubber. The reactivity of the amine end groups in polyamides has been exploited for the production of blends with non-polar elastomers, such as EPR and EPDM, by grafting the latter polymers with maleic anhydride, resulting in substantial improvements is impact strength over the blends with ethylene–acrylate copolymers.

The number of amide groups per unit length of a polymer chain also affects the amount of water that the polymer can absorb, which is generally high in comparison to other polymers. Nylon 11 and nylon 12 exhibit lower water absorption than nylon 6,6 and nylon 6. Grades in the form of blends are often produced as a means of reducing further the water absorption characteristics, as well as increasing the impact strength (as already indicated). Mouldings with very large cross-sections are produced in the form of castings by 'activated anionic polymerization' of caprolactam and can be obtained with very high molecular weights to achieve a very high resistance to abrasion and wear. The water absorption characteristics of nylons requires that the mouldings are 'normalized' to the equilibrium water absorption level in accordance with the expected level of humidity in the environment in which they are used. A comparison of the amount of water absorbed by the different types of nylons is shown in the plot against the relative humidity in the environment. (See Polyamide.)

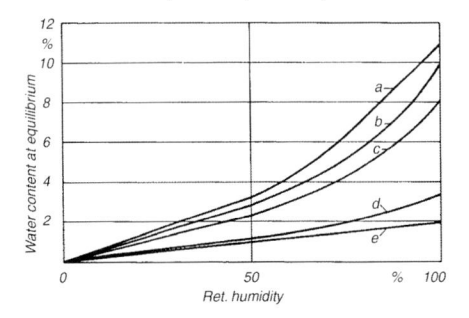

Water absorption of different types of nylons: (a) PA 6; (b) PA 6/6,6/6,10; (c) PA 6,6; (d) PA 6,10; (e) PA 11. Source: Domininghaus (1992).

29. Nylon Screw

An extruder screw comprising a long feed zone and a very short compression zone with a large compression ratio. This is due to the high melting point of nylons, particularly nylon 6,6, and the low viscosity of the polymer in its melt state.

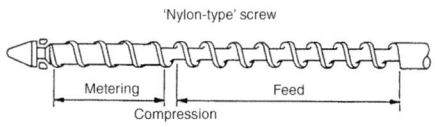

Typical features of a nylon screw used in extruders and injection moulding machines. Source: Unidentified original source.

O

1. Oil Absorption

An empirical measure of the surface area of powders, such as fillers, often using dibutyl phthalate (DBP) as the oil for the test. DBP displays good wetting characteristics owing to the high polarity of the phthalate group. A torque rheometer is often used for this test, which records the torque during mixing of the oil with the filler under cold conditions to form a paste. A maximum in the torque is registered when the oil completely fills the crevices between the particles. The oil absorption characteristics of the filler are expressed as millilitres of DBP per 100 g filler.

2. Oligomer

A short-chain polymer containing around 5–20 monomeric units.

3. Open Cell

Interconnected cells of a polymer foam. Open cells are widely found in foams produced from thermosetting systems. (See Foam.)

4. Optical Brightener

An additive used to mask yellow discolorations in polymers, which may arise through degradation reactions occurring during processing. These are fluorescent organic substances that absorb UV radiation at the far end of the visible spectrum (300–400 nm) and re-emit it at the lower end of the spectrum (450–550 nm). Typical optical brighteners used in polymer formulations are vinylene bisbenzoxazoles and benzosulfonamide derivatives of 4–naphthotriazolylstilbene.

5. Optical Microscopy

A microscopic examination method using incident visible light. (See Microscopy.)

6. Optical Path Difference

Also known as the relative retardation. (See Orientation and Refractive index.)

7. Optical Properties

Properties concerned with the response of a material towards visible light. The most widely measured optical properties for polymers are light transmission factor, refractive index, specular reflectance or gloss, haze and see-through clarity. (See Light transmission factor, Refractive index, Gloss, Haze and See-through clarity.)

8. Organic–Inorganic Hybrid

A term used to describe compositions consisting of two distinct nano-phases, where the inorganic phase is usually a metal oxide produced *in situ* by the sol–gel process. Often the organic and inorganic domains form co-continuous phases. A typical morphology of an epoxy–silica hybrid is shown in the micrograph.

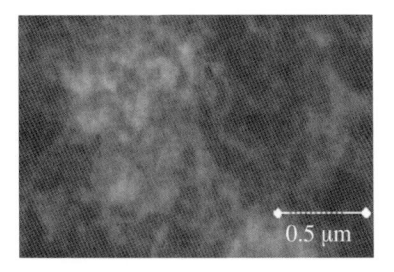

TEM micrograph of an organic–inorganic hybrid based on epoxy–silica. Source: Prezzi (2003).

Polymers in Industry from A–Z: A Concise Encyclopedia, First Edition. Leno Mascia.
© 2012 Wiley-VCH Verlag GmbH & Co. KGaA. Published 2012 by Wiley-VCH Verlag GmbH & Co. KGaA.

9. Organic-Modified Filler

A filler that has been treated with an organic additive either to form deposits on the surface of the particles or to penetrate into the galleries of the structural layers of the filler to produce the exfoliation of the constituent platelets. Fillers are coated with stearic acid or with an organic stearate to prevent them from becoming agglomerated when dispersed into a polymer. Sometimes the filler is coated with a functional organotrialkoxysilane as a means of increasing the adhesion between the filler particle and the polymer matrix.

10. Organosol

A plastisol containing an organic solvent (usually white spirit) as diluent. Not widely used in view of the toxicity implications arising from volatilization of the solvent. (See PVC and Plastisol.)

11. Orientation

A term used to describe the alignment of polymer molecules or reinforcing fibres within a composite, relative to a reference axis. Accordingly, orientation can be monoaxial or biaxial, depending on whether the molecules or fibres preferentially align along one direction or lay within a plane. The orientation of molecular chains takes place when the polymer is deformed in the rubbery state or in the 'plastic' state during cold drawing after the yield point. When orientation is induced in the rubbery state, the polymer has to be cooled down to 'freeze-in' the molecular alignment. This is achieved by the development of sufficiently strong intermolecular forces, via the reduction in intermolecular distances, which prevent the recoiling of the molecular chains into the random configuration, which is the thermodynamically stable state. This occurs naturally for deformations carried out in the 'plastic' state, owing to the lower temperature of the polymer environment during the deformations that induce molecular orientation.

12. Orientation Function

A parameter that quantifies the degree of alignment of polymer chains relative to reference axes. An orientation function is defined in terms of the deviation of the average vector formed by the projection of molecules on a specific axis relative to its random configuration. The monoaxial orientation function (f^*) for a specific direction, say direction x, can be described by a vector (representing the axis of symmetry of the average special configuration of polymer molecules) forming equal angles with respect to the other two directions, in this case y and z, giving

$$f^* = f(x/y) = f(x/z) = \tfrac{1}{2}(3\cos^2\alpha - 1)$$
$$\text{and}\quad f(y/z) = 0,$$

where α is the average angle formed by the projection vector. This means that the monoaxial orientation function f^* increases from 0 for the unoriented state, where the special configuration of polymer molecules is random (for which $\cos^2\alpha = 1/3$), to 1 for perfect monoaxial orientation (for which $\cos^2\alpha = 1$, i.e. for $\alpha = 0$).

The orientation state of crystalline polymers has to be represented by two orientation functions, one for the amorphous phase and one for the crystalline phase. For the latter, the reference vector, representing the spatial configuration of polymer molecules, is the crystal axis, which coincides with the direction of the chain folds.

Accordingly, the overall orientation can be taken as the algebraic average of the two orientation functions calculated by the law of mixtures based on the relative volume fractions of the two phases. Measurements of the orientation functions can be made by various methods. The most widely used method is based on measurements of the birefringence exhibited by the oriented sample. In this case the orientation function is equal to the ratio of the measured birefringence to the calculated maximum birefringence corresponding to the full alignment of the molecules along the reference axis, that is $\alpha = 0$. (See Birefringence.)

13. Osmometry

A method for determining the molecular weight of polymers from measurements of the osmotic pressure of solutions at various solute concentrations. (See Molecular weight and Osmotic pressure.)

14. Osmotic Pressure

The pressure generated in a pure solvent as a result of infusion of a solute from an adjacent solution through a membrane. Alternatively, the pressure can be created by the diffusion of solvent into the solution. Whether the solvent or the solutes migrate through the membrane depends on the nature of the system. For the case of polymer solutions, it is the solvent that permeates through the membrane, owing to the very small size of the solvent molecules relative to the size of polymer molecules. This phenomenon is caused by the thermodynamic drive towards concentration equilibrium, which allows the solvent to penetrate into the solution, thereby creating an expansion of the solution.

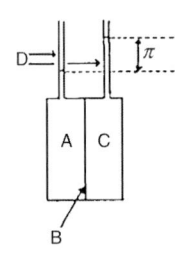

Schematic illustration of the generation of a pressure, π, in the solution of 'cell' C, (seen as a rise in the height of the liquid in the capillary), derived from the arrival of solvent from 'cell' A, diffusing through the membrane B.

This phenomenon has been exploited for the measurement of the number-average molecular weight of polymers, M_n, exploiting the unique characteristic of osmotic pressure as a colligative property, that is, a property that depends only on the 'number' of molecules (N) present. This is related to the molecular weight, M, and to the concentration of the polymer in the solution by the expression $N = N_A c/M$, where N_A is the Avogadro number (6.023×10^{18} molecules per gram of polymer). The actual procedure involves measuring the osmotic pressure resulting from solutions at different concentrations and plotting π/c against c. The value of π/c at $c=0$, obtained by extrapolation, corresponds to RT/M_n, where R is the universal gas constant and T is the absolute temperature. (See Colligative properties and Molecular weight.)

15. Oxirane Ring

Also known as epoxy group. (See Epoxy resin.)

16. Oxo-Biodegradable Polymer

A polymer capable of undergoing a controlled rapid UV-induced oxidative degradation and produce low-molecular-weight biodegradable species, through the formation of alcohol, ketone, aldehyde, ester and acid groups. Polymers containing C=C double bonds and tertiary hydrogen atoms along the molecular chains are particularly prone to oxidative degradation. The process can be accelerated by the incorporation of additives consisting of salts of transition metals. Polymers containing small amounts of carbon monoxide along the polymer chains, such as ethylene–CO copolymers, have also been produced to meet these specific requirements.

17. Oxymethylene Polymer

A polymer containing repeating methylene oxide along the molecular chains. These polymers are abbreviated to PMO, that is, poly(methylene oxide). They are known commercially as 'acetals' and are regarded as engineering polymers. (See Acetal.)

18. Ozone

Formed from the reaction of atomic oxygen with molecular oxygen, that is,

$$O + O_2 \rightleftharpoons O_3,$$

which is reversible. Within the conditions in which polymers are used, the atomic oxygen for the forward reaction originates by the splitting of molecular oxygen through the adsorption of light with wavelength less than 240 nm. This radiation can derive either from sunlight or by electrical (corona) discharges from electrical circuits. The reverse reaction takes place as a result of the absorption of light at higher wavelength, such as visible light and infrared at wavelengths up to 1200 nm. Therefore, the quantity of ozone actually formed depends on the prevailing conditions.

The presence of ozone in the atmosphere can accelerate considerably the degradation of polymers, particularly those containing double bonds along the polymer chains, such natural rubber and all diene-type elastomers. The atomic oxygen formed from the decomposition of ozone can also react with water to form hydrogen peroxide, which will rapidly decompose into two hydroxyl radicals and attack the polymer chains by abstracting a hydrogen atom, thereby starting the initiation of the polymer degradation reactions. (See Thermal degradation and UV degradation.) These reactions cause the breakdown of the original molecular structure of the polymer, that is,

$$H_2O_2 + h\nu \rightarrow 2HO^{\bullet}$$

then

$$HO^{\bullet} + \text{x}\cdots\text{CH=CH}\cdots \rightarrow \cdots\text{CH--C}\cdots + H_2O$$

P

1. Paint

Generic term for polymer solutions and emulsions used to deposit coatings.

2. Parallel-Plate Rheometer

A rheometer used for measuring the viscosity of polymer melts at low shear rates. It operates on the drag flow principle, whereby a shear rate (velocity gradient) is imposed on the melt placed between a stationary plate and a rotating disc (see diagram). It can be made to operate in a rotational mode or in a dynamic oscillatory fashion. The shear rate and shear stress vary from zero at the centre to a maximum at the edge of the circular plate.

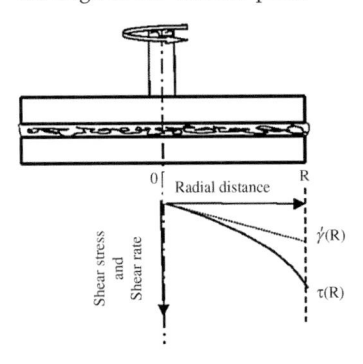

Schematic diagram of parallel plates and variation of shear rate ($\dot{\gamma}$) and shear stress (τ) with distance from centre (0) to edge of disc (R).

Owing to these variations, the viscosity, η, has to be estimated from the values of the shear stress and shear rate at the edge of the rotating plate, that is, $\eta = \tau(R)/\dot{\gamma}(R)$. In view of the non-Newtonian behaviour of polymer melts, the shear stress cannot increase linearly from the centre. However, for simplicity, this aspect is often ignored, and the viscosity is expressed as an 'apparent viscosity', based on the assumption of a linear change. The shear rate and shear stress are calculated from the peripheral velocity gradient and torque recorded on the spindle that drives the plate, that is,

$$\dot{\gamma}(R) = \frac{V(R)}{h} = \frac{2\pi \Omega R}{h} \text{ and } \tau(R) = \frac{3T}{2\pi R^3},$$

where $V(R)$ is the peripheral velocity, h is the gap distance, Ω is the angular velocity of the spindle, T is the torque and R is the radius of the rotating cone.

With some equipment it is also possible to measure the vertical thrust, from which it is possible to calculate the first normal stress difference N_1 and the first normal stress coefficient, ψ_1, that is, $\psi_1 = N_1/\dot{\gamma}^2$. (See Normal stress difference.) These latter measurements, however, are rarely made. For operations made in an oscillatory mode, the data are displayed in terms of 'real' shear modulus G' and 'imaginary' shear modulus G''. The latter corresponds to the 'real viscosity' η', calculated as the ratio G''/ω, where ω is the angular velocity of the oscillatory motion. The viscosity is calculated as a complex viscosity, η^*, comprising a real component η' and an imaginary component η'', that is, $\eta^* = \eta' - i\eta''$, where $i = \sqrt{-1}$ (the imaginary number). The imaginary term, η'', is directly related to G', that is, $\eta'' = G'/\omega$, which is a direct measure of a fundamental parameter for the melt elasticity characteristics of the polymer. (See Complex compliance and Viscosity.)

3. Parison

(See Blow moulding.)

4. Parkesine

The first tradename for cellulose nitrate products, given by the inventor Alexander

Parkes. The term 'Celluloid' was used later in the USA for similar materials.

5. Particulate Composite

Contain micrometre-sized particles (known as filler) for reinforcement. (See Composite and Filler.)

6. Paste

A term generally used to describe a very viscous suspension of PVC particles in plasticizer. (See PVC.)

7. Peel Strength

A parameter that characterizes the resistance of flexible adherend to debonding from the substrate. The peel strength is expressed as the recorded force divided by the thickness of the adherend. (See Fracture mechanics and Adhesive.)

8. Peel Test

Measures the force required to separate either a thin flexible adherend from a rigid substrate, the test being known as the 'L-peel test', or two thin flexible adherends, in which case the test is known as the 'T-peel test'. (See Adhesive.)

9. Pelletizer

The unit of a compounding line that produces pellets after the melt emerges in the form of a 'strand' from the die of the extruder. An example is shown.

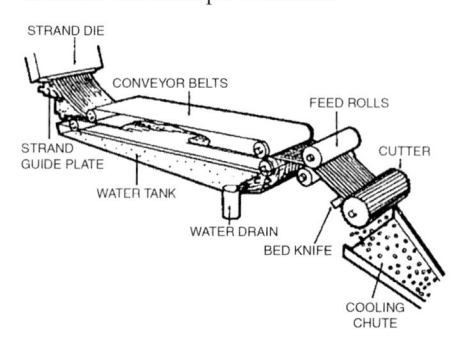

Example of a strand pelletizer. Source: Rosato (1998).

Pelletizers are also widely used for producing tablets, pellets or 'pre-weighed' feedstock of thermosetting moulding powders, as well as for compacting fine particulate fillers, such as carbon black. For these systems the pelletizers work on the cold-compaction principle using plunger–cavity components. The 'green strength' of the pellets is often increased with the use of binders deposited on the surface of the particles.

10. Pendulum Impact Test

A test in which the load is delivered at high speed by a mass placed at the extremity of a pendulum to a specific loading point of a supported or clamped specimen. The apparatus records the energy used in fracturing the specimen by registering the loss of potential energy (ΔU) from the reduction in the height reached by the striking mass after fracturing the specimen. The loss of energy of the pendulum is, therefore, given by the expression $\Delta U = mg(h_o - h_f)$, where m is the mass of the pendulum, and h_o and

h_f are the respective heights before and after fracture of the specimen.

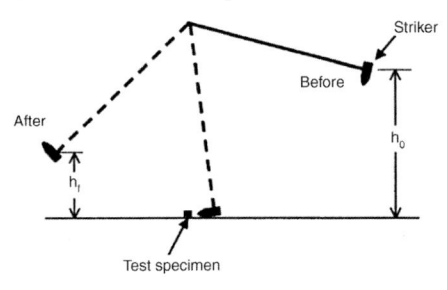

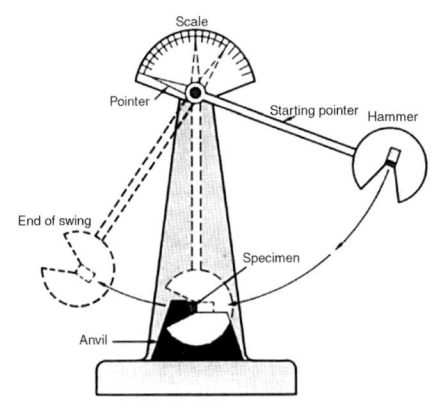

Principle of pendulum impact tests. Source: Unidentified original source.

The specimen is loaded either as a cantilever beam (Izod tests) or in a three-point bending mode (Charpy tests). In either case the load is applied in the direction that causes crack propagation from the initial artificial notch, as shown.

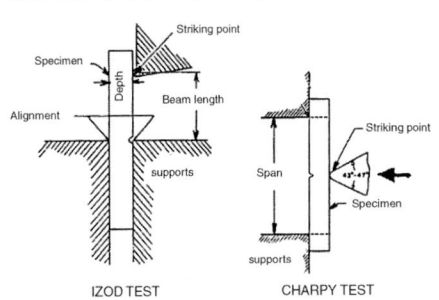

Geometry of specimens and loading modes in pendulum impact tests. Source: Unidentified original source.

11. Peptizer

An additive used to assist the breakdown of polymer chains during the mastication of raw rubber stock. Efficient peptizers consist of Fe, Co, V and other transition-metal complexes. However, these are not widely used, as they will also affect the thermal oxidative stability of the vulcanized rubber. The more commonly used peptizers are the traditional types, such as pentachlorothiophenol (PCTP) and its zinc salts (Zn PCTP).

12. Perfluoroether Polymer (PFA)

These are crystalline perfluoroalkoxy (PFA) copolymers of tetrafluoroethylene, which can be represented by the formula

$$\left[-CF_2-CF_2-CF-CF_2-CF_2-\right]\!\!-$$
$$|$$
$$O$$
$$|$$
$$R$$

where R is the fluoroalkyl group, C_nF_{2n+1}. The melting point of these polymers is around 300 °C, which is only slightly lower than that of PTFE. This allows PFA to be processed by conventional thermoplastic methods, owing also to the lower viscosity associated with the reduction in molecular weight. PFA retains many of the characteristics of PTFE, except for coefficient of friction.

13. Permanent Set

A term used to denote the amount by which a rubber sample fails to recover its original dimensions after removing the applied load. For instance, the difference between the length shortly after retraction and the original length, expressed as a percentage of the original length, is called the 'tension set'. The same calculation applied to tests in compression gives the 'compression set'. When the same reasoning is applied to the retraction of heat-shrinkable products after increasing the temperature above that used in the expansion process, this gives an estimate of what is sometimes known as 'amnesia'.

14. Permeability

A parameter (P) that describes the pressure-related rate of permeation of a gas or liquid through a medium, normally a sheet, film or membrane, that is, $F = P dp/dx$, where F is the mass diffusion rate and dp/dx is the pressure gradient through the thickness of the medium. Permeability is related to the diffusion coefficient (D) and the solubility (S) by the expression $P = DS$. It is a widely used parameter to describe the barrier properties of films in packaging, which are often multilayer structures that can be specifically designed to achieve the desired level of permeability. The overall permeability value (P_c) can be calculated from the permeability values of the individual layers by the application of the basic principles of the law of mixtures under isoflux conditions, that is, the rate of permeation is the same through each layer, which corresponds to a situation in which the total pressure gradient is the sum of the pressure gradients through each individual layer. For a three-layer film, for instance, this gives an expression for the permeability (P_c) in the form of

$$1/P_c = \tau_1/P_1 + \tau_2/P_2 + \tau_3/P_3,$$

where τ_i is the fractional thickness of each individual film, t_i, relative to the total thickness, t_{total}, that is, $\tau_i = t_i/t_{total}$, and the digits 1, 2 and 3 refer to each of the three layers. Note that the units of permeability are m³/m s Pa. Similarly to the diffusion coefficient, the variation of the permeability with temperature is described by the Arrhenius equation using the concept of activation energy as the fundamental parameter that characterizes the sensitivity of the permeability to changes in temperature. (See Activation energy.)

15. Permittivity

A property of a dielectric material (ε) that characterizes the capability of a dielectric to store charge in an electric field. Permittivity is defined as the ratio of the charge density Q/A (where Q is the charge accumulated in coulombs and A is the surface area) to the applied stress V/L (where V is the applied voltage and L is the thickness of the dielectric), that is, $\varepsilon = (Q/V) \times (L/A)$ in F/m, where Q/V corresponds to the capacitance. In an alternating field the permittivity is expressed as a complex parameter, that is $\varepsilon^* = \varepsilon' - i\varepsilon''$, where i is the complex number. The imaginary term ε'' corresponds to the deviation of a dielectric from a pure capacitor, corresponding to the amount of electrical energy dissipated as thermal energy. This characteristic can also be described in terms of the phase angle (known also as the loss angle) between the charge and loss vectors, that is, $\tan \delta = \varepsilon''/\varepsilon'$. These parameters vary with the frequency of the alternating field, and are modelled with an analogue consisting of a resistor and a capacitor. The variation of the related parameters with the product $\omega\tau$ (where ω is the frequency and τ is the relaxation time) is shown. (See Complex permittivity and Dielectric properties.)

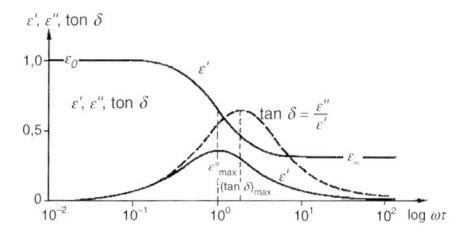

Variation of permittivity parameters of a dielectric with $\omega\tau$. Source: Hoffmann et al. (1977).

16. Peroxide

An agent used as curative for unsaturated elastomers and as initiator for free-radical polymerization of monomers and resins. Peroxy R–OO–R' compounds undergo homolytic scission when heated or exposed to UV irradiation, producing two free radicals, R–O• and R'–O•. These are highly reactive towards double bonds and will start the initiation stage of polymerization and cross-linking reactions. The decomposition characteristics of peroxides are described by their 'half-life', t_{hl}, and a nominal 'decomposition temperature', T_d. The t_{hl} value corresponds to the time taken for the peroxide to reach 50% conversion into free radicals, while the T_d value is the approximate temperature at which the peroxide undergoes the maximum decomposition rate. Typical peroxides used as polymerization initiators and curatives for elastomers are shown.

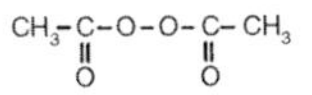

Diacetylperoxide

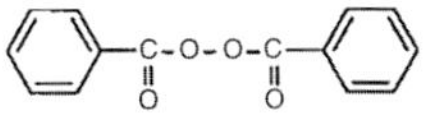

Dibenzoylperoxide

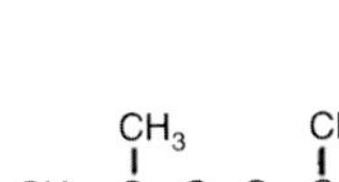

Di-tert-butylperoxide

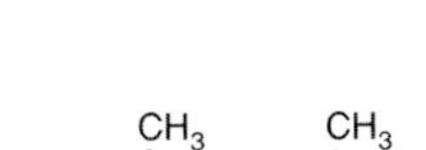

Dicumylperoxide

There are also systems available containing multiple peroxide groups, which therefore provide a high yield of peroxy radicals. Examples of these are shown.

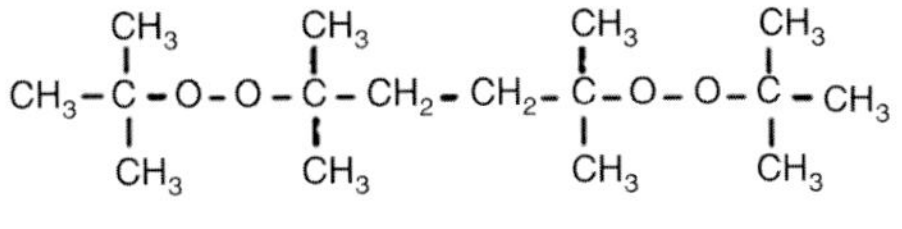

2,5-Bis-(tert-butylperoxy)-2,5-dimethyl hexane

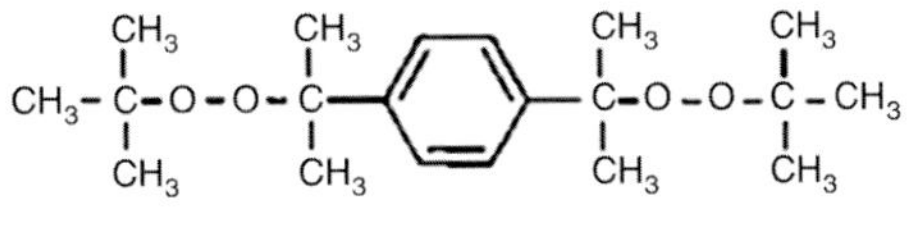

1,4-Bis-(tert-butylperoxyisopropyl)-benzene

A widely used peroxide for low-temperature curing reactions, as in the case of unsaturated polyesters for use in composites, is methyl ethyl ketone (MEK) peroxide, which is believed to contain a mixture of monomeric and dimeric compounds, represented by the chemical structure shown.

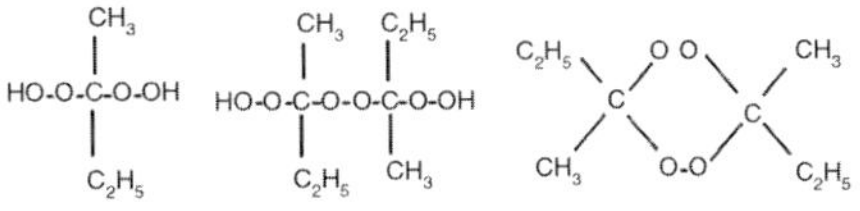

Monomeric   Linear dimer   Cyclic dimer

Formulae of commercial grades of MEK peroxide.

17. Peroxide Decomposer

Known also as a secondary stabilizer, these are used in combination with antioxidants for the stabilization of polymers against degradation by thermal or UV-induced oxidation reactions. (See Stabilizer and Antioxidant.)

18. Phase Angle

(See Loss angle.)

19. Phase Inversion

A phenomenon observed during the mass polymerization of certain reactive mixtures, such as the curing of CTBN-toughened epoxy resins and in the production of high-impact polystyrene (HIPS) and acrylonitrile–butadiene–styrene (ABS) terpolymer alloys. The phenomenon is driven by the tendency of a system to minimize its internal energy through thermodynamic considerations of the physical stability of miscible mixtures. During the course of polymerization involving two components, the originally miscible system can undergo phase separation through loss of miscibility as the molecular weight increases. This is also driven by the system seeking to minimize its energy through a reduction in viscosity, achievable via the precipitation of polymer particles from the miscible mixture. If at a later stage these conditions arise again, the precipitation of one component may bring about the redissolution of the other, driven again by the resulting reduction in viscosity. These two events are identified in the diagram, related to the production of styrene–butadiene block copolymers.

telechelic epoxy end groups of the liquid rubber. (See Epoxy resin.)

20. Phenol Formaldehyde

(See Phenolic.)

21. Phenolic

A generic term for resins based on phenol formaldehyde. Those known as 'novolac' contain around 5–10 phenol groups joined by methylene groups. Others, known as 'resole', consist of 2–4 methylolphenol

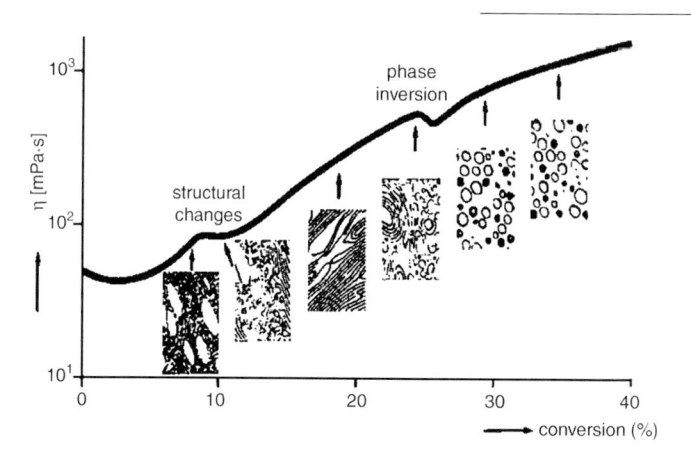

Data recorded during the production of a styrene–butadiene diblock copolymer through the polymerization of styrene units linked to a pre-formed polybutadiene. Source: Echte et al. (1981).

A situation similar to this can arise in the curing of epoxy resins containing a pre-reacted liquid rubber modifier. In this case, however, phase inversion will take place within the phase-separated particles after the surrounding epoxy matrix has reached gelation conditions. At this point the remaining hardener in the epoxy matrix can still diffuse into non-cross-linked precipitated particles, thermodynamically driven by the reaction possibilities with the

units with methylene and dimethylene ether bridges.

Chemical structure of a typical novolac phenolic resin.

Chemical structure of a typical resole phenolic resin.

Novolacs are mainly used for the production of moulding powders mixed with stoichiometric amounts of a hardener consisting of cyclic hexamethylenetetramine, $(CH_2)_6N_4$, known as HEXA, and wood flour as reinforcing filler. Curing takes place through hydrolysis of HEXA, to produce formaldehyde and ammonia, that is,

$$(CH_2)_6N_4 + H_2O \rightarrow 6CH_2O + 4NH_3$$

The CH_2O in turn enters the network through addition reactions on the benzene ring of the phenolic groups to form methylol groups. Sometimes paraformaldehyde (a low-molecular-weight polymeric form of formaldehyde) is used as hardener. Curing carried out at high temperatures during moulding (from around 150 °C) tends to decompose the less stable dimethylene ether and dimethylene amine bridges into more stable methylene links, producing a tighter network. Undoubtedly some condensation reactions take place also between methylol groups in the phenolic resin and the –OH groups in the cellulose units of wood flour. The gaseous products formed during curing tend to be absorbed by the wood flour, although some may remain dissolved in the actual phenolic network, owing to the hydrophilic and acidic nature of the phenolic groups.

Resoles are used in coatings, composites (paper and cotton fabric reinforcement) and adhesive formulations, often as water solutions. For these systems curing takes place at low temperatures through direct condensation reactions of the methylol groups, giving rise to considerable formation of dimethylene ether linkages, unless the products are post-cured at high temperatures. Strong acids are used, for example, phosphoric acid or p-toluenesulfonic acid, to provide a catalytic action for the cross-linking reactions. For coating applications, resoles are often etherified with methanol, ethanol or butanol, to produce resins that are more soluble in less polar solvents and to enhance their compatibility with other resins (e.g. epoxy, polyester and rosin resins) in mixed systems or with reactive plasticizers (e.g. liquid polybutadiene). The curing reactions occur via the formation of quinone methide units, some of which remain permanently in the structure, giving rise to the formation of dark-coloured products, for example.

Reaction mechanism for the formation of cross-links during curing phenolic resins.

Reaction mechanism for the grafting of polybutadiene chains on phenolic resins during curing.

Moulding compounds may contain wood flour as filler or other natural products, such as olive kernel flour. Special products may also contain inorganic fillers and fibres, including mica flakes (for electrical grades) and graphite for components subjected to friction, such as bushings. In addition to resin and hardener, the formulations contain a lubricant, an accelerator and sometimes a plasticizer. NBR elastomer particles are sometimes used to improve the impact resistance. Particularly relevant in this respect is the ability of phenolic compounds to cross-link diene elastomers. (See Vulcanization.) This practice is widely used for formulations used in structural adhesives. Phenolic resins are also widely used for high-performance composites based on yarn and fabric reinforcement, as cotton, aramid and glass fibre fabrics.

22. Phenolic Antioxidant

Antioxidants based on *ortho*- and *para*-hindered phenols. (See Antioxidant and Stabilizer.)

23. Phenoxy Resin

An end-capped high-molecular-weight epoxy resin. A glassy polymer with T_g around 80 °C, used primarily in hot-melt adhesives and as toughening agents in epoxy powder coatings.

24. Phillips Process

A polymerization method for the production of high-density polyethylene (first developed at Phillips Petroleum), which uses a chromium-based catalyst deposited on porous, high-surface-area metal oxide particles. The calcination of the particles at around 500–600 °C brings the chromium to its hexavalent form, which is responsible for chemisorption and subsequent initiation reactions for the polymerization of the ethylene monomer. (See Ethylene polymer, Ziegler catalyst and Metallocene catalyst.)

25. Phosphazene Elastomer

Based on polymers containing phosphorus and nitrogen atoms along the backbone chains, with alkoxy side groups containing fluorine atoms, as indicated by the formula on the right.

Two polyphosphazenes.

Polyphosphazenes are cross-linked via a free-radical mechanism with peroxides or high-energy radiation. They have a T_g in the region of -80 °C and are mainly used for their very high resistance to oils.

26. Photodegradation

Degradation induced by UV light. (See UV degradation.)

27. Photoelasticity

A methodology used for stress analysis, using a model structural component produced from a glassy polymer, such as polycarbonate or a cured epoxy resin. The strain resulting from the application of a stress brings about a small alignment of molecular chains, which is sufficient to create a certain amount of anisotropy that can be quantified by measuring the related birefringence. The photoelastic characteristics of a birefringent material can be expressed in terms of a 'stress optical coefficient' (C) defined as

$$n_1 - n_2 = C(\sigma_1 - \sigma_2) = 2C\tau_{max},$$

where ($n_1 - n_2$) is the difference in refractive index in two reference directions, σ denotes the principal stress and τ is the shear stress. The isochromatic fringes observed in the product represent the contour levels of maximum shear stress, which identify the areas of stress concentration and make it possible to quantify the stress distribution in the product.

28. Photoinitiator

A polymerization or curing initiator activated by UV light. (See Initiator and Free radical.) Photoinitiators absorb light in the UV–visible range (250–450 nm) and convert the absorbed radiation energy into chemical energy. This produces reactive intermediates, such as free radicals and reactive cations, which initiate the polymerization or curing reactions of functional monomers and oligomers. Photoinitiators for radical polymerization can act through bond cleavage, usually α-cleavage of an aromatic ketone group present in the photoinitiator (Norrish I cleavage) or by intermolecular hydrogen abstraction from a hydrogen donor by the photoinitiator (Norrish II cleavage). (See Norrish I and Norrish II.)

Typical initiator systems of Norrish I type are dimethoxyphenylacetophenone (DMPA) and diethoxyacetophenone (DEAP).

DMPA

DEAP

Initiators of Norrish II type are usually benzophenone (BP) with a tertiary amine (TA).

BP + TA

These initiators are less sensitive to air inhibition than the Norrish I types. (See Photopolymerization.) Typical cationic photoinitiators are -onium salts, that is, diaryliodonium (DAI) or triarylsulfonium (TAS) salts, with PF_6^- and SbF_6^- counter-ions.

DAI   TAS

When irradiated with UV light, these salts produce strong protonic acids, which initiate the cationic polymerization of systems such as epoxy resins and vinyl ethers. Some photoinitiators operate in combination with a photosensitizer, particularly when visible light is used to initiate the polymerization reactions. The photosensitizer absorbs energy from a light source and transfers it to the co-initiator, producing free radicals.

A typical example is the use of photoinitiator, 2–methyl-1–[4–(methylthio)-phenyl]-2–morpholinopropane-1–one (TPMK).

TPMK

This absorbs light in the 275–325 nm range and is used in conjunction with thioxanthone (TX) as photosensitizer, which has its main absorption at 380–420 nm.

TX

(Information provided by A. Priola and M. Sangermano, Politecnico di Torino, 2011.)

29. Photooxidation

(See UV degradation.)

30. Photopolymerization

Polymerization induced by radiation, usually UV. Photopolymerization can take place by a free-radical or a cationic mechanism depending on the chemical nature of the reactive species. Photopolymerization is the basis of important commercial processes, with a wide range of applications. Free-radical photopolymerization is carried out with unsaturated monomers with the aid of photoinitiators capable of absorbing light at certain frequencies and producing free radicals, which induce polymerization in the same way as for thermally induced polymerization. (See Photoinitiator.) The most important monomers used for free-radical photopolymerization are acrylates, methacrylates and unsaturated polyesters. Cationic photopolymerization can be performed using mainly epoxy monomers and vinyl ethers. Usually polyfunctional monomers or oligomers are employed to produce highly cross-linked polymer networks. It is noteworthy that oxygen in the atmosphere inhibits photopolymerization by the free-radical mechanism. Bimolecular radical photoinitiators are, however, less sensitive to air inhibition. (See Photoinitiator.) Conversely, cationic photopolymerization is not susceptible to oxygen inhibition.

31. Photoresist Polymer

Polymers used in the fabrication of integrated circuits using UV light to induce cleavage of the polymer chains so that the polymer becomes soluble in solvents. A typical example is shown for photoresists based on polyimides. Through the absorption of UV light, the strain in the four-membered ring connecting the two imide groups causes it to split.

Structure of a typical polyimide used as a photoresist.

Splitting of polyimide chains through UV absorption.

32. Physical Ageing

Denotes the slight increase in density of glassy polymers, both linear and cross-linked types, when exposed for long periods to temperatures below the glass transition temperature (T_g). This is entirely a physical process that arises purely from the thermodynamic drive of the polymer to acquire the minimum internal energy. In the case of crystalline polymers, this is achieved through crystallization, including secondary crystallization processes taking place during ageing at temperatures between T_g and the melting point of the polymer. For glassy polymers, the system seeks to achieve the most stable state (lowest energy state) consistent with the lowest level of free volumes, which can be brought about through the closer packing of molecular chains. This is manifested as a suppression of relaxations in both the β and α transitions, as shown in the diagram for rigid PVC samples.

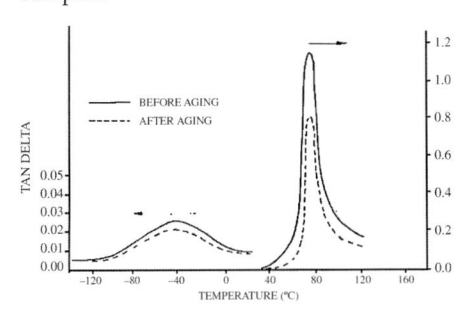

Effect of physical ageing on the low-temperature β transition and high-temperature α transition for rigid PVC. Source: Mascia and Margetts (1987).

The rate at which physical ageing takes place depends on the difference between the T_g of the polymer and the ambient temperature, and goes through a maximum at a temperature just below the T_g. In the processing of linear glassy polymers, the melt is usually cooled at a fast rate, which brings the polymer into its glassy state before reaching the minimum internal energy. Hence, the system will seek to achieve this state through an 'ageing' process. In the case of cross-linked glassy polymers, the 'curing' reactions that take place after gelation require sufficiently large free volumes to allow the reactive groups to interact through chain rotations. In so doing, an unstable state is set up, which is prone to physical ageing similarly to glassy linear polymers. This is particularly applicable for systems cured at temperatures lower than the T_g of the final product, that is, the so-called 'cold-cured' or 'ambient-cured' systems. Physical ageing can be considered as a reversible process insofar as it can be erased by heating the polymer to temperatures just above T_g and cooling it down slowly to ambient temperatures. The internal energy interpretation of physical ageing is supported by thermal analysis through observations made in a DSC heating scan. Upon reaching temperatures just above the T_g of the polymer, a small endothermic peak is developed immediately after the rapid decrease in the heat flow taking place during the glass transition. The effect of physical ageing on properties reflects the typical changes expected from an increase in density, such as increased brittleness of the polymer.

33. Physical Blowing Agent

(See Blowing agent.)

34. Piezoelectric Polymer

A polymer that exhibits piezoelectric behaviour, that is, a characteristic of certain materials to generate a small current under the influence of mechanical stresses. Among all the commercially available polymers, only poly(vinylidene fluoride) (PVDF) and, to a much lesser extent, nylon 11 (PA 11) are capable of exhibiting piezoelectric behaviour. In the case of PVDF, there are large dipoles arising from the vast difference in electron density between the CH_2 and CF_2 in the repeating units owing to the polymorphic nature of the crystals. The piezoelectric behaviour is acquired after drawing a film to align the polymer chains and subsequently applying a very large electrical stress (around 5000 kV/cm) to store charge on the dipoles. In this way, when a mechanical stress is exerted onto the film, the position of the dipoles is altered, causing the generation of an electric current through charge transfer. Main applications include microphones, earphones, loudspeakers and burglar alarms. (See Poly(vinylidene fluoride).)

35. Pigment

An additive used to impart colour and opacity to a polymer. Pigments can be either inorganic or organic. The latter are produced from dyes that are made insoluble with binders or carriers. Because of the interaction between the light scattering and light absorption characteristics of pigments, the prediction of colour obtained from a mixture of pigments is more complex than the simple additive and subtractive colour mixing relationships available for dyes. (See Colour matching.) Invariably, white colours are obtained with the use of titania pigments (rutile and anatase) and black with carbon blacks. Other inorganic pigments include cadmium sulfides and lead chromates for yellows, and chromium oxide for greens. A pearl appearance is often obtained with the use of metal flakes, such as aluminium flakes and copper–zinc alloy flakes. Examples of organic pigments are shown.

Three organic pigments: (top) Yellow 151, 13980, Monoazo; (middle) Blue 15:3, 47160, Phthalocyanine; and (bottom) Red 122, 73915, Quinacridone.

36. Pinhole

Small holes appearing in films or coatings that result from the presence of particle impurities during production, which may create an interfacial tear, leaving behind a hole.

37. Pinking

A coloration developed in polymer products after exposure to UV light. This arises from the fluorescent characteristics of some amine additives, such as the hindered amine light stabilizers (HALS) used as UV stabilizers.

38. Plane Strain

A term used widely in fracture mechanics to denote conditions by which the strains resulting from applied stresses are confined to the plane, that is, there is no change in thickness and, therefore, the strain in the thickness direction is zero. This situation arises in areas surrounding the crack tip of a thick specimen, owing to the absence of stress on the free surfaces of the crack. (See Fracture mechanics.)

39. Plane Stress

A term used widely in fracture mechanics to denote conditions by which the stresses around the crack tip are planar. This implies that there no stresses acting in the thickness direction, hence plane stress conditions are stated as $\sigma_z = 0$. A similar description is used within the context of yield criteria. This situation arises when the specimens are very thin, so that the stress-free state of the two outer surfaces of the specimen predominates through the entire thickness. A gradual transition from plane stress to plane strain conditions takes place with increasing thickness of the specimen. (See Fracture mechanics and Yield criteria.)

40. Plasma

(See Cold plasma.)

41. Plastic

A term introduced in the late nineteenth century to describe man-made rigid organic materials, later to be recognized as polymers. Plastics are classified as 'thermoplastics' when they are produced from linear polymers, and are capable of exhibiting a reversible melt state, which allows them to be shaped repeatedly through melting and cooling. Plastics based on cross-linked, or network, macromolecular systems are known as 'thermosets', owing to the irreversible nature of the melting process used to shape manufactured products. Prior to becoming thermosets, the organic components are multifunctional reactive oligomeric compounds, known as 'thermosetting resins', which can be made to flow and can be shaped at temperatures above their 'softening point'.

42. Plastic Deformation

Deformation occurring through yielding, a phenomenon normally associated with ductile behaviour of materials. (See Yield criteria.)

43. Plastication

A term used to describe the melting of polymer granules along the transition zone of an extruder or injection moulding machine, as shown in the diagram.

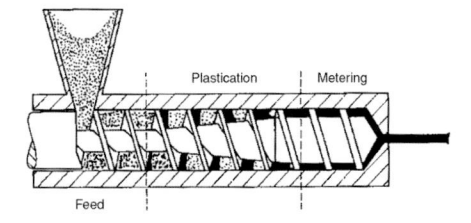

Schematic description of the three events taking place along the channels of the screw of an extruder. Source: Unidentified original source.

44. Plasticization

A phenomenon or mechanism to describe the lowering of the glass transition temperature (T_g) of a polymer, usually a glassy polymer. (See Plasticizer.)

- **Internal plasticization** This is brought about by modification of the chemical structure of the polymer either by creating greater chain flexibility within the backbone of polymer chains, or through the introduction of side groups that reduce the strength of intermolecular forces and increase the free volume. The principle can be illustrated by examining the reduction in T_g with increasing length of the alkyl ester group in polyacrylates, from methyl ($T_g = 6\,°C$), ethyl ($T_g = -24\,°C$), propyl ($T_g = -45\,°C$) to butyl ($T_g = -55\,°C$). In the case of poly(vinyl chloride) (PVC), for instance, internal plasticization is achieved in copolymer grades with vinyl acetate (VA), which decreases the T_g according to the comonomer content from 82 °C down to about 75 °C (with around 20–25% VA). It is noted that the reduction in T_g in copolymer grades of PVC is not as large as in the polyacrylate examples, owing to the lower incidence of longer side groups introduced via the incorporation of VA units. For thermoset polymers, plasticization can be brought about through chemical modifications resulting in a reduction in the cross-linking (network) density. This is usually achieved by the incorporation of a component with a lower functionality, which is often known as a reactive plasticizer.
- **External plasticization** A reduction in the T_g of an existing polymer can also be obtained by external plasticization, consisting of the addition of a miscible monomeric or oligomeric compound (known as a plasticizer) as a means of reducing the strength of the molecular attractions between polymer chains.

In the case of crystalline polymers, plasticization is confined primarily to the amorphous phase, but some of the plasticizer can penetrate into the lamellar constituents of the crystals. This increases the amount of defects, thereby reducing the temperature for the onset of melting of the crystals. When the degree of crystallinity is low, and the quantity of plasticizer added is very large, the crystals may accommodate a considerable amount of plasticizer, so that the overall melting process will take place at much lower temperatures. This is a technique used to produce the so-called 'thermo-reversible gels'.

45. Plasticizer

An additive, usually liquid, consisting of high-molecular-weight monomeric compounds or low-molecular-weight polymers, which is mixed with a polymer (usually a glassy polymer) to produce a miscible mixture exhibiting a lower glass transition temperature than the original polymer. The presence of plasticizer molecules dispersed between polymer chains reduces the overall intermolecular forces by a dipole screening mechanism, or increases the 'free volume' through the creation of obstacles to molecular packing. Dipole screening not only decreases the T_g but also produces a broadening of the modulus–temperature curve, and provides a higher level of flexibility at low temperatures.

Plasticizers have been classified in a number of ways, using different criteria. A widely used classification is according to the plasticization efficiency, defined in terms of the reduction of glass transition temperature achieved for a given weight fraction of plasticizer, which is determined primarily by increase in free volume. Another classification is made on the basis of the level of miscibility with the host polymer. Accordingly, a 'primary plasticizer' is one that is miscible with the polymer over the entire concentration range, while a 'secondary plasticizer' is only miscible at low concentrations. Plasticizers are also distinguished as 'non-migratory' types if they exhibit a high resistance either to migration into adjacent materials or to extraction by liquid environments. These are usually polymeric in nature and exhibit slow diffusion owing to the large size of the molecules. Polymers that are available as plasticized grades, in order of plasticizer consumption, are PVC (>80%), nitrile rubber, neoprene rubber, cellulosics and polyamides. The following are the main types of plasticizers available, in order of consumption.

- **Dialkyl phthalates**

  where R is either octyl (DOP), isooctyl (DIOP) or isododecyl (DIDP)
- **Dialkyl aliphatic esters**
  $RO-CO-(CH_2)_n O-CO-R$
  where R is octyl, and $n = 2$ for adipate (DOA) or $n = 8$ sebacate (DOS)
- **Triaryl phosphate**

  where R_1, R_2 and R_3 are typically phenyl (TPP), cresyl (TCP) or a mixture of the two groups
- **Poly(propylene adipate)** (See Polyester)
- **Sulfonamides**

Their main characteristics and uses are shown in the table.

46. Plastisol

A term for PVC paste.

47. Plastograph

A laboratory apparatus with the geometrical features of an internal mixer (Banbury type), also known as a torque rheometer, used to study the fusion and gelation characteristics of polymers and elastomers. The apparatus records the temperature of the mixture and the torque exerted by the rotors to maintain a constant speed in a mixing run. (See Mixer and Torque rheometer.)

48. Plate-Out

An industrial term (jargon) used to describe the formation of 'hard' deposits of additives or impurities present in a polymer formulation on the metal surface of the processing equipment. These can mar the surface of manufactured products, such as calendered sheets. Plate-out depositions are particularly likely to take place if any of the additives are miscible with the external lubricant.

Plasticizer nature	Plasticizer type	Plasticizer names	Polymers plasticized
Phthalates	General purpose, Primary types	dioctyl (DOP), diisododecyl (DIDP), dibutyl (DBP)	PVC, nitrile rubber, neoprene rubber, cellulosics, poly(vinyl acetate)
Dialkyl esters	Low temperature, Secondary types	dioctyl adipate (DOA), dioctyl sebacate (DOS),	PVC
Phosphates	Flame retardant, Primary types	triphenyl (TPP), tricresyl (TCP)	PVC, cellulosics, nitrile rubber, neoprene rubber
Polyesters	Non-migratory, Secondary types	poly(propylene adipate), (MW = 1500–3000)	PVC
Sulfonamides	Secondary types	N-ethyl SA, o- and p-toluene SA, formaldehyde SA resin	polyamides

49. Plating

A term used to describe the deposition of a metallic layer on the surface of an article by an electrolytic deposition method.

50. Plug-Assisted Vacuum Forming

A technique used in thermoforming of rigid polymer sheets. (See Thermoforming.)

51. Plug Flow

A type of flow with a flat velocity gradient profile, originating from slip phenomena at the walls of the flow channel and/or resulting from very pronounced pseudoplastic behaviour of the polymer melt, that is, a low power-law index. (See Non-Newtonian behaviour.)

52. PMR

Abbreviation for a technology (developed at NASA-Langley) known as 'polymerization of monomer reactants', used to describe systems consisting of carbon fibres impregnated in a stoichiometric mixture of an aromatic anhydride or a methyl ester (usually perfluoroisopropylidene diphthalic anhydride) with an aromatic diamine as a means of producing the required polyimide through *in-situ* polymerization during the manufacture of composites.

Perfluoroisopropylidene diphthalic anhydride (6FDA).

Other systems used for adhesives consist of mixtures of a polymeric aromatic diamine with norbor-5-ene-2,5-dicarboxylic methyl monoester, cured in two stages. The first stage produces a reactive end-capped prepolymer, such as

which cures further through free-radical polymerization of the end groups.

53. Poiseuille Equation

An equation first derived by Poiseuille for the volumetric flow rate (Q) of a Newtonian liquid through a circular channel, that is

$$Q = \pi R^4 \Delta P / 8\eta L,$$

where R is the radius, ΔP is the pressure drop along the length L of the channel and η is the viscosity of the fluid.

54. Poisson Ratio

The ratio of the lateral strain to the longitudinal strain resulting from uniaxial tension or compression deformations. The definition is illustrated in the diagram.

Force

Longitudinal strain
$\varepsilon_1 = (L_f - L_0)/L_0$

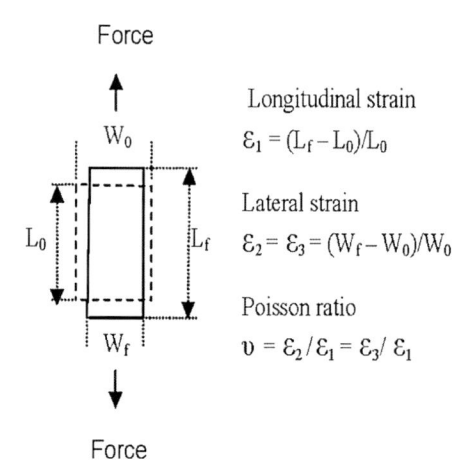

Lateral strain
$\varepsilon_2 = \varepsilon_3 = (W_f - W_0)/W_0$

Poisson ratio
$\upsilon = \varepsilon_2/\varepsilon_1 = \varepsilon_3/\varepsilon_1$

Force

Illustration of the concept of Poisson ratio.

Note that the Poisson ratio varies from 0.5 for materials that undergo no volume change, such as rubber, to about 0.3–0.35 for a glassy polymer, due to volumetric changes, that is, expansion in tension and contraction in compression. (See Modulus.)

55. Polarity

Denotes the presence and strength of dipoles in a solvent, auxiliary component or polymer.

56. Polarization

A phenomenon and a parameter describing the response of a material to excitations by an electromagnetic field. The total polarization is the sum of various terms, respectively, electronic, atomic, orientation and interfacial. The contribution of each polarization term to the permittivity of the material is related to the frequency of the applied stimulus. (See Permittivity.) At low frequencies, as in the case of an AC voltage, the predominant polarizations are interfacial polarization, resulting from the response of slow-moving charges at interfaces with filler particles or reinforcing fibres, and dipolar polarization, resulting from the movement of permanent dipoles in the structure. The type of polarization that takes place at different frequencies is illustrated.

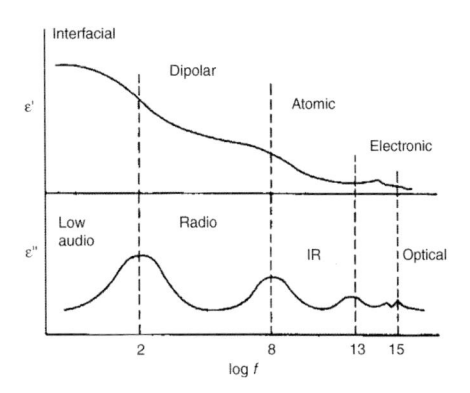

Variation of permittivity with frequency of applied stimulus. Source: Unidentified original source.

57. Polyacetylene

A polymer with the structure $-(CH=CH)_n-CH=CH-$ produced by the Ziegler catalysis of acetylene. The conjugated double bonds form the basis for developing an intrinsically high electronic conductivity by a variety of doping methods. (See Conductive polymer.)

58. Poly(Amic Acid)

A polymer precursor for the production of polyimides, normally available in N-methylpyrrolidone (NMP) or dimethylformamide (DMF) solution to prevent imidization reactions occurring during storage. (See Polyimide.) NMP and DMF molecules form physical bridges through hydrogen bonding. Internal imidization reactions occur until the temperature is increased to about 150 °C. At this temperature the solvent–amic acid bridges break down and conversion to polyimide begins to take place via internal condensation reactions (imidization), as shown.

$$\left\{ \begin{array}{c} -NH-\overset{O}{\underset{}{C}} \\ HOC \\ \overset{}{\underset{O}{}} \end{array} Ar \begin{array}{c} \overset{O}{\underset{}{C}}-NH-Ar'-NH- \\ COH \\ \overset{}{\underset{O}{}} \end{array} \right\}_n \longrightarrow \left\{ \begin{array}{c} \overset{O}{\underset{}{C}} \\ N \\ \overset{}{\underset{C}{}} \\ \overset{}{\underset{O}{}} \end{array} Ar \begin{array}{c} \overset{O}{\underset{}{C}} \\ \\ \overset{}{\underset{C}{}} \\ \overset{}{\underset{O}{}} \end{array} Ar' - N \right\}_n$$

Polyamic acid solution in NMP — Imidization by heat — Polyimide (insoluble in any solvent)

A widely used poly(amic acid) for the production of films, wire coatings and interlayer insulation in microelectronics is produced from the reaction of pyromellitic dianhydride (PMDA) and oxydianiline (ODA), which can be represented by

$$\left[\text{structure of PMDA-ODA polyamic acid} \right]_n$$

59. Polyamide

Polymers produced by the condensation reactions of a dicarboxylic acid and a diamine or via internal condensation reactions of an α,ω–amino acid, both through the intermediate formation of a salt in an aqueous medium. The monomers can be either aliphatic or aromatic in nature. Aliphatic polyamides can also be obtained by ring opening polymerization of a cyclic amide in the presence of water, or by either anionic or cationic polymerization. They are generally referred to as 'nylons'. (See Nylon.) Also available are aromatic polyamides in which the high chain stiffness, coupled with the strong intermolecular forces resulting from the amide groups, has been exploited for the production of polymers with a high melting point, in the region of 500 °C. These can only be processed by solution processing techniques in concentrated sulfuric acid for the production of high-strength fibres or papers. (See Aramid and Nomex.)

60. Polybenzimidazole (PBI)

A heat-resistant polymer used for high-temperature applications. The typical structure of PBI is as shown:

$$\left[\text{PBI structure} \right]_n$$

61. Polybutadiene

(See Diene elastomer.)

62. Poly(But-1-ene) (PB)

A polyolefin represented by the formula shown, where the CH_2–CH_3 side group appears on the same side along the polymer chain providing an isotactic structure.

$$-\left[CH_2CH \right]_n$$
$$\quad\quad\quad | $$
$$\quad\quad\quad CH_2$$
$$\quad\quad\quad | $$
$$\quad\quad\quad CH_3$$

It can crystallize in two forms, respectively, type II, which is a soft rubber-like polymer that is obtained on cooling from the melt state, and type I, resulting from a transformation of type II after about one week standing at room temperature.

The transformation takes place at maximum rate at about 30 °C. Type I has a higher melting point, in the region of 125–130 °C, and also a higher density in the region of 0.91–0.92 g/cm^3. It has a high resistance to creep, manifested by a small rate of increase in strain with time under stress, which makes it attractive for tubings and pipes. Some copolymers with a lower melting point are also available.

63. Poly(Butylene Terephthalate) (PBT)

(See Polyester.)

64. Polycarbonate (PC)

A glassy engineering thermoplastic polymer with a T_g in the region of 150 °C, represented by the formula shown.

$$\left[-O-\underset{}{\bigcirc}-\underset{CH_3}{\overset{CH_3}{\underset{|}{\overset{|}{C}}}}-\underset{}{\bigcirc}-O-\underset{\overset{\|}{O}}{C} - \right]_n$$

It is widely used for applications requiring a combination of desirable characteristics, such as rigidity, creep resistance, toughness and transparency, as well as high resistance to UV light. Special branched grades are used for the production of DVDs owing to their ability to exhibit very low warping in injection moulding. This arises from the absence of molecular orientation, which results in a uniform shrinkage during cooling in the mould. The lack of orientation also results in a lack of internal stress and the absence of birefringence patterns under polarized light, which makes it an excellent candidate for optical devices.

Polycarbonate is miscible with a variety of polymers, notable among which is poly(butylene terephthalate). Mixing the two together with the addition of semi-miscible acrylate elastomer, which forms particulate rubbery inclusions, results in a well-balanced combination of useful properties, such as retention of modulus and strength at high temperature, toughness at low temperatures, oil/fluid resistance and good thermal oxidative and UV stability. Widely used in the automotive industry are blends with acrylonitrile–butadiene–styrene (ABS) terpolymer alloys, utilizing the miscibility of the styrene–acrylonitrile (SAN) polymer component with polycarbonate and allowing the immiscible butadiene–acrylonitrile elastomer component to form rubbery inclusions to increase the impact strength. The micrograph shows the presence of spherical rubbery inclusions derived from ABS and also a zone surrounding these particles consisting of a semi-miscible mixture of polycarbonate and the SAN component, chemically bonded to the rubber particle.

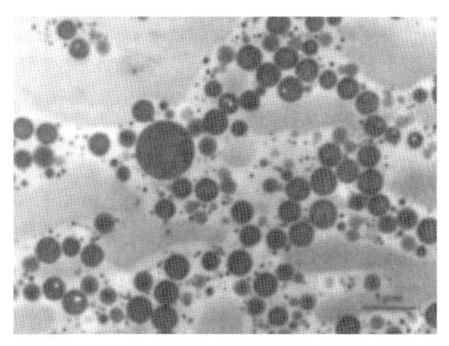

Electron micrograph of a blend of polycarbonate and ABS. Source: Herpels (1989).

65. Poly(Carborane–Siloxane)

Linear polymers containing carbon, silicon and boron in the backbone chains, with useful elastomeric properties, which can be used continuously at temperatures up to 300 °C. They can be formulated, processed and vulcanized in the same way as conventional silicone elastomers. The chemical structure of the base polymer is shown.

66. Polychloroprene

(See Diene elastomer.)

67. Polychlorotrifluoroethylene (PCTFE)

A polymer with a melting point around 215 °C, a T_g in the region of 45 °C and a density of $2.1\,\mathrm{g/cm^3}$. PCTFE has an exceptional thermal oxidation resistance and extremely good barrier properties towards water vapour. The chemical structure can be represented by the formula $-[CClFCF_2]_n-$ but it is also available as an alternating copolymer with ethylene. Although it was intended to be a processable polymer equivalent to PTFE, it falls short of these expectations in terms of coefficient of friction and chemical resistance. The polarity arising from the substantial difference in electron density between the Cl and F atoms attached to the same C atom also brings about considerable deterioration in electrical properties relative to PTFE.

68. Polyelectrolyte

Polymers containing ionic side groups in the molecular chains. These can form salts and can perform like any electrolyte capable of producing ionic conduction through dissociation and movement of the counterions, if sufficiently small in size, hopping from one ionic group to another. Examples of cationic and anionic polyelectrolytes are shown.

Cationic polyelectrolyte: poly-(4-vinylpyridine) quaternized with butyl bromide.

Anionic polyelectrolyte: sodium polyacrylate.

Both are examples of linear polyelectrolytes that will dissolve in water. There are also examples of polyelectrolytes in the form of cross-linked networks, which will swell in water to produce 'gels' and produce a mechanism for the exchange of ions with inorganic electrolyte solutions, as illustrated.

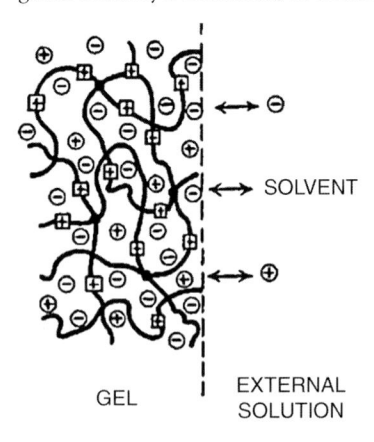

Swollen ionic 'gel' in equilibrium with an electrolyte solution. Source: Flory (1953).

The free anions can associate with the cations fixed on the network and also move freely from one cationic site to another through dynamic associations. In this way the swollen ionic gel acts as its own membrane, preventing the fixed cations from diffusing into the surrounding solution.

This produces a mechanism for the exchange of anions from another solution. (See Ion exchange resin and Membrane.)

Polymer electrolytes without solvents, known as solid polymer electrolytes, are used in lithium batteries. These are based on poly(ethylene oxide) containing dissolved lithium salts, such as lithium perchlorate, $LiClO_4$, and lithium bis(trifluoromethylsulfonyl)imide, $Li[N(SO_2CF_3)_2]$ (commonly known as TFSI). The ionic conductivity arise from the ability of Li^+ ions to form coordination bonds with the closely spaced ether groups along the polymer chains, thereby producing paths for the Li^+ to hop along while maintaining charge neutrality with the neighbouring anions. Special copolymer grades of poly(ethylene oxide) (PEO) are usually used in order to destroy the crystallinity of standard PEO as a means of enhancing the mobility of Li^+, thereby increasing the conductivity.

69. Polyester

Polymers containing repeating ester groups in the backbone chains. These are produced both as thermosetting resins (for alkyds and unsaturated polyesters) and as linear thermoplastic polymers (aliphatic, aromatic or mixed aliphatic–aromatic). The linear aliphatic polymers are used primarily as hot-melt adhesives owing to their quite low melting point (e.g. poly(ethylene succinate), $T_m = 108\,°C$) or as polymeric plasticizers (e.g. poly(ethylene glutarate), $T_m = 10\,°C$, or poly(ethylene adipate), $T_m = 50\,°C$). The difference in melting point arises from the different number of CH_2 groups of the diacid used, which is two for the succinate, three for the glutarate and four for the adipate. These follow the trend that smaller and even numbers of CH_2 give higher melting points than larger and odd numbers, owing to the lower packing density of the latter molecules in the respective crystal lattices.

The mixed aliphatic–aromatic polymers, in particular poly(butylene terephthalate) (PBT; $T_m = 222\,°C$; $T_g = 40–45\,°C$) and poly(ethylene terephthalate) (PET; $T_m = 265\,°C$, $T_g = 72–78\,°C$), are used primarily for fibres, plastics and films. PET is commercially used in much larger quantities than PBT primarily on account of its lower cost. PBT crystallizes at a much faster rate than PET and is much tougher in the unoriented crystalline state. For this reason, PBT is used widely for injection moulding, particularly as a glass-fibre-reinforced grade. PET is used mostly for the production of fibres, biaxially drawn films and injection blow moulding of bottles and containers. For the latter application, advantage is taken of the ability of PET to be moulded into amorphous pre-forms that can be stretched at around $100–110\,°C$ to undergo stress-induced crystallization, which brings about enormous increases in strength and toughness, as well as a high level of optical transparency. Poly(ethylene naphthanate) (PEN) has been introduced in recent years as it has a T_g in the region of $125\,°C$, owing to the high chain stiffness brought about by the bulky naphthanate rings. The main use of PEN is for the production of biaxially drawn films for motor insulation. Thermoplastic aliphatic–aromatic polyesters have been used also for the production of a variety of engineering polymer alloys for use in the automotive industry and electrical and electronic components. The main advantages of using polyesters for the production of alloys for automotive applications, for instance, are their high melting point, which makes possible their use in high-temperature paint stoving operations, as well as their resistance to oils.

High-T_g amorphous polyesters are produced, with the used cycloaliphatic glycols, such as dimethylolcyclohexane, utilizing the rigidity provided by the aromatic and cyclohexane rings in the backbone of the polymer chains. Even higher T_g values are exhibited by polyesters made via polycondensation of diphenols with a mixture of terephthalic acid and phthalic anhydride, often referred to as polyarylates. In these systems, the lack of symmetry of the constituent units of the polymer chains prevents the development of crystallinity, as indicated by the example shown.

$$\left[O\!-\!\!\left\langle\!\!\bigcirc\!\!\right\rangle\!\!-\!\!\underset{\underset{CH_3}{|}}{\overset{\overset{CH_3}{|}}{C}}\!\!-\!\!\left\langle\!\!\bigcirc\!\!\right\rangle\!\!-\!O\!-\!\underset{\overset{\|}{O}}{C}\!\!-\!\!\left\langle\!\!\bigcirc\!\!\right\rangle\!\!-\!\underset{\overset{\|}{O}}{C}\right]_n$$

Linear polyesters are also synthesized as multi-block copolymers for the production of thermoplastic elastomers, where the hard (crystalline) domains of poly(butylene terephthalate) have a melting point (T_m) within the range of about 170–210 °C. The soft domains consist of low-molecular-weight segments of poly(tetramethylene oxide) with a glass transition (T_g) in the region of −80 °C. (See Thermoplastic elastomer.)

70. Poly(Ether Imide) (PEI)

(See Polyimide.)

71. Polyethylene

(See Ethylene polymer.)

71.1 Ethylene Copolymer

(See Ethylene polymer.)

71.2 Ethylene–Methacrylic Acid Ionomeric Copolymer

(See Ionomer.)

72. Poly(Ethylene Terephthalate) (PET)

(See Polyester.)

73. Polyhydroxyalkanoate (PHA)

A family of biopolymers synthesized by many bacteria. The most common is poly(3–hydroxybutyrate) (PHB). In order to overcome the excessive brittleness inherent to PHB, due to the high degree of crystallinity, copolymers are usually produced to introduce long-chain side groups, such as poly(hydroxybutyrate-*co*-hydroxyvalerate) or poly(hydroxybutyrate-*co*-hydroxyhexanoate). (See Biopolymer.)

74. Polyimide (PI)

A class of polymers well known for their good properties at high temperature and for their very high resistance to thermal oxidation and to absorption of solvents. However, they are susceptible to hydrolytic degradation under acidic conditions. There are two main types of polyimides widely used commercially, as follows.

- **Curable systems** These are produced either from poly(amic acid) solutions or from reactive monomers, and are used for coatings, as matrices for composites or in the production of films. (See Poly(amic acid).) The chemical structures of the two polyimides most widely used for coatings and microelectronics interlayers, or in the form of films, are shown. The first, derived from poly(amic acid) solution, has a T_g in the region of 280 °C, whereas the other, obtained from diphenyl dianhydride, has a T_g value that can be greater than 300 °C. Both PIs have a service temperature rating greater than 250 °C.

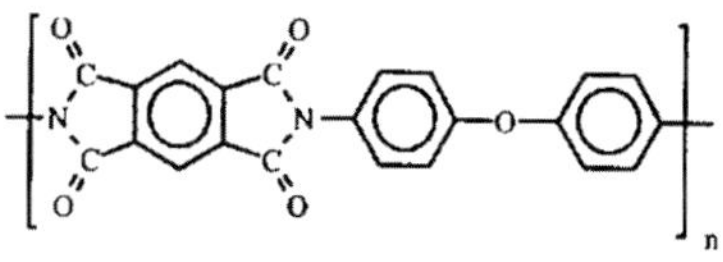

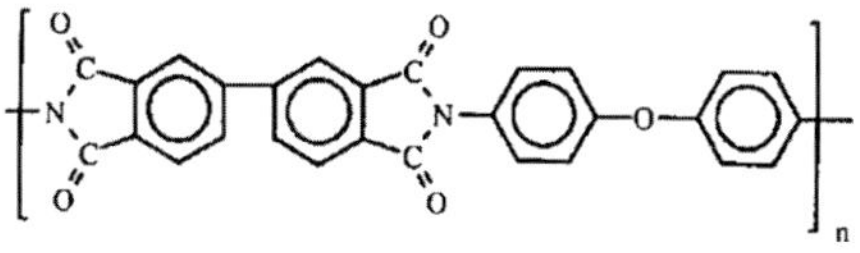

- **Thermoplastics systems** These are used for the manufacture of moulded or extruded products. The main polymer in this class that is available commercially is poly(ether imide) (PEI), which is a glassy polymer with a T_g in the region of 230 °C and a service temperature rating around 180 °C. Its chemical structure is shown.

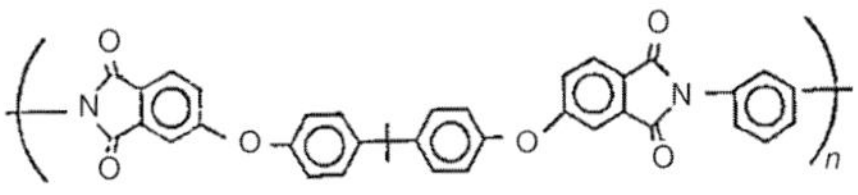

Structure of the thermoplastic, poly(ether imide).

75. Polyisobutylene

(See Butyl rubber.)

76. Polyisocyanurate

(See Urethane polymer and resin.)

77. Polyisoprene

(See Diene elastomer.)

78. Polyketone

These are aromatic ether ketones with a high melting point and an exceptionally high thermal oxidative stability, which allows them to be processed by conventional thermoplastics processing techniques, albeit at higher temperatures. Adding these characteristics to the high ductility even in the crystalline state makes poly(aryl ether ketone)s unique among all the processable aromatic polymers. The crystallinity prevents solvent absorption and embrittlement through crazing phenomena. The two main systems available commercially are poly(aryl ether ketone)s (PEK) and poly(aryl ether ether ketone)s (PEEK) represented by the formulae shown.

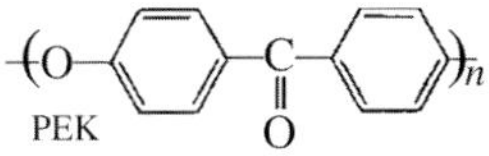

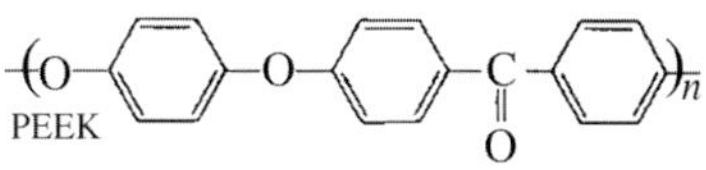

The properties are remarkably similar, with PEK exhibiting a slightly higher T_g and higher melting point. The values are respectively in the region of 145 °C and 335 °C, with densities of 1.265 g/cm^3 for the amorphous polymers, obtained by melt quenching, and about 1.32 g/cm^3, depending on the degree of crystallinity.

A major use of poly(aryl ether ketone)s is for matrices for advanced carbon-fibre composites.

79. Poly(Lactic Acid) (PLA)

A biopolymer obtained by condensation reaction of lactic acid. (See Polylactide and Biopolymer.)

80. Polylactide

Corresponds to poly(lactic acid) produced from ring opening polymerization of lactide. Poly(lactic acid) and polylactide are chemically the same (known as PLA) and can also readily depolymerize thorough hydrolysis down to the original monomer. The polymerization and depolymerization reactions are shown.

Because two stereoisomeric forms of lactic acid exist, there are three forms of PLA: poly(L–lactic acid) (PLLA), corresponding to the syndiotactic structure; poly(D–lactic acid) (PDLA), equivalent to the isotactic structure; and poly(meso-lactic acid), also known as poly(D,L–lactic acid) (PDLLA), as it contains equimolecular amounts of D- and L–lactic acid units. While PLLA and PDLA are crystalline, with similar characteristics, the PDLLA form is amorphous. Polymerization conditions can be adjusted to produce a variety of PLA, containing different ratios of L to D monomeric units, which gives polymers with different levels of crystallinity and properties. Both the melting point (T_m) and glass transition temperature (T_g) of the polymer vary according to the degree of crystallinity. Reported values for T_m are between 120 and 200 °C, which vary not only according to composition but also depending on the thermal history of the sample.

81. Polymer

A term first introduced in the early part of the nineteenth century by the Swedish scientist Berzelius for siloxane compounds containing several repeating units. The term became more widely used following the work of the German scientist Staudinger, when he illustrated that natural rubber, as well as synthetic products derived from the polymerization of monomers, contained many thousand repeating units. He also devised a method for measuring the molecular weight of polymers by the well-known solution viscosity technique. Initially, the term 'high polymers' was used to emphasize the very high-molecular-weight feature of these organic compounds.

82. Polymer Alloy

A term used to describe 'well-compatibilized polymer blends'.

83. Polymer Blend

A mixture of two or more polymers. The properties of polymers can be improved by blending with auxiliary components, such as plasticizers, or with other polymers to obtain the so-called 'polymer blends' or 'polymer alloys'.

- **Miscible blends** are mixtures of two or more polymers that are miscible at molecular level and will exhibit properties that are intermediate between those of the polymer components. These include properties such as glass transition temperature, light transmission and gas/vapour permeability. These types of blends are widely found in the fields of adhesives and coatings.

- **Heterophase polymer blends** are mixtures of two or three polymers that are either totally immiscible or semi-miscible, so that they form two distinct phases, as shown.

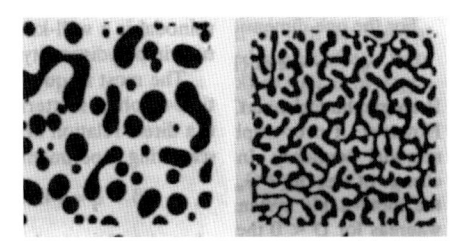

Particulate dispersed blend (left) and co-continuous phase morphology blend (right).

The morphology of polymer blends can be either a particulate type (left), where the minor component becomes the dispersed phase, or in the form of co-continuous phase systems (right). In particulate polymer blends, normally the continuous phase is a glassy polymer and the dispersed phase is a rubbery polymer. Either component can be a linear or cross-linked polymer. These blends are normally produced to increase the fracture toughness of glassy polymers. Typical systems are high-impact polystyrene (HIPS) and acrylonitrile–butadiene–styrene (ABS) terpolymer blend.

84. Polymerization

The process or chemical reactions that lead to the formation of polymers from the constituent monomer components.

Polymerization processes can be classified in a number of ways, but more usually into condensation and addition polymerization. (See Condensation polymerization and Addition polymerization.) In the first case, single monomer units or a mixture of two or more monomer units add to one another with the loss of water, or other species, derived from the combination of end groups in the monomer units involved in the reaction, for example,

$$n\ HO-M-H \rightarrow -(M)_n- + n\ H_2O$$

or

$$n\ HO-M-H + n\ HO-CM-H \rightarrow -(M-CM)_n- + n\ H_2O$$

In addition polymerization, monomer units produce polymer chains by the direct addition of monomer units to one another. Addition polymerization can be further divided into free-radical polymerization and ionic polymerization. (See Free-radical polymerization, Anionic polymerization and Cationic polymerization.)

Polymerization can also be divided into homopolymerization, when only one type of monomer unit is used for the formation of polymer chains, and copolymerization, when two or more different monomer units are used to produce polymer chains. (See Polymer and Copolymer.) Free-radical polymerization can be carried out in different ways: (i) from suspension or emulsion of monomer in water, (ii) in solution using either a solvent that dissolves the polymer or one that allows the polymer to separate out as a dispersion in solvent, and (iii) mass or bulk polymerization, where the polymer is either soluble in the monomer or phase-separates at a certain level of conversion.

85. Poly(4-Methylpent-1-ene) (PMP)

An isotactic polyolefin represented by the formula:

$$-[CH_2-CH]_n-$$
$$|$$
$$CH_2$$
$$|$$
$$CH(CH_3)_2$$

It was the first crystalline polymer that was found to be completely transparent, owing to the similar density of the crystalline and amorphous regions (in the region of $0.83\,g/cm^3$). The bulky side groups, $CH_2CH(CH_3)_2$, along the chains provide considerable rotational hindrance, which results in a polymer with a very high melting point at around 245 °C.

86. Polyol

A general term to describe hydroxyl-terminated oligomers, usually polyether types, known also as polyglycols. The more common polyglycols are poly(ethylene glycol) (PEG) and poly(propylene glycol) (PPG). The structure of these is shown.

$$HOCH_2CH_2-[-OCH_2CH_2-]_n-OH \quad HOCHCH_2-[-OCH_2CH-]_n-OH$$
$$||$$
$$CH_3 CH_3$$
$$PEG PPG$$

There are also polyester types, such as poly(ethylene adipate) and other mixed aromatic–aliphatic polyesters. Multifunctional polyols (trifunctional types in particular) are available to produce network structures, such as those in polyurethanes.

87. Polyolefin

A polymer produced from olefin monomers. The more common types are polyethylene and polypropylene, but there are several other types of considerable importance.

88. Polypeptide

A polymer composed of protein-type amino acid units linked by peptide (amide) bonds. (See Biopolymer.)

89. Poly(Phenylene Oxide) (PPO)

This polymer is obtained by a unique polymerization method based on the catalytic oxidation of 2,6-dimethylphenol, according to the scheme

PPO is an amorphous polymer with T_g in the region of 190–210 °C and a high melt viscosity. The polymer is susceptible to rapid decomposition at temperatures above T_g owing to the formation of conjugated methide groups, followed by a rapid oxidation of the methyl groups.

Since PPO is miscible with polystyrene, blends are produced as a means of reducing the processing temperature and the melt viscosity. Although this considerably reduces the T_g, the blends can still be used up to temperatures of about 120–130 °C. More usually commercial blends are produced using mixtures with high-impact polystyrene (HIPS) to benefit from the enhanced toughness resulting from the rubbery polybutadiene inclusions. More thermally stable blends are produced with the use of thermoplastic elastomers based on styrene block copolymers. Other blends with low-melting-point polyamides are also produced to improve the solvent and oil resistance for use in the automotive and electrical and electronics industry. In these systems, compatibilization is achieved through the formation of graft copolymers via free-radical decomposition reactions. (See Polymer blend.)

90. Poly(Phenylene Sulfide) (PPS)

A crystalline aromatic polymer represented by the structure

The quoted melting point is 288 °C and the T_g is 93 °C. The polymer is extremely brittle and is, therefore, used mainly as a fibre-reinforced grade for load-bearing applications in compression, or as the matrix for advanced composites, owing to the retention of properties up to high temperatures and the high resistance to solvents and chemicals, arising from the high level of crystallinity.

91. Polypropylene (PP)

A polyolefin represented by the formula shown, where the CH_3 group appears on the same side along the polymer chain, providing an isotactic configuration.

$$\left[CH_2CH\text{—}CH_3 \right]_n$$

It is produced by Ziegler–Natta and metallocene catalysis and is available as random and block copolymer grades with ethylene. Through many ethylene–propylene combinations within the blocks, and molecular-weight diversity, it is possible to tailor the structure to achieve a wide variation of optical, mechanical and rheological properties. The melting point of PP is in the range of 170 °C for homopolymers, down to 135 °C for copolymers, while the T_g remains at around −15 °C.

PP is a polymer with a reasonably high rigidity, having a Young's modulus in the range 1–2 GPa, depending on density, which varies from 0.90–0.91 g/cm³ according to the degree of crystallinity. The lack of polarity has two contrasting effects, providing excellent electrical properties but poor resistance to mineral oils. The presence of tertiary C−H groups along the polymer chain makes it vulnerable to attack by free radicals, leading to poor thermal and UV stability. On the other hand, this feature has provided the opportunity to deliberately induce chain scission in the chains to produce the so-called 'controlled rheology' grades with low molecular weight and low polydispersity, which results in a very low melt viscosity. This is a desirable characteristic for the production of fibres and flat films, as well as coatings. Unlike all polyethylene grades, PP does not suffer from environmental stress cracking problems. (See Polyolefin.)

92. Polypyrrole

(See Conductive polymer.)

93. Polysaccharide

(See Biopolymer.)

94. Polystyrene

(See Styrene polymer.)

95. Polysulfone

These are glassy aromatic polymers with a very high T_g. The two main types found commercially are as follows:

- **Poly(aryl sulfone)** designated as PSU, with a T_g in the region of 190 °C, and represented by the formula

$$\left[\bigcirc\text{—}\underset{CH_3}{\overset{CH_3}{C}}\text{—}\bigcirc\text{—}O\text{—}\bigcirc\text{—}SO_2\text{—}\bigcirc\text{—}O \right]_n$$

Structure of a poly(aryl sulfone).

- **Poly(ether sulfone)** with a T_g around 230 °C, and represented by the formula

$$\left[\bigcirc\text{—}O\text{—}\bigcirc\text{—}SO_2 \right]_n$$

Structure of a poly(ether sulfone).

Even higher T_g values have been achieved by introducing very rigid biphenyl groups into the polymer chains. The absence of hydrocarbon segments in the polymer confers a very high thermal oxidative stability. Even for the case of poly(aryl sulfone)s, the CH_3 groups are quite resistant to thermal oxidation. They have a very poor resistance to UV light owing to very strong absorption by the SO_2 groups. Another major difficulty with these polymers is their susceptibility to brittle failure occurring through crazing, which is exacerbated in the presence of solvents.

96. Polytetrafluoroethylene (PTFE)

A unique polymer produced by conventional suspension and emulsion polymerization with free-radical initiators, resulting in polymers with exceptionally high molecular weights due to the lack of chain termination reactions. At the same time, the absence of chain transfer reactions during polymerization allows the formation of linear chains free of any branches. These conditions result in a highly crystalline polymer with a very high melting point (327 °C), a very high stability to thermal and UV degradation, and an excellent resistance to solvents. The complete lack of net dipoles along the chains, due to internal cancellation of local C–F dipoles acting in opposite directions, confers on the polymer excellent dielectric characteristics. Coupled with lack of reaction possibilities at high temperatures to produce graphitic (char-like) residues, the polymer also exhibits an extremely high resistance to tracking failures at high voltages, particularly in the presence of environmental pollution, which is the main weakness of most polymer insulators. The presence of four 'heavy' fluorine atoms in each repeating unit, together with the fairly high degree of crystallinity, results in a polymer with a high density, in the range 2.14–2.20 g/cm^3.

One of the most outstanding features of PTFE is its very low coefficient of friction. With values as low as 0.04, it has often been said to be equivalent to 'wet ice on ice'. PTFE is a soft and ductile polymer, which maintains a high ductility even at very low temperatures. In general, it suffers from a low yield strength and a poor resistance to creep, which has prevented it from being used more widely in structural applications. A major factor that has limited the uses of PTFE is its extremely high melt viscosity, which requires high pressures and the generation of high shear stresses to promote flow. This can result in large rises in temperature, which can lead to rapid depolymerization reactions at temperatures above 400 °C, with the formation of highly toxic volatiles. The diagram, representing the mechanical spectrum of the deformational behaviour of PTFE, shows that the modulus values at high temperatures are quite low and are associated with two secondary transitions.

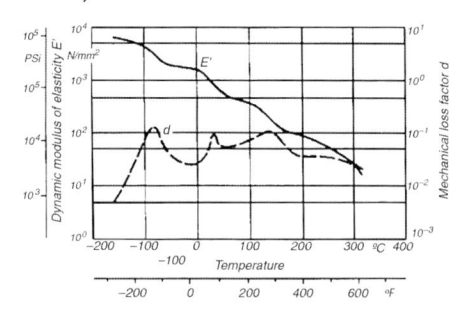

Dynamic modulus and mechanical loss factor of PTFE. Source: Domininghaus (1992).

These secondary transitions bring about phase transformations, which is a rare occurrence in polymers. The transition at 19 °C is associated with the transformation of the triclinic crystals into a less-ordered hexagonal packing. This is manifested as sudden peaks in the coefficient of thermal expansion. The transition at very low temperatures is a characteristic transition related to molecular relaxations within the amorphous phase (glass transition), while the transition at 127 °C is again associated with the amorphous phase, but is an actual physical transition of physical states, from amorphous solid to supercooled liquid state. PTFE is available in many different grades, many of which contain inorganic fillers or fibre reinforcement to enhance the mechanical properties. Fillers include (i) molybdenum sulfide and graphite as solid lubricants to increase the wear resistance, (ii) carbon fibres and glass fibres for reinforcement, and (iii) bronze particles to improve several characteristics, such as creep and wear resistance, as well as increasing the thermal conductivity to reduce rises in temperature arising from frictional contacts.

97. Poly(Tetrafluoroethylene–Ethylene) Copolymer (PETFE)

A copolymer containing 25% ethylene, characterized by a very high melting point (about 280 °C), an exceptional thermal oxidative stability (service temperatures up to about 200 °C) and high resistance to solvents. These characteristics are close to those exhibited by PTFE, with T_m around 330 °C and a service temperature up to 250 °C. Unlike PTFE, the PETFE copolymers can be processed by conventional polymer processing methods and have much better mechanical properties than PTFE, except for the coefficient of friction.

98. Poly(Tetrafluoroethylene–Hexafluoropropylene) Copolymer (FEP)

A copolymer containing small amounts of hexafluoropropylene in the chains to disrupt the chain regularity and create less perfect crystals as a means of reducing the melting point to temperatures below that of PTFE. A comparison is shown of the structures of hexafluoropropylene monomer units (left) and tetrafluoroethylene units (right).

The compact nature of the pendent CF_3 group does not create large changes in the packing of the molecular chains, so that the polymer retains its ability to form crystals and the density does not decrease very much, that is, from about 2.2 for PTFE to 2.15 g/cm^3 for FEP. At the same time the melting point decreases sufficiently to enable the polymer to be processed by the conventional methods used for thermoplastics, albeit at much higher temperatures. As a result of the decrease in degree of crystallinity, however, the polymer suffers from deterioration in mechanical properties, particularly with respect to the coefficient of friction, which is much higher than that of PTFE.

99. Polythiophene

A researched intrinsically conductive polymer, represented by the formula

100. Poly(Vinyl Butyral)

(See Vinyl polymer.)

101. Poly(Vinyl Carbazole)

(See Vinyl polymer.)

102. Poly(Vinylidene Chloride) (PVDC)

A crystalline polymer with a melting point in the range 198–205 °C. The chemical structure can be represented with the general formula

$$-[CH_2CCl_2]_n-$$

PVDC has a low permeability to gases and liquid additives used for aromatization and flavouring of food, which makes it very suitable as a coating or lining material for film packaging and containers. Due to the susceptibility to decomposition through dehydrochlorination, followed by rapid thermal oxidation, the melt processable grades are in the form of random copolymers with ethyl acrylate or acrylonitrile monomers, as a means of reducing the degree of crystallinity and melting point. Copolymers with vinyl chloride or vinyl acetate are used the form of water emulsions as surface coatings for paper, polymer films and containers.

103. Poly(Vinylidene Fluoride) (PVDF or PVF$_2$)

A crystalline high-molecular-weight polymer exerting polymorphism due to the presence of a substantial amount of head-to-head sequences of monomeric units in the polymer chains, which gives rise to different crystal structures with melting points in the range of 154–184 °C, T_g around −40 °C and a density of 1.78 g/cm^3. The mechanical properties, for example, modulus, yield strength and ductility, are similar to those of HDPE. However, PVDF has a much greater resistance to solvents and a much higher stability towards thermal oxidation and UV-induced degradation, owing to the presence of CF$_2$ groups and the total absence of double bonds and tertiary CH groups in the molecular chains. The general molecular structure can be represented by the formula

$$-[CH_2CF_2]_n-$$

The large difference in electron density between two adjacent CH$_2$ and CF$_2$ groups along the polymer chain creates a very large dipole, which is responsible for the very large dielectric constant, the highest of all commercial thermoplastics. For this reason, PVDF can only be used as a secondary insulation or as a protective component in wire and cables. The large dipoles, however, confer on PVDF a unique characteristic that has been exploited for the production of piezoelectric devices. There are available also a large number of copolymers with varying amounts of hexafluoropropylene units, to obtain more flexible products, due to the reduction in crystallinity, while maintaining the high resistance to thermal and UV degradation.

104. Post-Curing

A treatment of cross-linked polymer products carried out at high temperatures to optimize the desired properties.

105. Pot Life

A practical term used to describe the available time for the processing of a thermosetting resin system. A certain amount of reactions in cross-linkable systems can be tolerated before the viscosity increases to levels that render the system unprocessable. Pot life is related to the more precisely defined concept of gel time. (See Gel time.)

106. Powder Coating

A process for producing coatings from 'dry' powders. A widely used powder coating process is 'electrostatic spraying', which takes place in successive stages, comprising: (i) electrification of the powder by passing it through the nozzle of a 'gun' subject to a high-voltage electric field; (ii) dispersion of the charged particles through electrostatic repulsions; and (iii) deposition of a thin layer of powder on the substrate via electrostatic attractions, which are connecting the conductive object to earth. These three stages of the process for producing the coatings from dry powder are illustrated.

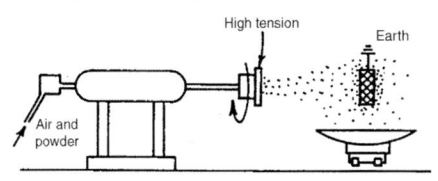

Schematic diagram of electrostatic spray coating set-up. Source: Mascia (1989).

A uniform coating is achieved as a result of the powder coating layer, formed on the substrate, which begins to repel the incoming charged particles when a certain thickness is reached. This is due to the insulating effect produced, which prevents the incoming particles from being discharged to earth. In the final stage the coated objected is transferred to an oven for sintering. (See Powder sintering.)

Another important powder coating process is 'fluidized bed coating', which involves the immersion of a heated object into a fluid bed of suspended powder, which melts and sinters the powder coating. A second heating stage may be required to complete the thermofusion operation. A variant to the conventional fluidized bed coating described above is a technique known as 'electrostatic fluidized bed coating', which charges the particles as a means of obtaining better control of the thickness of the coating. At the same time this makes it possible to use an earthed metal article to attract the charged particles, so that the deposited powder coating can then be sintered as a separate step.

107. Powder Sintering

A technique mostly used for the processing of PTFE, which consists in producing a preform by compaction in cold conditions and then raising the temperature above the melting point of the polymer for a sufficient length of time to allow molecular diffusion across the grains. Upon cooling, crystallization takes place over the entire mass, which erases the previous interfacial boundaries. This can be done also as a continuous process by ram extrusion in line with a sintering oven set-up. (See Polytetrafluoroethylene.)

108. Power Law

A response that, when raised to the power n, becomes a linear function of the excitation. A typical example is the relationship between shear stress, τ, and shear rate, $\dot{\gamma}$, for the flow of polymer melts, that is, $\tau = k\dot{\gamma}^n$. This implies that a log–log plot of the shear stress against shear rate produces a straight line with gradient equal to n. (See Power-law index and Non-Newtonian liquid.)

109. Power-Law Index

The exponent n of the power-law relationship between shear stress, τ, and apparent shear rate, $\dot{\gamma}$, for the flow of a polymer melt, that is, $\tau = k\dot{\gamma}^n$. The value of n decreases from 1 (for Newtonian behaviour) to around 0.3–0.4 for most polymer melts. The value of n is obtained from a log–log plot of shear stress against apparent shear rate derived from rheological measurements, as shown in the diagram.

Typical log–log plot of shear stress against apparent shear rate at three different temperatures, $T_1 < T_2 < T_3$.

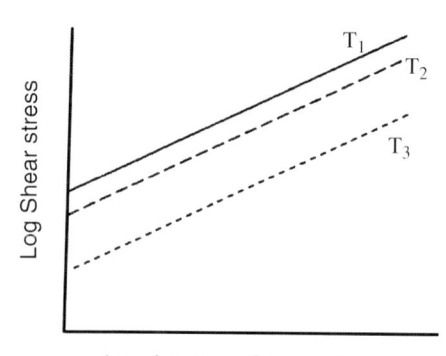

The diagram shows that the power-law index is not affected by changes in temperature within the range of temperatures used for the common processing operations, such as injection moulding and extrusion. (See Capillary rheometer, Rabinowitsch equation and Viscosity.)

110. Pre-form

A small thick-walled precursor for the production of bottles by injection blow moulding.

Typical pre-form used for the production of PET bottles.

111. Prepreg

A term (jargon) for resin-impregnated reinforcing fibres or fabrics used for the manufacture of structures by compression moulding. During the production of pre-pregs, usually the resin is partially reacted with the hardener in order to increase viscosity by increasing the molecular weight of the oligomeric components of the resin.

112. Pressure-Sensitive Adhesive (PSA)

(See Adhesive.)

113. Primary Plasticizer

(See Plasticizer.)

114. Primary Stabilizer

(See Stabilizer.)

115. Primary Transition

(See Transition and Glass transition temperature.)

116. Primer

A term used to describe a coating deposited on the substrate before the main coating layer is applied. A primer coating can have many functions depending on the system considered. It can simply act as an interfacial layer to enhance the adhesion of the primary coating with the substrate or it can have a functional role, such as corrosion protection through the slow release of inhibitors. For the case of porous substrates, the primer can used to seal the surface in order to prevent the primary coating from penetrating deeply.

117. Processing

A term used to describe any operation on the polymer involved in the manufacture of products.

118. Processing Aid

An additive or auxiliary component of a formulation used to enhance the processing characteristics of a polymer or elastomer.

119. Processing Stabilizer

(See Stabilizer.)

120. Propagation Reaction

The reaction step taking place immediately after the initiation step. The propagation step is followed by a termination step, often via chain transfer reactions. The processes involving propagation reactions are polymerization and thermal degradation reactions, such as thermal and UV-assisted oxidation reactions.

121. Proton Exchange Membrane

(See Nafion.)

122. Pseudoplastic

A term describing the behaviour of polymer melts by which the viscosity decreases with increasing shear rate during flow. This arises from a nonlinear relationship between shear stress and shear rate, which is usually modelled by a power-law expression or the Carreau equation. (See Non-Newtonian behaviour.)

123. PTC Polymer

The abbreviation PTC stands for 'positive temperature coefficient', which refers to the rapid increase in resistivity (i.e. positive gradient) experienced by certain polymer products above a specific temperature. The phenomenon is reversible, so that the polymer regains its high conductivity when is cooled down to the original temperature. The sudden change in the current density through the material at some specific temperature, therefore, produces a 'switch' effect in the circuit when a PTC polymer is used to separate the conducting elements. Since the switch effect is reversible, a PTC polymer can be used as a temperature regulator in heaters.

The PTC characteristics are obtained by the incorporation of a conductive filler, usually carbon black, in a crystalline polymer in amounts just enough to reach the percolation threshold for the formation of conductive channels for the flow of the current. The threshold conditions for the percolation of the conductive particles are reached at fairly low concentrations, as these are confined entirely within the amorphous regions. The switching to a non-conductive polymer takes place when the melting point of the polymer is reached. The large thermal expansion caused by the melting of the crystals breaks up the contact between the carbon black particles, thereby interrupting the flow of current, as shown schematically.

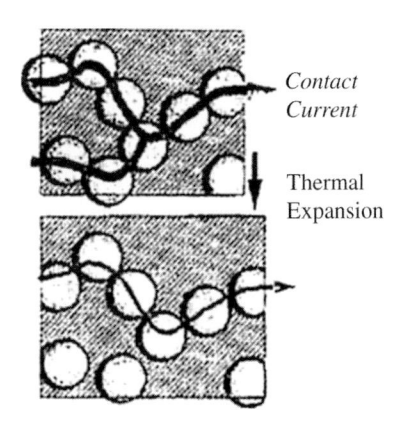

Illustration of the breaking up of the contact current mechanism due to thermal expansion during the melting of polymer crystals. Source: Unidentified original source.

Further expansion of the matrix when the melting of the crystals is complete causes the total disruption of the conductive paths. In order to achieve reversibility of the switching characteristics, the polymer has to be cross-linked as a means of achieving stable morphology through the prevention of the migration and possible agglomeration of the conductive particles. At the same time, to ensure that the threshold conditions are reached at the lowest possible concentration of conductive particles, it is preferable to cross-link the polymer at low temperatures, using high-energy radiation techniques, such as electron beaming.

124. Pultrusion

(See Composite manufacture.)

125. Purging

A term used to describe the removal of residual polymer from the barrel of an extruder or injection moulding machine after completing a particular processing operation or changing to a new polymer. Purging is usually carried out by passing through the new polymer until the barrel is completely free of the previously used polymer. This often involves the use of an intermediate polymer that is more miscible with the first polymer and has a higher melt viscosity, in order to allow the removal of residues in 'stagnation spots' of the flow channels.

126. PVC

Abbreviation for poly(vinyl chloride). (See Vinyl polymer.) Available in two forms: rigid grades, known as PVC-U, do not contain plasticizers but may be in the form of blends; and plasticized grades, known as PVC-P. The latter are available as compounds, dry blends (jargon specific to PVC powders) or pastes, also known as plastisols. Flexible PVC grades are rubbery and, therefore, contain a large amount of plasticizer to bring the T_g down to values below $-20\,°C$. The amount of plasticizer used is usually higher than 20% and can be as high as 80% for very flexible formulations. The latter is made possible by the presence of crystal domains that are swollen by the plasticizer, without bringing them into solution, thereby creating a strong thermo-reversible gel. (See Gel.) The use of small amounts of plasticizer to reduce primarily the melt viscosity is not generally recommended, as it may give rise to the 'antiplasticization' phenomenon, which brings about considerable embrittlement in the final product. The plasticizers used are normally: diphthalate esters of long-chain alcohols (e.g. octyl, nonyl or dodecyl phthalates); tricresyl phosphate; and secondary plasticizers, usually used in combination with the above primary plasticizers, namely, dioctyl adipate and dioctyl sebacate, as well as low-molecular-weight aliphatic polyesters derived from the condensation reactions of both adipic acid and sebacic acid with ethylene glycol. (See Plasticizer.)

In all formulations, whether rigid (PVC-U) or flexible (PVC-P) types, in addition to property modifiers, such as fillers, fire retardants and antistatic agents, there are a number of other essential additives that have to be incorporated into the mixes as a means of assisting processing operation and to improve the service performance of products. Predominant among these additives are the thermal stabilizers, which have been specially developed to enable PVC to be processed at the highest possible temperature. Other important additives for PVC-U formulations are lubricants and processing aids.

PVC grades are also widely used in the form of melt blends as a means of improving the processing characteristics of the polymer or the mechanical properties of the products. Examples of blends of PVC with small amounts of other polymers as a method for improving the processing characteristics follow:

a) the addition of small amounts high-molecular-weight poly(methyl methacrylate) (PMMA), which functions as a processing aid by speeding up the fusion of PVC particles in extrusion operations, because PMMA is fully miscible with PVC and, therefore, the optical transmission characteristics are not impaired;

b) the addition of ethyl acrylate–methyl methacrylate copolymers, chlorinated polyethylene (CPE) or ethylene–vinyl acetate (EVA) to increase the melt strength in extrusion and to enhance the drawdown characteristics of foils and sheets for thermoforming.

Another important reason for producing PVC blends is to improve the fracture toughness and impact strength. Typical is the addition of 5–15% of specially formulated acrylonitrile–butadiene–styrene (ABS) or methyl methacrylate–butadiene–styrene (MBS) terpolymer alloys, exploiting the miscibility of the styrene–acrylonitrile copolymer component of ABS and the PMMA component of MBS. This assists the dispersion of butadiene–styrene rubber particles and provides a strong bond with the glassy PVC matrix, through the miscible PVC–PMMA interphases. Well known also are blends of PVC and nitrile rubber (NBR) to improve the resistance to oils and fluids, as well as increasing the flexibility of products, such as hoses, belting and wires and cables.

126.1 Dry blend

Dry blends are PVC porous powders that have been made to adsorb large quantities of a plasticizer through high-speed mixing at high temperature. In this way the plasticizer fills up the micropores within particles about 100–150 µm diameter. These are produced by specially devised polymerization techniques. One of these is known as 'microsuspension', and another is two-stage bulk polymerization. The latter takes advantage of the fact that PVC is not soluble in its monomer and therefore will precipitate out at about 5–10% conversion. The suspension obtained in the first high-speed reactor is transferred to a second reactor where more monomer and initiator are added to continue the polymerization reactions. The process is stopped at about 80% conversion to outgas the unreacted monomer.

126.2 PVC paste

These consist of emulsion PVC particles (20–150 µm) dispersed in plasticizer. When the temperature is increased, the plasticizer first migrates into the spaces between the primary particles, 0.1–0.2 µm diameter, to gel the paste (i.e. a soft solid state). The structure of a typical emulsion particle used for pastes is shown. After gelation, the plasticizer diffuses into the molecular structure to produce the plasticized polymer.

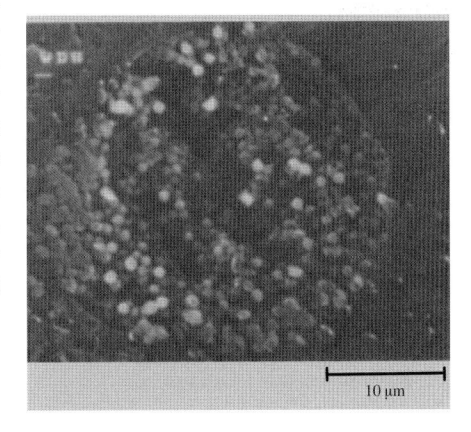

Micrograph of a PVC emulsion particle. Source: Wilson (1995).

126.3 Heat stabilization of PVC

This requires special consideration owing to the unique degradation mechanism of vinyl polymers, as shown.

$$-CH_2-CH-CH_2-CH-CH_2- \xrightarrow{energy} -CH_2-CH-CH_2-CH-CH_2-$$
$$|| |$$
$$ClCl Cl\quad +Cl*$$

$$\downarrow$$

$$-CH_2-CH-CH_2-CH=CH- \xleftarrow{\text{(catalysed by HCl)}} -CH_2-CH-CH_2-CH=CH-$$
$$+Cl* Cl\quad +HCl$$

(weak allylic bond)

$$\downarrow$$

$$-CH_2-CH=CH-CH=CH- \quad +HCl \quad \text{etc.}$$

} conjugated unsaturation responsible for the early development of the polymer discolouration

with the use of additives consisting of basic lead salts and weak basic soaps, such as basic lead carbonate ($PbO \cdot PbCO_3$), tribasic lead sulfate ($3PbO \cdot PbSO_4 \cdot H_2O$) and dibasic lead phosphite ($2PbO \cdot PbHPO_3 \cdot \frac{1}{2}H_2O$).

From the reaction scheme, it can be deduced that stabilizers for PVC have to fulfil the following functions:

a) absorb and neutralize the HCl evolved as a means of arresting the autocatalytic chain reactions, as well as preventing the corrosion of processing equipment;

b) prevent oxidative reactions at the 'weak' allyl groups formed in the first step of the evolution of HCl, as well as disrupt the conjugated double bonds, which give rise to discoloration; and

c) substitute active Cl atoms in the allyl position after the formation of HCl.

Originally the stabilization of PVC addressed primarily the first requirement, Stabilization through displacement of labile Cl atoms from the chains has been achieved with the use of miscible metal carboxylates, such as mixed cadmium barium laureates

$$Cd[OCOC_{11}H_{22}]_2 Ba[OCOOCOC_{11}H_{22}]_2$$

The disruptions of the conjugated double bonds formed in the chains from the unzipping elimination of HCl can be accomplished with the use of mercaptides and maleate esters, through the mechanism shown.

$$\sim\!\!\sim\!CH=CH\!\sim\!\!\sim + RSH \dashrightarrow \sim\!\!\sim\!CH_2-CH\!\sim\!\!\sim$$
$$|$$
$$SR$$

$$\sim\!\!\sim\!CH=CH-CH=CH\!\sim\!\!\sim + \underset{CH}{\overset{CH}{\underset{\diagdown}{\diagup}}}\!\!\!\!\begin{array}{l}COOR\text{ (or M)}\\ \\ COOR\text{ (or M)}\end{array} \longrightarrow \underset{\underset{\text{(or M)}}{COOR\ COOR}}{\bigcirc}$$

The mercaptides can also assist the stabilization of PVC by acting as free-radical quenchers in the simultaneous oxidation reactions. The more recent stabilizers are organo-tin and organo-phosphite compounds that act primarily through the displacement of the labile Cl atom, according to the scheme shown.

~~~CH-CH$_2$~~~ + Bu\Sn/OCOR / Bu/ \OCOR ---> -CH-CH$_2$- with Bu, Cl\Sn/OCOR, Bu/ \OCOR

~~~CH-CH$_2$~~~ + P(OR)$_3$ ---> ~~~CH-CH$_2$~~~
+ •Cl   O=P(OR)$_2$ + RCl

The more widely available systems are dibutyl-tin maleate and dibutyl-tin mercaptides, respectively, represented with the chemical formulae shown.

$$\left[(C_4H_9)_2\,SnOCOCH{=}CH\atop COO \right]_n$$

$(C_4H_9)_2\,Sn(SC_{12}H_{25})_2$

The majority of PVC formulations make use of mixtures of stabilizers, as they act synergistically, which is to say that the combined effect of two or more stabilizers is much greater than is achievable with the use of single stabilizers at equal levels of addition. It has also been found that the addition of epoxy compounds, such as epoxidized soya bean oil, can enhance further the synergistic action of the combination of conventional stabilizers. (See Vinyl polymer.)

127. PVC Paste

(See PVC.)

128. PVT Diagram

The term refers to plots of the specific volume of polymer melts against temperature at different levels of pressure. The data are used to produce a pressure profile for an injection moulding cycle as a means of controlling the volumetric shrinkage of an injection-moulded part, as shown in the diagram for a particular grade of polystyrene.

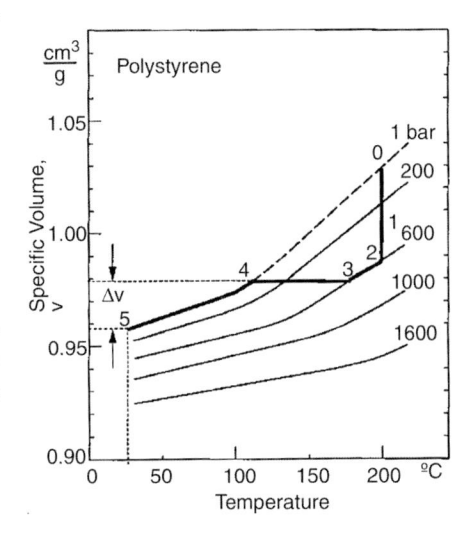

Plots of the specific volume of a polystyrene melt against temperature at different levels of pressure, and programmed pressure profile during cooling. Source: Osswald (1998).

The number sequence from 1 to 4 indicates the programming of the hold-on pressure on the polymer melt. The trace from 4 to 5 represents the natural path of the volumetric shrinkage of the polymer in the cavity due to further cooling, which causes the moulded part to shrink away from the walls of the cavities, thereby allowing the pressure to fall off to reach atmospheric conditions. The shrinkage taking place at atmospheric pressure corresponds to the actual 'mould shrinkage' experienced by the moulded part. From a close analysis of the PVT curves, it can be inferred that, if the injection pressure were to be increased to 1000 bar, the extent of mould shrinkage would decrease to zero, so that the dimensions of the moulded part become exactly the same as those of the cavity of the mould.

In the diagram are reported the PVT data for a polyamide 6,6 (PA 6,6), which is a crystalline polymer and undergoes a much higher level of natural shrinkage than polystyrene (amorphous polymer). The data suggest that the application of a high pressure would not be possible to compensate for the high shrinkage resulting from the crystallization of the polymer. This means that while for glassy amorphous polymer the linear mould shrinkage is usually around 0.5–0.8%, for crystalline polymers the shrinkage values are in the region of 1–2%.

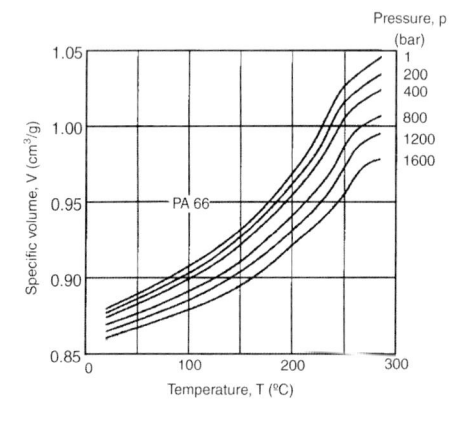

Plots of the specific volume of a polyamide 6,6 melt against temperature at different levels of pressure. Source: Osswald (1998.)

129. Pyrolysis

The breaking down of polymer chains into low-molecular-weight species when exposed to high temperatures. The mechanism and type of products formed depend on the nature of the polymer and the environmental conditions, particularly on whether pyrolysis is carried out in an inert environment, such as nitrogen, or an oxidizing environment, such as oxygen or air. The thermal decomposition of some polymers occurs via a depolymerization mechanism producing the original monomer. For some polymers, such a polytetrafluoroethylene (PTFE), poly(methylene oxide) (PMO) and poly(methyl methacrylate) (PMMA), pyrolysis can produce a 100% yield of monomer. In other cases, such as polystyrene (PS) and poly(ethylene oxide) (PEO), the monomer yield is much less and is largely dependent on the environmental conditions. The pyrolysis of vinyl polymers, such as PVC and poly(vinyl acetate), in an inert atmosphere, takes place by the formation of the conjugated unsaturation along the chains, as follows:

$$-(CH_2-CHX)_n-CH_2CHX- \longrightarrow -(CH=CH)_n-CH=CH- + (n+1)HX$$

Other polymers, such as cellulosics, polyacrylonitrile (PAN) and phenolics produce carbonaceous residues and a variety of volatile products.

Q

1. Q Meter

An apparatus used to measure the permittivity and loss factor of a dielectric at frequencies up to about 70 MHz.

2. Quartz

The crystalline version of silica, SiO_2. It finds very little use in polymer compounds.

3. Quencher

An additive used to interact with activated species to bring them down to a lower energy state, thereby making them less reactive. Quenchers are widely used for stabilization of polymers against UV-induced degradation. The absorption of UV light brings a molecule (A) to an excited state (A*), known also as the triplet state. The collision of an excited molecule with a quencher (Q) brings it back to the ground energy state, known also as the singlet state, through the release of the absorbed energy as IR radiation, that is,

$$A + h\nu(UV) \rightarrow A^*$$

$$A^* + Q \rightarrow A + Q + h\nu(IR)$$

Quenchers used in combination with primary UV stabilizers are usually nickel complexes. There are, however, toxicity implications to be considered.

4. Quinone Structure

This is formed in phenol derivatives, such as those used to produce phenolic resins and antioxidants, as a result of reactions with oxygen. They are responsible for the development of discoloration with a yellowish or even brown tint. (See Phenolic antioxidant.)

R

1. Rabinowitsch Equation

An equation used to calculate the true shear rate at the wall ($\dot{\gamma}_w$) of a capillary rheometer in studies of the flow behaviour of polymer melts, based on a power-law behaviour,

$$\dot{\gamma}_w = \frac{3n+1}{4n}\left(\frac{4Q}{\pi R^3}\right),$$

where n is the power-law index, Q is the volumetric flow rate and R is the radius of the capillary. (See Non-Newtonian behaviour and Power law.)

2. Radiation

Can be defined as the energy emitted from an electronic or magnetic field and transmitted in the form of waves according to Planck's law, that is, $E = h\nu$, where E is the energy and ν is the frequency of the radiation waves, and h is known as Planck's constant (with the value 6.63×10^{-34} J s). The frequency is equal to the velocity (c) divided by the wavelength of the radiation (λ), that is, $\nu = c/\lambda$. The energy levels emitted by different types of radiation and the associated wavelength of the radiation are shown in the table.

The energy for laser radiation has not been given, as it depends on the wavelength, which varies according to the source (1 mW–10 W). The actual energy level of laser radiation is not much different from that of visible light, but it is highly concentrated on very small areas (10 µm²–1 mm²). The power associated with laser radiation is therefore very high, 10^7–10^8 W/m², while that of sunlight reaching Earth is about 10^3 W/m². From these data it is easy to see why laser light can be used to melt and etch the surface of materials.

Type	Wavelength (nm)	Energy (kJ/mol)
Far ultraviolet	100	1196
Vacuum ultraviolet	200	598
Mercury lamp	254	471
Solar cut-off	295	406
Mid-range ultraviolet	350	341
End of ultraviolet range	390	306
Blue-green light	500	239
Red light	700	171
Near infrared	1000	120
Infrared	5000	24
Hard X-rays, soft gamma-rays	0.05	2.4×10^6
Hard gamma-rays	0.005	2.4×10^7

Source of data: Wypych (2008).

3. Radiation Processing

A term used to induce reactions in oligomers (resins) or linear polymers to produce cross-linked molecular structures or to functionalize the surface of existing polymers through grafting reactions. Usually the mechanism is a free-radical type but cationic initiators have also been used for the cross-linking of specific epoxide oligomers. Radiation sources used vary from UV (100–1000 nm wavelength) to electron beaming (10^{-4}–10^{-1} nm). Ionizing radiation emitted from electron beams is very energetic (0.5–2.5 MeV, with a power of up to 20 kW) and can penetrate depths to 3–4 mm, whereas UV light can only be used effectively to induce reactions to a maximum of about 0.5 mm from the surface. The absorption of radiation brings the molecules to their excited state, which can in turn cause bond cleavage to produce free radicals or ions. (See Photopolymerization.)

The radiation dose is measured in either kilograms (kGy) or megarads (Mrad): 1 Gy = 1 J/kg = 1 W s/kg = 100 rad, hence 1 kGy = 10^{-1} Mrad. The irradiation of polymers from high-energy sources can result in cross-linking as well as chain scission or both reactions simultaneously, depending on the structure of the polymer and whether additives are present. These are classified as 'pro-rads' if they enhance the yield of cross-linked structures (multifunctional monomers), or 'anti-rads' if they absorb radiation and, therefore, act as screens for the polymer chains. The latter additives are usually aromatic amines or sulfides, and phenols or quinines. The most widely used pro-rads are trimethylolpropane trimethacrylate (TMPTMA), triallyl cyanurate (TAC) and triallyl isocyanurate (TAIC). The polymers that cross-link more readily are those containing unsaturation along the chains or as side groups. The ability of linear polyethylenes to be cross-linked readily by radiation methods is due to the presence of structural defects, consisting primarily of vinyl and vinylidene groups. Vinyl polymers can also cross-link readily due to the weakness of the tertiary C—H bonds, which undergo homolytic scission to produce free radicals. Conversely, the complete absence of unsaturation or tertiary C—H groups in the polymer chains results mainly in chain scission reactions. Typical among these are poly(methyl methacrylate) and polytetrafluoroethylene. Polymers such as polystyrene, polypropylene and polyamides will undergo extensive chain scission alongside some degree of cross-linking.

4. Radical

A term used to denote chemical species containing an unpaired electron, called 'free radical', which results from the breaking of covalent bonds. A typical example is the homolytic scission of a peroxide initiator for the free-radical polymerization of unsaturated monomers, that is,

$$HO-OH + heat/UV \rightarrow 2{}^{\bullet}OH$$

5. Radical Scavenger

An additive that reacts with free radicals present on polymer chains and prevents the onset or continuation of propagation reactions, which could lead to the degradation of properties. The most common type of scavengers are also known as primary stabilizers and are based on hindered phenols or hindered amines.

6. Radius of Gyration

A theoretical concept in polymer science denoting the root-mean-square distance of a particular atom or group in a polymer chain from the centre of gravity. This parameter is used to measure the effective size of polymer molecules.

7. Raman Spectroscopy

A vibrational spectroscopy technique based on the use of intense monochromatic radiation in the visible region to promote the transition in energy levels through vibrations of primary bonds. A small fraction of the incident radiation is scattered at a different wavelength. Such a change in wavelength is known as the 'Raman shift', which forms the basis for the characterization of the molecular structure of polymers. (See Infrared spectroscopy.)

8. Random Copolymer

A term used to describe a copolymer in which the two monomeric components are distributed at random along the polymer chain. This type of copolymer is also known as a statistical copolymer, identified as –(A-*stat*-B)$_n$– for a random copolymer containing A and B units. Most copolymers used for commercial products are random types. (See Block copolymer.)

9. Rapid Prototyping

A series of different processing methods used to make prototypes without having to machine or join together small units. It is also used for producing items in the manufacture of special articles that cannot be economically produced by conventional moulding methods. (See Stereolithography.)

10. Rayon

A fibre based on regenerated cellulose. (See Cellophane.)

11. Reaction Moulding

(See Reaction processing.)

12. Reaction Processing

A generic term for manufacturing or compounding processes involving chemical reactions between the components. All moulding processes involving the formation of cross-links, therefore, fall in this category.

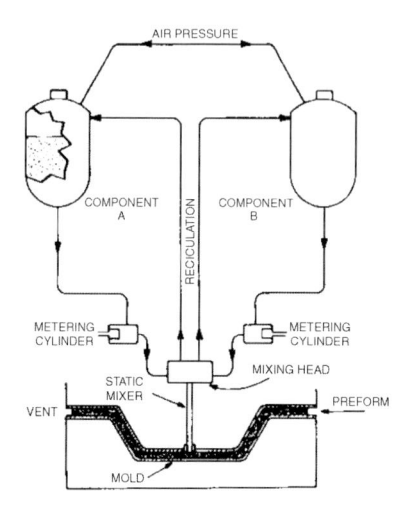

Example of structural reaction injection moulding: a thermosetting liquid mixture is injected into a cavity containing a glass-fibre pre-form. Source: Muccio (1994).

Compounding techniques are sometimes used to induce some chemical modifications of the polymer chains, particularly for the purpose of introducing functional groups. A typical example is the grafting of acid groups along the molecular chains of a polyolefin through grafting reactions with maleic anhydride or acrylic acid.

13. Reactive Diluent

A miscible monofunctional component of a resin mixture, used to reduce the viscosity and to decrease the glass transition temperature (T_g) of the cured products. (See Epoxy resin.)

14. Reactivity Ratio

A concept used in free-radical copolymerization to control the composition and random distribution of the two monomer components. The definition is based on the rate constants k for the addition of a monomer molecule to a growing chain containing the reactive radical. These are denoted as monomer reactivity ratios r_1 and r_2, defined as $r_1 = k_{11}/k_{12}$ and $r_2 = k_{22}/k_{21}$, where the various rate constants are for the various possible reactions, that is, k_{11} for the addition of growing chain $M_1^\bullet$ onto monomer M_1, k_{12} for the addition of growing chain $M_1^\bullet$ onto monomer M_2, k_{22} for the addition of growing chain $M_2^\bullet$ onto monomer M_2 and k_{21} for the addition of growing chain $M_2^\bullet$ onto monomer M_1.

15. Reciprocating Screw

The screw of an injection moulding machine, which slides back while it rotates to 'plasticate' the polymer granules and deliver a predetermined amount of melt to the front. Then the screw is pushed forwards at a high speed to inject the melt through the nozzle into the cavities of the mould, via the sprue and runners. (See Injection moulding.)

16. Recoverable Strain

The magnitude of the total strain that can be recovered after removing the stress in a creep experiment. For linear polymers, the strain can recover completely if the applied stress has not reached, or closely approached, the conditions causing failure by yielding.

17. Recovery

A term used to denote the period that follows the removal of the stress in creep experiments. During this period, the strain decreases and can recover totally after a long period, usually about 10–20 times longer than the creep period. (See Creep.)

18. Recycling

A term that denotes the reuse of polymers previously manufactured. Recycling can be divided into three categories:

- **primary recycling** if the recycled polymer is used to produce the same article;
- **secondary recycling** if the polymer is re-used for the manufacture of articles different from those used previously; and
- **tertiary recycling** when the polymer is used for other purposes, such as incineration to produce heat, or being pyrolysed to produce chemicals.

19. Reduced Time

A concept related to creep experiments to normalize the time after removing the load, known as the recovery time (t_1), to the time under load, known as the creep time (t). The reduced time t_R is, therefore, defined as

$$t_R = (t-t_1)/t_1.$$

20. Reflection Factor

The ratio of the flux of reflected light (Φ_r) from the surface of a sheet or film to the flux of incident light (Φ_i), that is, $R = \Phi_r/\Phi_i$, where $\Phi = I/A$, I is the intensity of light and A is the surface area from which the light is reflected.

21. Refractive Index

A parameter that denotes the characteristics of a material related to the change in the velocity of the transmitted light. The diagram shows that the incident light is partly reflected from the upper surface and partly from the surface at the exit side.

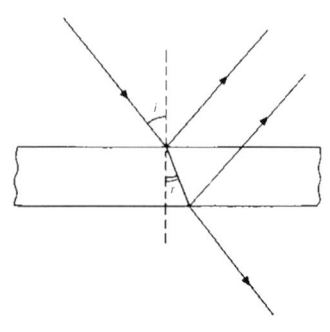

Reflection and refraction of light in passing through a transparent medium.

The transmitted beam of light is 'refracted' through a reduction in the path angle determined by the nature of the material, Thus, the refractive index $n(\lambda)$ is defined as

$$n(\lambda) = \frac{\sin i}{\sin r},$$

where i is the angle of incidence and r is the angle of refraction. The refractive index $n(\lambda)$ decreases with increasing wavelength of light. For PMMA, for instance, the value of $n(\lambda)$ is 1.50 at 450 nm and about 1.49 at 650 nm wavelength. The refractive index decreases also with increasing temperature, with the greatest change taking place upon reaching the glass transition temperature.

22. Regulator

Additives that are more reactive than the growing polymer chains during the course of polymerization so that small amounts can be used to stop the growth of polymer chains in order to reduce and control the molecular weight of the polymer produced. Typical chain regulators are carbon tetrachloride and pentaphenylethane.

23. Reinforcement

A method for increasing the modulus and strength of a material, brought about by the incorporation of high-modulus and/or high-strength fibres, such as glass fibres and carbon fibres. (See Composite and Fibre reinforcement theory.)

24. Reinforcement Factor

The ratio of the value of the modulus or strength of a composite to the corresponding value of the matrix. The reinforcing factor gives a measure of the efficiency of the reinforcing agent used in producing a composite.

25. Relative Permittivity

A dimensionless parameter that relates the permittivity of a dielectric (ε_d) to that of air (ε_{air}), which is approximately equal to the permittivity of vacuum. The relative permittivity of a dielectric is also known as the dielectric constant, K, and is defined as $K = \varepsilon_d / \varepsilon_{air}$.

The value of the relative permittivity of most polymers varies from about 2 to 5. The lowest values are obtained with non-polar polymers, such as polyethylene types, polytetrafluoroethylene and polypropylene. Even lower values are achievable with cellular products, owing to the presence of air in the cells. Higher values are obtained through the absorption of water. The polymer with the highest intrinsic permittivity is poly(vinylidene difluoride), with a K value of approximately 8. (See Permittivity.)

26. Relative Viscosity

The ratio of the viscosity of a solution to the viscosity of the solvent. A concept used in

27. Relaxation Modulus

Modulus value obtained from experiments carried out at constant strain, resulting in the decay of stress with time, known as stress relaxation. (See Viscoelastic behaviour and Standard linear solid.)

28. Relaxation Time

A concept used in viscoelasticity theory. (See Maxwell model and Standard linear solid.)

29. Relaxed Modulus

The value of the relaxation modulus for time tending to infinity. (See Standard linear solid.)

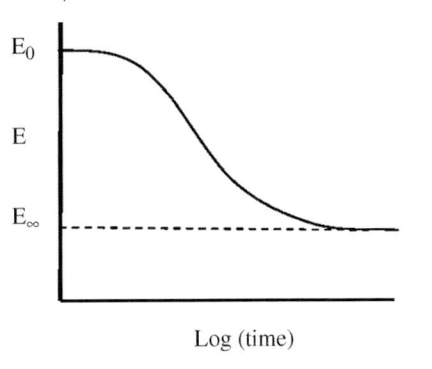

Variation of relaxation modulus with time: E_0 is the instantaneous modulus ($t=0$); E_∞ is the relaxed modulus ($t=\infty$).

30. Reptation

A term that describes the mechanism of the movements of polymer molecules relative to each other in order to produce flow when the temperature is increased above the rubbery state and, therefore, reaches the melt state. The constraints due to their coiled configuration and entanglements induce the molecules to dissipate their internal energy through serpentine motions confined within a curvilinear cylinder path, which causes them to slip past each other.

Schematic illustration of the serpentine motions (reptation) of polymer molecules in the melt state. Source: Elias (1993).

31. Residual Stress

(See Internal stress.)

32. Resin Transfer Moulding (RTM)

A manufacturing method for composites, consisting of a reinforcing fibre pre-form placed in the cavity of a mould. A resin is injected into the interstices between the fibres and the removal of air is assisted by the application of vacuum. (See Composite.)

33. Resistivity

(See Surface resistivity and Volume resistivity.)

34. Resole

(See Phenolic.)

35. Retardation Time

A concept used in viscoelasticity theory. (See Kelvin–Voigt model and Standard linear solid.)

36. Retarder

An additive used in rubber compounds to slow down the rate of cure, hence increasing the 'scorch time'. Typical retarders are diphenyl nitrosoamine, salicylic acid, benzoic acid and phthalic anhydride.

37. Rheology

A branch of science that is concerned with the non-Newtonian behaviour of liquids. (See Non-Newtonian behaviour.)

38. Rheometer

An apparatus used to study the rheological properties of polymer melts.

39. Rheopectic

A term that denotes the rheological behaviour of certain fluids whereby the viscosity decreases with the duration of the imposed shear rate.

40. Rockwell Hardness

Hardness value for rigid polymers obtained from measurements involving the penetration of a steel ball into sample supported on a massive hard base. (See Hardness.)

41. Rosin

A naturally occurring resin also known as colophony, found in trees, mostly pine species. Rosin essentially consist of a series of so-called resin acids, which contain unsaturation and carboxylic acid functionalities. The main acid components of rosin are abietic acid, neoabietic acid and dehydroabietic acid. Rosin is usually modified through reactions with the functional groups into useful resins for coatings. Notable among these are the products obtained by the reaction of rosin with maleic anhydride. The acid and anhydride functional groups can be reacted with polyhydric alcohols to produce the so-called maleic resins, which are essentially a type of alkyd resin.

42. Rotational Moulding

Also known as rotomoulding, is a process for the production of hollow mouldings at atmospheric pressure from polymers in the form of powder. The powder is fed into a preheated mould, which rotates, at low speed, in one or two directions in order to spread the polymer along the wall of the mould, where it melts and sinters. A typical set-up for the rotation of the mould is shown.

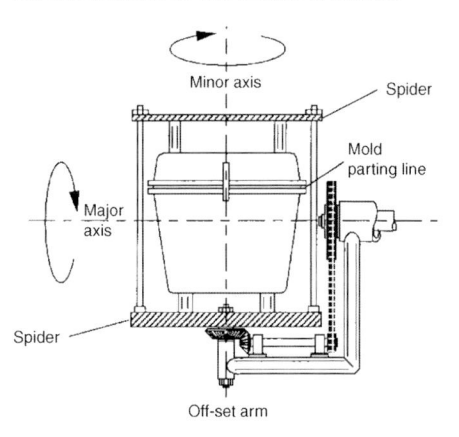

Typical set-up for the rotation of the mould in rotomoulding. Source: Osswald (1998).

During flow, the melt is subjected only to gravitational forces, as the centrifugal forces are negligible, because the rotational speed is low. Note that there is virtually no flow of the melt, as it adheres to the surface of the mould and remains in place due to the relatively high viscosity. Nevertheless the temperature has to be very high in order to cause a rapid melting of the powder and also to effectively remove the voids between particles through surface tension effects.

The melting and fusion of the particles are, in effect, events similar to those experienced in a 'free' sintering process, hence the parameter that controls the extent of flow of the melt and the removal of the air entrapped between particles is the ratio of surface energy (γ) to melt viscosity (η). In order to obtain the highest possible γ/η ratio, the process relies on the contrasting sensitivity of these two properties of polymers to increases in temperature, which is very low for surface energy and very high for melt viscosity. Before the powder melts and sticks to the surface of the mould, the spreading of the powder takes place by rolling movements under gravitational forces. As the shear stresses arising from gravitational forces on the melt are very low, there are severe limitations for the filling of small cavities in a mould, so that the topographic features of the article must be free of intricate asperities.

To complete a moulding cycle, the mould passes through four stages: (i) the article from the previous run is removed; (ii) then fresh powder is fed into the mould; (iii) the mould is heated by external sources; and (iv) this is followed by a cooling stage using the spray of fine jets of cold water.

43. Rubber

A term derived from the erasing characteristics of the constituent polymer, which exists in a 'rubbery state' at ambient temperatures. The constituent polymers of rubber products are also known as elastomers.

44. Rubber Elasticity

A term used to denote the total reversibility of the high elongations displayed by polymers in their rubbery state, as shown in the diagram.

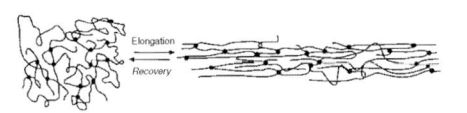

Pictorial description of the stretching and recovery of the cross-linked polymer chains of an elastomer. Source: Unidentified original source.

The term has acquired special significance in the mechanics of materials, as it has required the modification of the classical theory of elasticity based on infinitesimal deformations. To emphasize the difference from classical elastic behaviour, rubber elasticity is sometimes referred to as rubber-like elasticity. Rubber elasticity has acquired a special role also in thermodynamics, as it has made it possible to calculate the change in entropy with the stretching out of polymeric chains. From the second law of thermodynamics,

$$dA = -S\,dT + f\,dL - P\,dV,$$

where A is the free energy, S is the entropy, T is the absolute temperature, f is the applied force, L is the length of the specimen, P is the external pressure and V is the volume of the stretched specimen. Through appropriate theoretical considerations, and knowing that there is no change in volume, it is deduced that the change in entropy with increase in the length of the specimen is equal to the change in force per increase in temperature, that is,

$$(\delta S/\delta L)_T = -(\delta f/\delta T)_L.$$

This relationship implies that the force acting on a stretched rubber specimen increases linearly with temperature if the length is kept constant, a behaviour that arises entirely through the reduction in the entropy of the stretched polymer chains. Mechanics considerations require a redefinition of strain, from the classical elasticity, that is, $\varepsilon = (dL/L)_{dL \to 0}$. This is accomplished by considering the extension ratio (λ) as the parameter related to the applied stress, that is, $\lambda = L/L_0$, where L_0 is the initial length of the specimen when the stress, σ, is zero. From fundamental considerations, an expression is derived for the relationship between stress and extension ratio, known as the Mooney–Rivlin equation, that is,

$$\sigma = (C + C'/\lambda + C''\lambda^2)(\lambda - 1/\lambda^2).$$

The terms C' and C'' are characteristic constants of the rubber, depending on the chemical constitution of the polymer, the cross-linking density, and type and level of reinforcement. From the above relationship, it is deduced that the gradient of the plot of $\sigma/(\lambda - 1/\lambda^2)$ against $1/\lambda$ in the linear region corresponds to the shear modulus of the rubber. The diagram shows that there is a strong deviation from linearity at large extension ratios, owing to strain hardening phenomena, such as stress-induced crystallization.

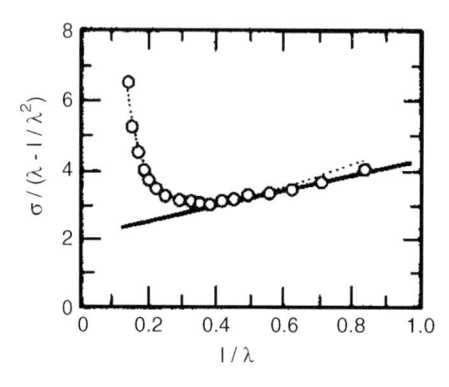

Mooney–Rivlin plot for rubber elastic behaviour. Source: Sato (1969).

Since the concept of Young's modulus used for small strains is not valid for large rubber extensions, in practice the terms M_{100} and M_{300} are used instead, which correspond to the value of the nominal stress (i.e. force/original cross-sectional area) divided by the nominal strain, $(100\Delta L/L)$, at 100% and 300% extension, respectively. These are empirical values that are useful for quality control and materials specifications.

45. Rubbery State

The state of a polymer above the glass transition temperature. (See Transition and Deformational behaviour.)

46. Runner

The channel that connects the spruc to the gates of an injection mould, as shown. (See Injection moulding, Mould and Gate.)

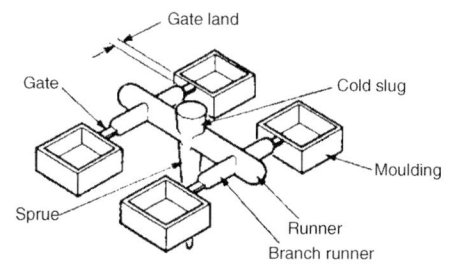

Example of channels feeding the melt into the cavities of an injection mould in sequence: sprue, runner and gate. Source: McCrum et al. (1988).

47. Rutile

(See Titania.)

S

1. Sag

A term used to denote the thinning of the polymer melt emerging from an extrusion die, or a polymer sheet in thermoforming prior to the application of vacuum or draping the sheet on to the mould.

2. Scanning Electron Microscopy (SEM)

A technique capable of providing much larger magnifications than light microscopy (more than 1000 times). The method is based on the use of a beam of high-energy electrons, produced from a filament and accelerated by a high voltage, to produce secondary electrons and/or backscattered electrons, as well as X-rays, when focused on a sample.

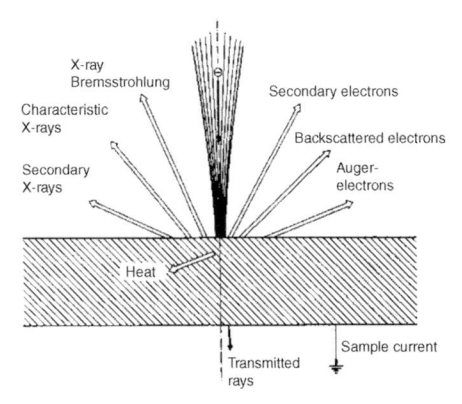

Schematic representation of events occurring when a sample is bombarded with high-energy electrons. Source: Unidentified original source.

The backscattered electrons are primary beam electrons (high-energy electrons) that have been elastically scattered by the nuclei in the sample and escape from the surface. Thus they can be used to obtain an elemental composition contrast in the sample. Secondary electrons are emitted with low energy from the upper surface layers of the sample (a few nanometres deep), which produces topographic images of the sample surface.

Non-conductive materials, such as polymers, have to be coated with a thin layer of highly conductive and inert metal, such as gold, using a sputtering technique. The sample is usually fractured after refrigeration in liquid nitrogen so that the fracture plane reproduces all the asperities arising from the heterogeneities in the structure. Chemical etching or chemical staining are sometimes used to enhance the topographic contrast resulting from the presence of two phases, even when the chemical composition is uniform and the heterogeneity arises from the existence of crystalline and amorphous domains of the same polymer. Chemical staining is carried out in samples where one of the components contains domains of a diene rubber, such as ABS or HIPS. The technique takes advantage of the ability of osmium tetroxide to react with aliphatic double bonds so that only the domains containing the diene elastomer will be 'stained' with heavy-metal ions. The reaction scheme for the staining process is shown.

Chemical staining of diene elastomer domains in polymer blends.

3. Scorch Time

The time at which the value of the torque reaches a certain pre-specified value in evaluating the curing characteristics of a rubber mix using a cure-meter, such as the Mooney viscometer. This parameter gives an indication of the length of time that a rubber mix maintains its processing characteristics at a specific temperature. (See Cure-meter.)

4. Screw–Die Interaction

A term used to describe the relationship between flow rate and pressure for the metering zone of the screw and that for the die of the extruder as a means of obtaining the operating conditions for an extrusion operation. The net flow rate of the melt through the metering zone can be written as (drag flow – pressure flow), that is, $Q_{screw} = Q_D - Q_P$, where $Q_D = \alpha N$ and $Q_P = \beta \Delta P/\eta$. The two constants α and β are known as the 'screw characteristics' insofar as they represent the geometric features of the screw, and N is the screw speed in revolutions per second. The flow rate through the die can be written as $Q_{die} = K\Delta P/\eta$, where ΔP is the head pressure and η is the viscosity of the melt for both equations. The operating conditions are determined from the flow balance $Q_{screw} = Q_{die}$, as there are no leaks of the melt in other parts of the extruder. The main difficulties in the solution of these equations for polymer melts arise from the shear stress and temperature dependences of the term η, resulting from their non-Newtonian behaviour, which cause considerable deviations from linearity in the flow rate (Q) against pressure (ΔP) plots for both the screw and the die sections, as shown in the diagram.

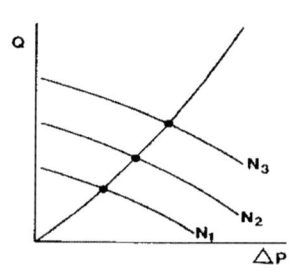

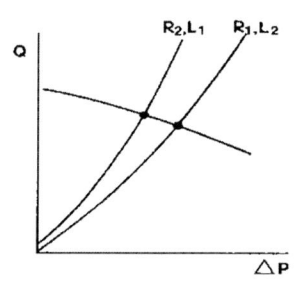

Plots of flow rate versus pressure drop for (a) the screw metering zone and (b) the die of an extruder, where N is the number of screw revolutions per second, R is the nominal radius of the die and L is the length of the die lips. Source: Birley et al. (1991).

The curves in the upper diagram are for three different screw speeds, while the curves at the bottom are for two different die constants, K. The operating conditions correspond to intercepts of the two plots, which give the values of the flow rate and the peak pressure between the die and the metering section of the screw at different screw speeds (N).

5. Secant Modulus

The ratio of stress to a pre-specified strain (say 1%) from a force–deformation curve obtained in tensile tests. (See Young's modulus.)

6. Secondary Crystallization

The term refers to the continuation of the crystallization process taking place at low temperatures. Secondary crystallization takes place primarily in the form of a lamella thickening process, but can also be associated with the formation of additional crystals of very small lateral dimensions, with specific melting characteristics. Some specific features of the secondary crystallization process include a linear increase in both the melting temperature of secondary crystals and the glass transition temperature with the logarithm of the crystallization time. (See Crystallization.)

7. Secondary Plasticizer

An auxiliary plasticizer. (See Plasticizer.)

8. See-Through Clarity

A term used in the packaging industry to denote the characteristics of a film that allows fine details of an object placed at some distance from the film to be seen. This property is related to the presence of large surface irregularities, such as 'sharkskin' and 'orange peel', which cause forward scattering of light at very small angles (between 1° and 1.5°). (See Optical properties.)

9. Seed Polymerization

A term sometimes used for an emulsion polymerization that starts from an existing polymer emulsion. A new monomer is polymerized on the surface of existing particles to produce an outer shell or a chemically bonded (compatibilized) mixture of two polymers.

10. Self-Extinguishing

A term used to describe the fire retardant characteristics of polymers, whose behaviour is characterized by their capability of allowing a propagating flame to auto-extinguish.

11. Semi-IPN

An interpenetrating polymer network (IPN) structure consisting of a linear polymer and a cross-linked polymer. (See Interpenetrating polymer network.)

12. Shape Memory Polymer

(See Heat-shrinkable product.)

13. Sharkskin

A type of defect found on the surface of extruded products, formed as the melt emerges from the die. The typical appearance of a polymer filament with sharkskin surface defects is shown in the photograph.

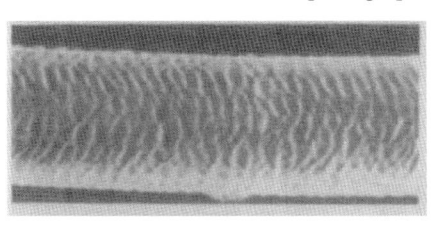

Sharkskin defects on the surface of a filament emerging from the die of a capillary rheometer.

Smoother defects, known as 'orange peel', are sometimes found on the surface of extruded products, particularly if these have been stretched out, as in blow moulded containers or tubular films. These types of defects arise as a result of the acceleration of the outer polymer layers at the exit die in order to catch up with the velocity in the centre due to the change from the parabolic flow profile within the die to a constant velocity within the filament after the die. This brings about a form of melt fracture localized at the surface due to the rapid stretching of the polymer outer layers. (See Melt fracture and Melt strength.)

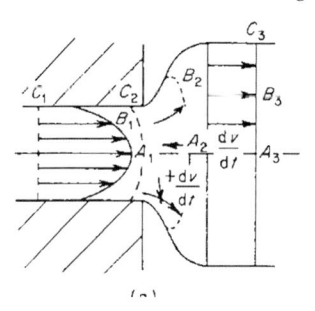

Mechanism for the formation of sharkskin defects in an extrudate at the die exit. Source: Mascia (1989).

14. Shear Modulus

The coefficient that relates the shear stress, τ, to the resultant shear strain, γ, that is, $G = \tau/\gamma$. (See Modulus.)

15. Shear Rate

A concept used in rheology to denote the velocity gradient perpendicular to the flow direction. (See Rheology and Viscosity.)

16. Shear Stress

(See Stress.)

17. Shelf-Life

A term used to describe the storage stability of a cross-linkable system, expressed in terms of maximum time that can elapse before being used.

18. Shift Factor

Also known as the 'time–temperature shift factor', this is a term to denote the amount of displacement of a creep or stress relaxation curve, for experiments carried out at a given temperature, along the time axis to overlap with the corresponding curve obtained for experiments at a different temperature, as a means of producing a master curve. The master curve represents an extrapolation of the data to low temperatures from data obtained at higher temperatures. (See Master curve and Time–temperature superposition.)

19. Shish-Kebab Crystal

Type of structure that is intermediate between a fringed micelle and chain folding. (See Crystallinity.)

20. Shore Hardness

A measure of the hardness of soft or rubbery polymers. (See Hardness.)

21. Side Group

A group attached to the backbone of a polymer chain.

22. Silane

A generic name for alkoxysilane compounds, normally in relation to coupling agents for composites. (See Coupling agent.)

23. Silica

A filler represented by the formula SiO_2, usually with very small particle dimensions. (See Filler.)

24. Silicone Fluid

Essentially low-molecular-weight polydimethylsiloxanes (PDMS) or polymethylphenylsiloxanes (PMPS), whose constituent units are represented by the chemical formulae shown.

PDMS   PMPS

Both have a wide service temperature range, from about -60 to $+200\,°C$. The phenylmethyl silicones have lower compressibility and a higher thermal oxidative stability, as well as being more resistant to high-energy radiation. The molecular weight varies widely depending on the required viscosity, which covers the range from about 5 cSt to 2×10^6 cSt. The range of uses is illustrated in the diagram.

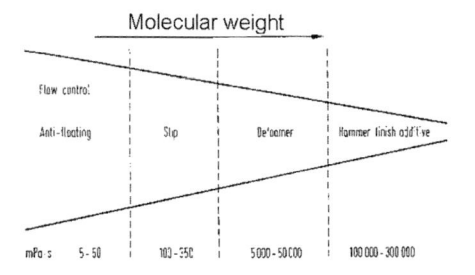

Applications of silicone fluids in relation to viscosity and molecular weight. Source: Bieleman (1996).

25. Silicone Rubber

A rubber containing silicon and oxygen atoms in the backbone of the constituent polymer chains, often identified with the letter Q. The most important silicone elastomers are based on high-molecular-weight polydimethylsiloxane (PDMS, also referred to as MQ) or polymethylphenylsiloxane (PMPS, and also PMQ). Cross-linking reactions can take place either via alkoxy or acetoxy functional groups, usually for room-temperature curing, or by free-radical reactions via pendent vinyl groups (known also as VMQ). The former mechanism relies on the hydrolysis of the alkoxy or acetoxy silane groups along the backbone chains, induced by moisture in the atmosphere, followed by condensation reactions, both catalysed by acid additives, such as acetic acid or organo-tin compounds. These reactions are normally carried out at room temperature, known as room-temperature vulcanization (RTV), producing either an alcohol or acetic acid as by-products, as shown in the reaction scheme.

Reaction scheme for RTV condensation reactions of silicone elastomers. Source: Pocius (2002).

The limitations of one-component systems requiring moisture for the initiation of the polymerization reactions are overcome by two-component systems. Curing of the latter systems takes place via direct condensation reactions between hydroxyl groups, catalysed by organo-tin compounds, such as dibutyl tin dilaurate (DBTL) and stannous octoate (STO). Both one- and two-component systems are mainly used as sealants and adhesives. The free-radical cross-linking reactions, on the other hand, are catalysed by peroxides and are mainly used for high-temperature cure as in moulded products. They contain about 0.5–1.0% pendent vinyl groups along the chain, introduced through copolymerization with the required amounts of methylvinylsiloxane monomer. For sealant and liner formulations, modern silicones use cure reactions based on platinum-catalysed vinyl addition of silane hydride to a vinylsilane, controlled by the addition of an inhibitor, as shown in the reaction scheme.

In effect the inhibitor retards the reaction, through complexation reactions with the catalyst, thereby prolonging the storage life of the uncured system. The inhibitor–catalyst complex breaks down when the system is heated, releasing the catalyst for the reaction.

Reaction scheme for the silane hydride cross-linking of vinyl-functionalized silicone elastomer. Source: Pocius (2002).

Silicone elastomers have a wide service temperature range and a high thermal oxidation stability, as well as excellent dielectric properties and a high resistance to tracking. Owing to their very low T_g (around $-120\,°C$) they retain their rubber-like characteristics down to very low temperatures. They suffer, however, from low resistance to solvents and poor mechanical strength. Unlike most other elastomers, PDMS systems cannot be reinforced with carbon black, but only with fumed silica. Better resistance to oils, and to non-polar solvents in general, is achieved by the fluorosilicone corresponding to polymethyltrifluoropropylsiloxane, known also as FMQ (FVMQ for the vinyl-functionalized system), which is prepared and cured in the same way as conventional PDMS. The chemical structure of FMQ is shown.

The presence of the fluorinated side groups, however, increases to some extent the T_g and impairs the mechanical properties relative to PDMS systems.

26. Sink Mark

(See Moulding defect.)

27. Size (or Sizing)

A term for coatings deposited on fibres during manufacture, sometimes referred to as 'finishing'. Starch has traditionally been used to treat the surface of cotton fibres to prevent them from fluffing and has also been used initially in the production of glass fibres for the same reason. The hydroxyl groups in starch provide a natural affinity for the SiOH groups on the surface of glass fibres. Later starch was replaced with poly (vinyl acetate), deposited from diluted emulsions. Over the last 10–20 years, there have been considerable new developments in the type of polymer emulsions and/or dispersions used for the 'sizing' of fibres in order to 'tailor' the interphase for specific fibre–matrix combinations. The emulsions used for coating fibres contain a number of additives, such as lubricants, antistatic agents, pH controllers, plasticizers and silane coupling agents. The term 'size' or 'sizing' is used to describe the overall composition of the coating on the fibres, which contains about 80% polymer and 20% additives. The amount of size used on fibres is about 0.5–2.0%, giving a coating thickness in the region of 0.1–0.5 μm. The micrograph reveals the type of morphological heterogeneity in size coatings on the surface of glass fibres that may arise from the presence of the multitude of additives and, possibly, also from the poor fusion of the polymer particles deposited from the emulsion.

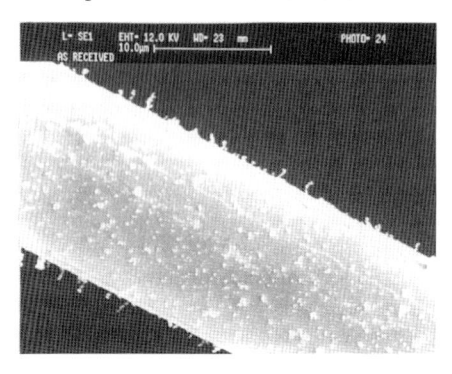

SEM micrograph of a glass fibre taken from a commercial roving. Source: Demirer (2000).

Sizes have to be miscible with the resin or polymer used as the matrix for composites so that chemical bonding can take place through reactions with the silane coupling agents anchored on the surface of glass fibres. (See Coupling agent.)

28. Size Exclusion Chromatography (SEC)

(See Gel permeation chromatography.)

29. Sizing Die

An auxiliary die placed after the extrusion die, whose functions are (i) to prevent the collapse of the melt at the die exit before the extrudate is cooled, and (ii) to provide the final dimensions to a tubing or a pipe. Two typical sizing die systems are shown in the diagrams.

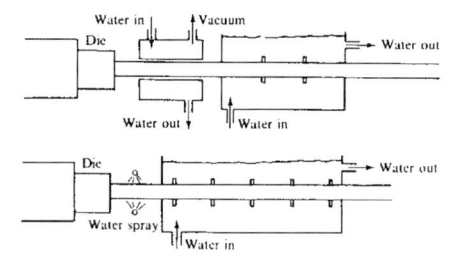

Typical sizing dies for tubings and pipes. Source: Rosato (1998).

The system shown at the top operates through a perforated brass jacket that allows the water to get to the surface of the pipe and act as a coolant and lubricant. Vacuum is applied to draw the surface of the pipe against the cooling jacket, thereby ensuring that the pipe does not flatten under gravity forces. In the adjacent cooling tank there are also calibrating rings to monitor the outer diameter of the extruded pipe. The take-off unit is often used to draw down the extrudate to the right dimensions. The system at the bottom operates by direct cooling of the extrudate by spraying fine jets of cold water immediately it emerges from the extrusion die.

30. Slip–Stick Effect

A phenomenon associated with the occurrence of 'melt fracture' of polymer melts above a critical shear stress value. (See Melt fracture.)

31. Slit Rheometer

A pressure rheometer for studying the rheological properties of polymer melts, using a slit die instead of the usual capillary die. A slit die makes it possible to insert pressure transducers along the channel length for recording the pressure at different positions along the flow path before the melt emerges from the die. From the linear plot of pressure against distance from the entry, one obtains (i) the gradient dP/dL, which is required to calculate the shear stress at the wall, that is, $\tau_w = (h/2)\,dP/dL$, and (ii) the residual pressure at the die exit (P_{exit}), which is associated with die swelling and corresponds to the first normal stress difference parameter N_1, related to melt elasticity. (See Normal stress difference.)

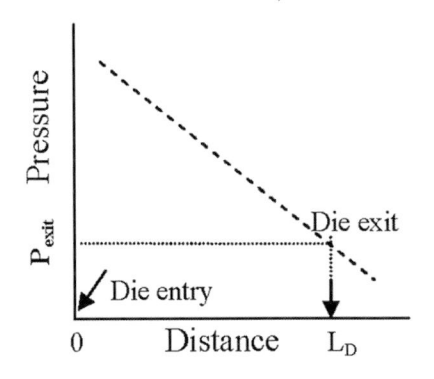

Plot of pressure against distance along the length L_D of the die of a slit rheometer.

The shear rate at the wall ($\dot{\gamma}_w$) is obtained from the expression $\dot{\gamma}_w = 6Q/Wh^2$, where Q is the volumetric flow rate, and W and h are the width and height of the channel. From the data it is possible to obtain the melt viscosity, $\eta = \tau_w/\dot{\gamma}_w$, and the first normal stress coefficient, $\psi_1 = N_1/\dot{\gamma}_w^2$, at different shear rates, obtained through changes in the volumetric flow rate. Often the slit die is fitted to a small extruder so that measurements can be made under conditions closer to those found in polymer processing operations. (See Capillary rheometer and Melt elasticity.)

32. Small-Angle X-Ray Scattering (SAXS)

A technique widely used for the characterization of nanosized domains in polymers, such as the lamellar structure of crystals and the geometric features of inorganic domains in nanocomposites. The principle of SAXS examinations is illustrated.

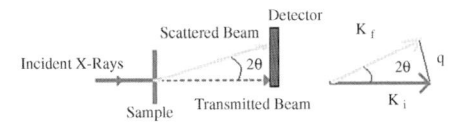

Schematic diagram of the principle of SAXS measurements. Source: Lavorgna (2009).

A selected wavevector of incident monochromatic radiation, K_i, is impinged onto the sample and the scattered intensity, K_f, is collected as a function of the scattering angle 2Θ. The relevant parameter that is required to analyse the interaction with the scattering domains is the vector q, representing the difference between the incident and scattered vectors, that is, $q = K_i - K_f$, obtained from the equation

$$q = (4\pi/\lambda)\sin\Theta.$$

From knowledge of the scattering vector q, it is possible to calculate the diffraction spacing (d) using the equation $d = 2\pi/q$. The scattered intensity $I(q)$ is the Fourier transform of the correlation function of the electron density $\rho(r)$, which corresponds to the probability of finding a scattering domain in position r in the sample if another scattering element is located at position 0. SAXS experiments are designed to measure $I(q)$ at very small scattering vectors and plots are obtained for $I(q)$ against scattering vector, q. An example of such plot is shown in the diagram for some experiments on silica-hybridized Nafion membranes, which were used to calculate the fractal size D_m of the silica domains.

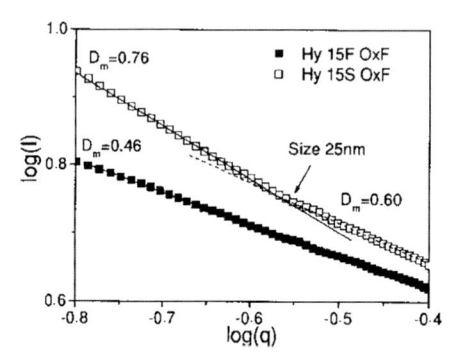

Example of a log–log plot of scattered intensity against vector q from SAXS experiments. Source: Lavorgna (2009).

Usually the data are normalized using the scattering invariant, INV, defined as the integral of the scattered intensity, that is,

$$\text{INV} = \int_0^\infty I(q)q^2 \, dq.$$

In the diagram are shown linear plots for the same samples used to calculate the separation distance between organic and inorganic domains, as a bimodal distribution with mean values, respectively, 2.1 and 3.7 nm.

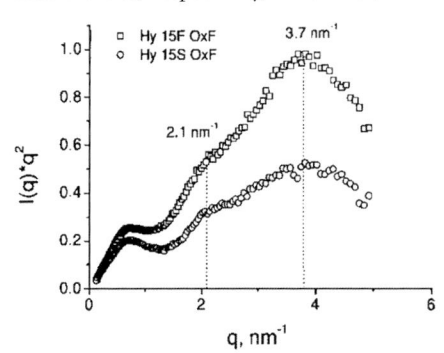

Example of plots of INV against vector q derived from SAXS experiments. Source: Lavorgna (2009).

33. Solar Radiation

Radiation from the Sun reaching the Earth. Only about 5% of all the radiation reaching the Earth gets through the atmosphere to the surface. Radiation with wavelength in the range 290–400 nm of the spectrum can be absorbed by many polymers and can induce oxidative degradation reactions.

34. Solid-State Polymerization

A term usually used to denote the increase in the molecular weight of poly(ethylene terephthalate) (PET) by advancing the condensation reactions, through thermal treatment of the granules under vacuum at temperatures just below the melting point ($T_m = 265\,°\text{C}$).

35. Solid-State Processing

Type of processing operation for thermoplastics carried out at temperatures either just above the T_g of a glassy polymer or between the T_g and T_m of a crystalline polymer. A typical example of solid-state processing is the stretch blow moulding of PET bottles and containers.

36. Solubility Coefficient

A parameter, S, that denotes the solubility of a gas or liquid in a solid relative to the applied pressure, that is, $S = C/P$, where C is the volume of gas or liquid dissolved in unit volume of polymer and P is the applied pressure. (See Permeability.)

37. Solubility Parameter

A parameter, δ, that describes the solubility (or miscibility) characteristics of polymers, additives or solvents. The definition is $\delta = E/v$, where E is the internal energy per unit volume and v is the volume of the molecule.

The solubility of additives in polymers, or the miscibility of two polymers in a mixture, is thermodynamically possible if the two components of the mixture have similar solubility parameters. The value of the solubility parameter is related to intermolecular interactions and corresponds to the sum of three terms, respectively, polar (δ_P), hydrogen bonding (δ_H) and non-polar or dispersive (δ_D).

38. Spandex Fibre

The term refers to elastomeric fibres where the fibre-forming component is a long-chain polymer consisting of more than 85% segmented polyurethane. The soft block is a macroglycol (linear polyol), while the hard block is formed from diphenyl-methane-4,4′-diisocyanate (MDI) and hydrazine (H_2NNH_2) or ethylene diamine ($H_2NCH_2CH_2NH_2$), according to the reaction scheme shown. (See Urethane polymer, Block copolymer and Fibre.)

HO ∼∼∼∼∼∼∼ OH   +   OCN—R—NCO ⟶
Macroglycol M.W. 2000   MDI

OCN—R—NHCOO ∼∼∼∼∼∼ OCONH—R—NCO $\xrightarrow{\quad H_2NNH_2 \quad}$
Prepolymer   Chain extender

—[—COO ∼∼∼∼∼∼ OCONH—R—NHCO—NHNH—CONH—R—NH—]$_n$
    Soft segment   Hard segment

R = —⟨phenyl⟩—CH$_2$—⟨phenyl⟩—

Reaction scheme for the synthesis of block copolymers for the production of spandex fibres.

39. Specific Impedance

A parameter that describes the capability of a dielectric to store electric charge in an alternating electric field. The complex specific impedance (Z_p^*) is related to the capacitance (C^*) and is defined as

$$Z_p^* = \frac{1}{\omega C^*(d/A)},$$

where $\omega = 2\pi f$ (f is the frequency), $C^* = Q^*/V$ (i.e. stored charge/applied voltage), d is the distance between the electrodes and A is the area of the electrodes. The impedance Z_p^* is related to the relative permittivity (ε_r^*) by the expression

$$Z_p^* = \frac{1}{\omega \varepsilon_r^* C_1}.$$

(See Permittivity.)

40. Spectrum

A term used to describe the detailed features of the variation of a physical parameter within a range of related causes or excitations. An example is the absorption spectrum of a polymer to display the details of the absorption of a certain type of radiation (e.g. infrared) as a function of wavelength or wavenumber.

41. Spherulite

A term used to describe the organization of crystals of polymers, emanating from a central nucleus, as shown in the micrographs.

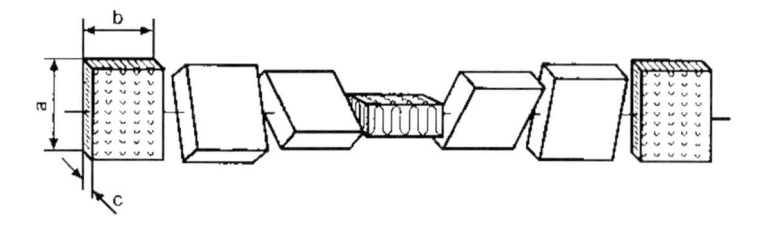

Lamellar helical stacking of folded crystal domains within a spherulite. Source: Ehrenstein (2001).

42. SPI Gel Time

A method used to determine the gel time of unsaturated polyester resins, devised by the Society of the Plastics Industry (SPI) in the USA. It consists of recording the time taken for the temperature of a specific amount of resin, contained in a test tube immersed in a water bath maintained at 50 °C, to rise to a pre-specified value as a result of the exothermicity of the cross-linking reactions.

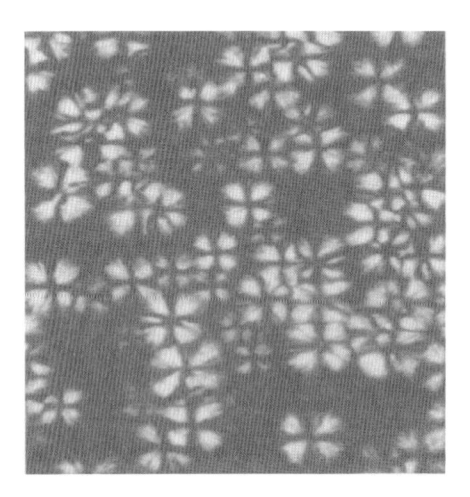

Typical appearance of spherulites in a crystalline polymer viewed in a polarizing microscope.

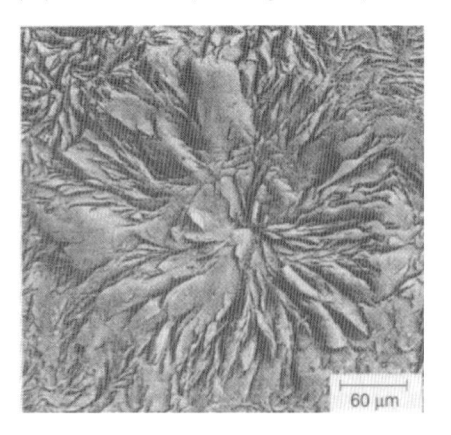

43. Spider Leg

These are the three sections of a tubing or pipe die that holds the central pin fixed on the main body of the die. (See Extrusion die.)

44. Spiral Flow Moulding

A technique used to evaluate the flow characteristics of polymers for injection moulding. The cavity of the mould is a graduated spiral that is fed from the central sprue and channels out to the open. The procedure involves measurements of the length of the spiral under different conditions to compare the moulding characteristics of different materials.

Optical micrograph of a spherulite of polyoxymethylene sample etched with 37% HCl solution. Source: Ehrenstein (2001).

The lamellae of folded-chain crystals emanate from a central nucleus in a spiral fashion as indicated schematically.

45. Sputtering

A technique used to deposit a thin layer of metal or silica on articles or films through vaporization induced by a high-temperature filament. (See Metallization.)

46. Stabilization

Although generally the term denotes the process or method that renders a system stable, within the polymer context it mainly refers to the use of stabilizers to prevent the occurrence of degradation reactions through the incorporation of additives, which are known as stabilizers. (See Thermal degradation, UV degradation, Stabilizer and Vinyl polymer.)

47. Stabilizer

An additive used to prevent or retard the degradation of polymers. Stabilizers function in a number of different ways according to the cause and mechanism of the degradation. In this respect they are broadly divided into:

a) **light or UV stabilizers** if they protect the polymer against the degradation effects of sunlight in combination with atmospheric oxygen; and

b) **processing stabilizers** when their intervention prevents degradation during manufacturing operations, involving the combined effect of high temperature and oxygen.

Stabilizers that protect the polymer by reacting with the free radicals produced from the interaction of oxygen with polymer chains are usually called 'antioxidants' or 'primary stabilizers'. The stabilizers that act as peroxide decomposers are referred to as 'secondary stabilizers', insofar as they are used in conjunction with a primary stabilizer to provide a synergistic effect. (See Antioxidant and UV stabilizer.) Special stabilizers can be used to protect polymers against degradation resulting through hydrolysis reactions, which cause chain scission and a reduction of molecular weight. (See Hydrolysis and Hydrolysis stabilizer.)

Stabilizers are sometimes classified in a manner that reflects both their chemical nature and the type of protection that they afford. For instance, HALS is the abbreviation for 'hindered amine light stabilizers'. The chemical classification of stabilizers is applied particularly to additives used in formulations of PVC and other vinyl polymers. In this case, stabilizers are classified as 'acid absorbers' if their function is to protect the equipment against the corrosive action of the acid formed, as well as minimizing the autocatalytic effect that they have on the actual formation of the acid from decomposition of the polymer chains. These stabilizers are further classified into lead, cadmium, barium, zinc or tin types to describe the chemical nature of the metal component of the compound used as additive. (See PVC and Stabilization.)

48. Standard Linear Solid (SLS)

A hybrid model that combines the relaxation characteristics of the Maxwell model and the creep behaviour of the Kelvin–Voigt model, as shown in the diagram.

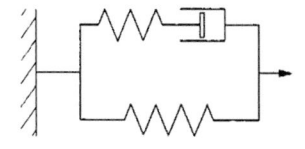

Standard linear solid (SLS) model.

The SLS model is more realistic than either the Maxwell model or the Kelvin–Voigt model for the linear viscoelastic behaviour of polymers insofar as it can exhibit both the stress relaxation and the creep behaviour of polymers. Note that the Kelvin–Voigt model does not exhibit stress relaxation characteristics. At the same time, the solution of the constitutive equation for the Maxwell model applied to creep situations (constant stress input) produces an unrealistic linear increase in strain from time zero and a residual constant strain when removing the stress after a creep period.

The physical interpretation of the SLS model with respect to stress relaxation situations is shown.

$$a_0\sigma + a_1 d\sigma/dt = b_0\varepsilon + b_1 d\varepsilon/dt,$$

where the constants a_0, a_1, b_0 and b_1 can be related to the two spring constants and the dashpot of the SLS model, resulting in the expression

$$\left(\frac{E_m + E_v}{E_m}\right)\sigma + \frac{\eta}{E_m}\frac{d\sigma}{dt} = \eta\frac{d\varepsilon}{Et} - E_v\varepsilon,$$

where E_m is the modulus of the Maxwell component, E_v is the modulus of Kelvin–Voigt component and η is the viscosity of the dashpot component. Solving the equation

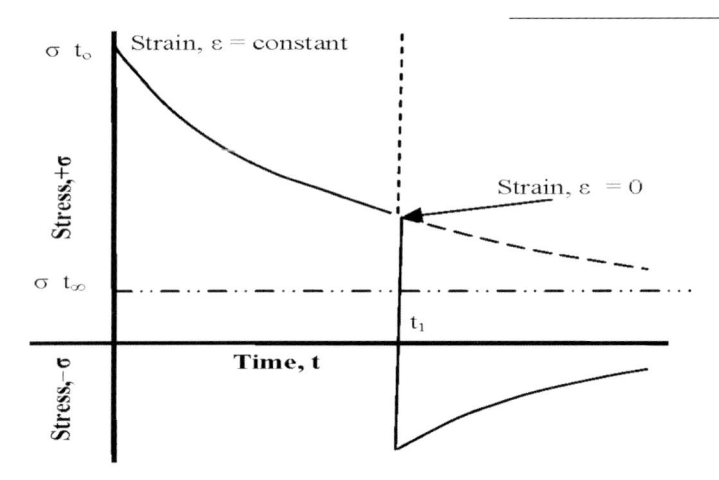

Stress relaxation characteristics of the standard linear solid (SLS) model.

The SLS model indicates that, if a strain is applied instantaneously at time zero, and held constant, there will be an exponential decay of the resulting stress, reaching a constant value at infinite time. In this respect the SLS model differs from the Maxwell model, which predicts complete relaxation of the stress. However, if the strain is forced to go to zero at time t_1, the resulting compressive stress will relax completely at infinite time.

The constitutive equation for the SLS model can be written as a generic expression for the variation of stress, σ, and strain, ε, with time, t, such as

for ε = constant (stress relaxation conditions) gives an expression for the variation of the relaxation modulus with time, $E(t)$, as

$$E(t) = E_\infty + (E_0 - E_\infty)\exp(-t/\lambda_R),$$

where E_∞ is the modulus at infinite time, known also as the relaxed modulus (E_R), E_0 is the modulus at time zero, known also as the instantaneous modulus, and $\lambda_R = \eta/(E_m + E_v)$ is the relaxation time.

Solving the constitutive equation for σ = constant (creep conditions), an equation for the compliance $D(t)$ is obtained in the form

$$D(t) = D_0 + (D_\infty - D_0)[1 - \exp(-t/\lambda_c)],$$

where D_0 is the instantaneous compliance, D_∞ is the compliance at infinite time, and $\lambda_c = \eta/E_v$ is the retardation time.

The physical interpretation of the SLS model for creep conditions is illustrated in the diagram in terms of the variation of the strain as a function of time during the creep period and after removing the stress (i.e. recovery period). The diagram shows that there is an instantaneous increase in strain (i.e. at time zero) followed by a retarded increase towards equilibrium at infinite time. If the stress is removed at a certain time, there will be an instantaneous recovery of the strain imposed at the start of the creep period, followed by an exponential retardation of the strain before the conditions of complete recovery are reached.

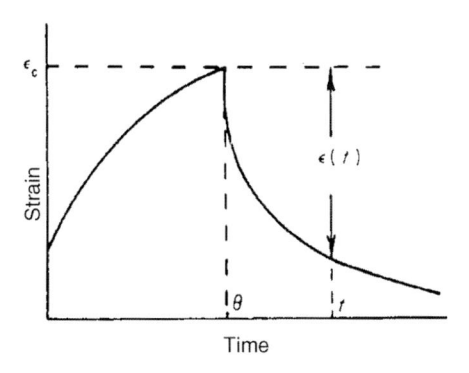

Evolution and recovery of strain for the standard linear solid (SLS) model under creep conditions.

49. Starch

A polysaccharide consisting of a mixture of amylase (linear polymer) and amylopectin (a highly branched polymer), as shown.

Chemical structure of (a) amylose and (b) amylopectin. Source: Rudnik (2008).

Natural starch contains 15–30% amylose with an average molecular weight (MW) that varies from 250 to 5000, and 70–85% amylopectin with MW varying within the range of 10 000 to 100 000. Heating starch in a closed vessel containing about 15% water to temperatures above 100 °C produces a rigid thermoplastic polymer (thermoplastic starch, TPS) with a glass transition temperature (T_g) in the region of 60 °C (after drying), which results from the destruction of the original crystalline structure. This restructured form of starch can be injection moulded into products with rigidity similar to that exhibited by polypropylene or HDPE, and can be plasticized using a variety of plasticizers, such as glycerol and poly(ethylene glycol). The TPS form of starch can be blended with other biodegradable polymers such as polycaprolactone, poly(lactic acid), poly(vinyl alcohol), poly(hydroxybutyrate-*co*-valerate) and poly(ester amide)s to tailor the properties to specific product requirements. (See Biopolymer.)

50. Staudinger

A German scientist who devised the first method for measuring the molecular weight of polymers. After this work, the term 'high polymers' became widely used to describe the long-chain macromolecular compounds, which are now known simply as 'polymers'.

51. Stearate

Usually as an inorganic salt of sodium, zinc and calcium types, used as external lubricants in polymer formulations.

52. Stearic Acid

An additive used in conjunction with zinc oxide to 'activate' the vulcanization of elastomers. Also used as a dispersing aid for fillers and as an internal lubricant in PVC-U formulations.

53. Stereolithograpy

A technique widely used for the rapid production of prototypes. In this process, usually a thermosetting resin, such as an epoxy resin containing specific photoinitiators, is used to build up three-dimensional shapes through sequential cross-linking reactions of successive thin layers of resins by means of a focused laser beam, usually operating in the UV range. The movement of the laser beam is programmed via a computer-aided design (CAD) system. The principle is illustrated in the diagram.

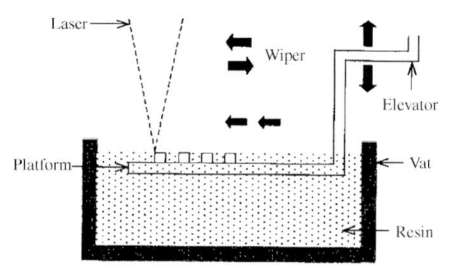

Principle of stereolithography prototyping. Source: Chua *et al.* (2003).

The same technique can also be used with thermoplastics capable of absorbing radiation within the limits of the power generated by the laser, to generate sufficient heat to melt a thin layer of polymer powder. Typical polymers used are polyamides, particularly nylon 11 or nylon 12, polycarbonate and thermoplastic elastomers.

54. Stiffness

A general term used to describe the resistance of a structure to bending deformations. It can be quantified through the definition of a stiffness coefficient (S) as the ratio of the applied load (P) and the resulting deflection (Δ), that is, $S = P/\Delta$. The reciprocal of the stiffness coefficient is known as the 'compliance'.

55. Stokes Equation

An equation used to determine the sedimentation time of particles suspended in a liquid medium through calculations of the sedimentation velocity (V), that is,

$$V = \frac{d^2(\rho-\rho_0)g}{18\eta},$$

where d is the particle diameter, ρ is the particle density, ρ_0 is the density of the medium, g is the gravity constant and η is the viscosity of the medium.

From an examination of this equation, for instance, one can predict the beneficial effects of reducing particle size and increasing the viscosity of the suspending medium with flocculants as a means of improving the stability of suspensions. It must be noted, however, that the Stokes equation assumes that there are no interactive forces acting between particles, which are determined by their zeta potential value. Consequently, the equation can only be used for dilute suspensions, that is, volumetric concentrations up to about 3%. Empirical modifications can be used for estimations related to suspensions with higher particle concentrations. For concentrations between 3% and 10%, the calculated velocity is corrected by a factor equal to $(1-\varphi_s)^{4.5}$, where φ_s is the volume fraction of the solid particles. Another limitation of the Stokes equation arises from deviations of the particles from the assumption of spherical geometry.

56. Strain

A concept used in engineering mechanics to take into account the dimensions of a component or specimen subjected to deformations. Accordingly, strain (ε) is defined as the change in linear dimension divided by the original dimension. Strains can be divided into:

a) **normal strain** (compression or tension) when the change in dimension takes place along the direction of the force, that is, $\varepsilon = dL/L$;

b) **shear strain** when the deformation causes a distortion of the geometry (e.g. a square section becomes rhomboidal) without changing the volume, that is, $\gamma = dX/L = \tan\alpha$, where dX is the change in dimension perpendicular to the length (L), so that α represents the deviation of the angle from the original 90° angle formed by the surfaces of the body deformed; and

c) **volumetric strain** (ε_V) when there is a change in dimensions in three perpendicular directions, which gives $\varepsilon_V = \varepsilon_1 + \varepsilon_2 + \varepsilon_3$.

57. Strain Rate

The rate of change of strain with time, a variable used in tensile tests carried out to measure the strength or evaluate the toughness of materials.

58. Stress

A concept used in both mechanical and electrical engineering to take into account the dimensions of a component or specimen over which a mechanical force acts or the distance over which a voltage is applied. A mechanical stress (σ) can be simply defined as the force (F) divided by the area (A), while an electrical stress, ξ, is the voltage (V) divided by the distance (L). In more rigorous terms, these should be written as the ratio of infinitesimal quantities, that is, $\sigma = dF/dA$ and $\xi = dV/dL$.

58.1 Mechanical Stress

There are three types of mechanical stresses: (i) normal stress (tensile or compressive) when the force acts perpendicular to a surface; (ii) shear stress when the force acts parallel to a surface; and (iii) hydrostatic stress (pressure p or hydrostatic tension σ_H) when three forces, equal in magnitude, act in three perpendicular directions over an area. The different types of stress that can act on a body are shown in the diagram.

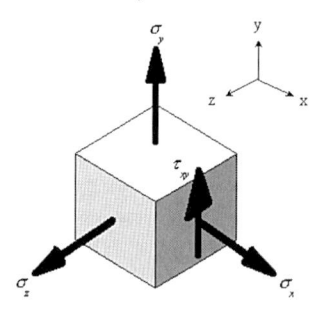

Identification of normal stresses and shear stresses that can act on a body.

Stress is a tensor quantity requiring two digits for identification: the second digit indicates the direction of the force, while the first denotes the area over which the force is applied. For instance, σ_{xx} denotes a normal stress acting in direction x on a plane perpendicular to direction x, which is often identified simply as σ_x. On the other hand, σ_{xy} denotes a force acting in direction y on a plane perpendicular to direction x. This means that σ_{yx} is a shear stress, which is more usually represented by the symbol τ_{xy}. If three stresses acting in perpendicular directions are different in magnitude, then the hydrostatic stress component is taken as the average value, that is, $\sigma_H = (\sigma_x + \sigma_y + \sigma_z)/3$.

58.2 Electrical Stress

The same arguments can be made with respect to electrical stresses.

59. Stress Concentration

A term used to describe the intensification of stresses (mechanical or electrical) arising when there are sharp discontinuities in the geometry of a product or specimen. These discontinuities can be in the form of V notches or sharp edges. Stress concentrations are experienced particularly in joints, irrespective of whether they are mechanical or electrical. An example of stress concentrations in electrical components is shown in the diagram for the case of a shielded cable at the point where the insulation has been cut to produce a joint.

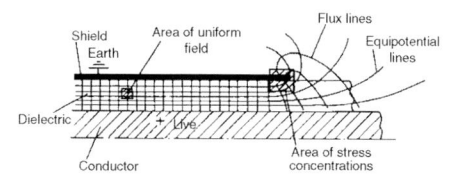

Stress concentrations in a cable where the insulation is stripped for the preparation of a joint. Source: Mascia (1989).

60. Stress Grading

A design aimed to attenuate the stress concentration, mechanical or electrical, arising when two adjacent components of a structure, or a circuit, are made from materials with vastly different Young's moduli (mechanical structures) or volume resistivities (electrical circuits). Stress grading is achieved by introducing an interlayer composed of a material with an intermediate modulus or resistivity.

61. Stress Optical Coefficient

(See Photoelasticity.)

62. Stress–Strain Curve

A plot of stress against strain data obtained from mechanical tests and usually used to calculate the Young's modulus of the material.

63. Stretch Blow Moulding

(See Injection blow moulding.)

64. Stripper Plate

A mechanism used to eject concave or hollow thin-walled mouldings from the cavities of an injection mould. When the mould opens, the stripper plate pushes the mouldings away from the respective cores by exerting forces along the ribs of individual mouldings. (See Ejection mechanism.)

65. Structural Foam

A foam with a solid skin obtained from rigid polymers. (See Foam.)

66. Styrene–Butadiene Rubber (SBR)

(See Diene elastomer.)

67. Styrene–Butadiene–Styrene (SBS) Thermoplastic Elastomer

(See Thermoplastic elastomer.)

68. Styrene–Isoprene–Styrene (SIS) Thermoplastic Elastomer

(See Thermoplastic elastomer.)

69. Styrene Polymer

A polymer containing a predominant number of styrene units in the polymer chains. These include both the homopolymer and copolymers, which are described below.

69.1 Polystyrene (PS)

Available commercially predominantly as an atactic homopolymer, where the benzene side groups are distributed at random in space. (See Tacticity.) PS is glassy polymer (amorphous, although often referred to as 'crystalline' due to its transparency). The chemical structure of PS is represented by the formula:

$$\left[CH_2 - CH(C_6H_5) \right]_n$$

The glass transition temperature is in the region of 100 °C and the density is 1.05 g/cm^3. It is produced predominantly by bulk or suspension free-radical polymerization. Apart from the main use for the production of injection-moulded articles, it is widely used for the manufacture of foams (known as expanded polystyrene, as it is produced from expandable beads containing pentane as blowing agent) and also for the production of biaxially oriented films.

69.2 Syndiotactic Polystyrene (SynPS)

Polystyrene can also be polymerized with a Natta catalyst to produce stereospecific polymers with a high melting point. In particular, the syndiotactic version has been commercialized in recent years for the production of films. SynPS has a melting point of 273 °C and a T_g around 100 °C.

69.3 Styrene Copolymer

A variety of random copolymers, produced by free-radical polymerization methods, are available in the plastics industry, as well as for coatings and binders. The comonomer chosen varies according to the desired change in T_g or to improve the chemical resistance, particularly in relation to the susceptibility to crazing. Copolymers with α-methylstyrene (styrene–α-methylstyrene, SMS) and acrylonitrile (styrene–acrylonitrile, SAN) have a T_g around 115–120 °C. The structure of SAN can be represented by the formula:

69 Styrene Polymer

$$\left[-CH_2-CH(CN)-CH_2-CH(C_6H_5)- \right]_n$$

SAN also provides a better resistance than PS to crazing in hydrocarbons and vegetable oils. Improved weathering resistance is achieved with the use of copolymers of methyl methacrylate without any significant changes in the T_g. Copolymers and terpolymers with other acrylate esters, on the other hand, are more widely used as water emulsions for coatings and binders, as these exhibit a considerably lower T_g.

69.4 Styrene–Maleic Anhydride Copolymer (SMA)

These may be considered as functionalized polystyrene grades insofar as they contain a small number of succinic anhydride units (5–10%) in the backbone chains, obtained by copolymerization of styrene with maleic anhydride. This method makes it possible to introduce a larger quantity of anhydride functional groups on the polystyrene chains than by post-polymerization grafting techniques. The presence of reactive anhydride groups in the chains can be exploited in a number of ways, particularly in improving the adhesion characteristics towards reinforcing fillers and fibres.

69.5 High-Impact Polystyrene (HIPS)

Represents a wide range of toughened grades of polystyrene containing particles of essentially polybutadiene (PB) rubber chemically bonded to the surrounding glassy polystyrene (PS) matrix. These are produced by dissolving the elastomer in styrene monomer and conducting a free-radical polymerization in bulk. The free radicals produced by the initiator can also attack the PB chains, thereby producing a quantity of grafted block copolymer acting as compatibilizer for the immiscible PS and PB homopolymers. By varying the molar ratio of reactants and the polymerization conditions, as well as the reaction procedure, it is possible to obtain a wide range of PS–PB alloys with different morphological structures, as shown in the micrographs.

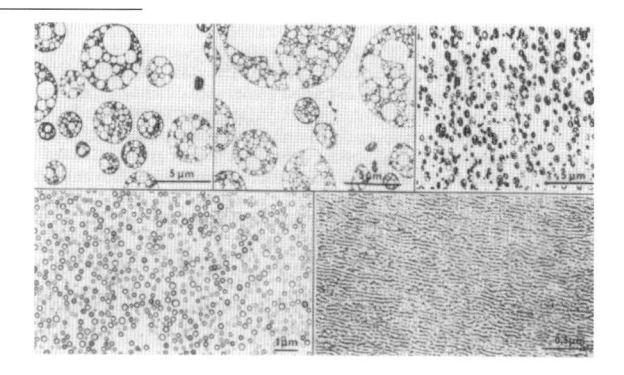

Morphological features of several types of HIPS. Source: Courtesy of BASF.

The details of the morphological structure can have a predominant effect on properties. For instance, the top left and top centre micrographs are typical HIPS obtained through conditions that lead to phase inversion during the course of polymerization. (See Phase inversion.) Typically, the dispersed rubber particles of these systems contain occluded sub-micrometre sized particles of glassy polystyrene. Note that, whereas the systems with large particles (micrographs at the top) are opaque in view of the large difference in refractive index between particles and matrix, the system at bottom left, containing fine discrete particles of PB, produces translucent mouldings. The HIPS shown in the bottom right micrograph was produced by an anionic solution polymerization method, producing a very fine lamellar rubber phase, resulting in products that are completely transparent. This is due to the fact that the dimensions of the dispersed scattering centre of the PB rubber particles are smaller than the wavelength of visible light. Owing to some inevitable plasticization effect provided by solubilized grafted PB chains in the surrounding PS matrix, all HIPS grades tend to exhibit a somewhat lower T_g than that of pure PS.

69.6 Acrylonitrile–Butadiene–Styrene (ABS) Terpolymer Alloy

Consists of compatibilized two-phase blends with an acrylonitrile–styrene copolymer as the main glassy component and a butadiene–acrylonitrile copolymer as the dispersed rubber component. The rubber particles contain inclusions of glassy polymer, which act as reinforcement to counteract the reduction in T_g brought about by the presence of miscible components of the main copolymers. Similarly to HIPS there is a wide range of different grades available commercially, varying in levels of rubber modification and method used to produce the blends. (See ABS.)

69.7 Acrylonitrile–Acrylate–Styrene (ASA) Terpolymer Alloy

Consist of compatibilized two-phase blends of a styrene–acrylonitrile copolymer with an acrylate rubber, produced in the same way as ABS. The replacement of the polybutadiene elastomer with an acrylate rubber, such as copolymers of butyl acrylate–ethyl acrylate, brings about large improvements in thermal oxidation stability and resistance to UV light, as well as in the resistance to mineral oil.

70. Surface Energy (Surface Tension)

Forces and energy on the surface of a liquid or solid, arising from the imbalance of molecular interactions between molecules in the bulk and those at the surface. The molecules exposed to the surface will attract or repel molecules from the environment. If the latter is air (or any gas), the number of molecules that can interact with those of the solid or liquid surface is very small, so that an imbalance of energy between bulk and surface still remains. In the case of liquids, the presence of surface forces is evidenced by the tendency of a free liquid droplet to assume a spherical geometry, which gives the lowest surface per unit volume. Surface energy plays a crucial role in the understanding of adhesion phenomena. The surface energy (γ) of water at room temperature is $72\,\text{mJ/m}^2$, which is quite high due to the strong hydrogen bonds between molecules in the bulk. The value decreases with the reduction in strength of intermolecular forces, so that for glycerol the value of γ becomes $63\,\text{mJ/m}^2$ and for a typical epoxy resin the value is $47\,\text{mJ/m}^2$. The values go right down when the intermolecular forces are weak, as in the case of hexane, where there are only van der Waals forces acting between molecules, giving a surface energy around $18\,\text{mJ/m}^2$.

The surface energy of the most common types of polymers are shown in ascending order in the table. For comparison the values for wood and iron are also reported.

For these, the γ values are very high, owing to the very large number of hydrogen bonds and covalent bonds in wood, and the strong ionic interactions in iron.

Substrate	Surface energy (mJ/m²)
Polytetrafluoroethylene (PTFE)	18.5
Polytrifluoroethylene	22
Poly(vinylidene fluoride) (PVDF)	25
Poly(vinyl fluoride) (PVF)	28
Polyethylene	31
Polypropylene	31
Polystyrene	33
Poly(vinyl alcohol) (PVA)	37
Poly(methyl methacrylate) (PMMA)	39
Poly(vinyl chloride) (TVC)	39
Poly(vinylidene chloride) (PVDC)	40
Poly(ethylene terephthalate) (PET)	43
Poly(hexamethylene adipamide) (nylon 6,6)	46
Epoxy resin (cured)	45
Wood	200–300
Iron	~2000

Source of data: Wake (1992).

71. Surface Resistivity

Denotes the resistivity of a dielectric, which characterizes its capability to 'resist' the flow of a current over the surface. Since 'resistivity' is defined as the ratio of electrical stress to the resulting current density, the related formula for surface resistivity (ρ_S) can be derived from first principles as follows. One has stress = voltage (V) per unit distance, and surface current density = current (I) per unit frontal width. Using the apparatus and specimen geometry shown, the final equation for ρ_S becomes

$$\rho_S = (V/g)(P/I) = (P/g)R_S,$$

where R_S is the surface resistance (V/I), P is the average perimeter of the guarded electrode ($2\pi D_{\text{average}}$) and g is the gap distance between the inner electrode and the guard ring. Note that the dimensions of surface resistivity are ohm/square ($\Omega/\square$), where square ($\square$) indicates that the property is related to the surface characteristics, so that it can be differentiated from volume resistivity, which is a bulk characteristic of the dielectric. The schematic diagram for the electrical circuit measurement of the surface resistivity is shown, where G represents the galvanometer for measuring the current (I) resulting from the applied voltage (V).

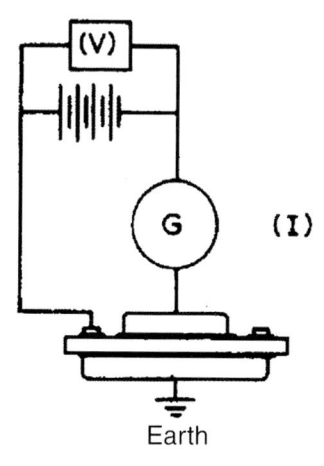

Circuit for measuring the current flowing across the gap between the inner electrode and the guard ring.

The geometric setup for the electrodes (top and bottom) and the guard ring is shown. Note that the function of the guard ring is to ensure that only the current flowing through the gap between the electrode and the guard ring is recorded by the galvanometer. The circuit arrangement shown allows any leakage of current flowing over the edge of the plaque to go to earth.

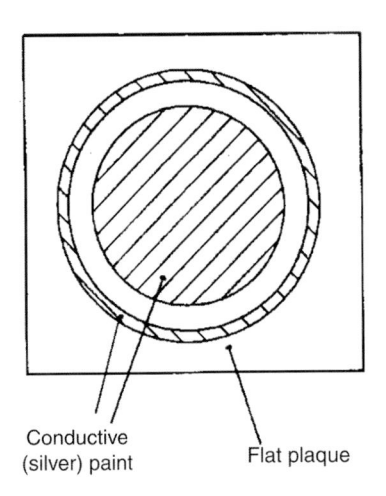

Top electrode and guard ring attached to the plaque used as specimen for measuring surface resistivity.

72. Surfactant

Agent used to reduce the surface energy of liquids or the interfacial energy between solid and liquid or between liquid and liquid. There are three types of surfactants:

a) **Anionic surfactants** consist of long aliphatic chains (hydrophobic) and $SO_3^-Na^+$ (hydrophilic) end groups. The most common type is dodecylbenzene sodium sulfonate.

b) **Cationic surfactants** are mostly ammonium quaternary salts (bromides or chlorides) of long-chain hydrocarbons, used primarily as emulsifying agents for emulsion polymerization and as exfoliating agents for nanoclays. An example is $C_xH_yNH_4^+Br^-$, where $x = 12–18$ and $y = 25–37$.

c) **Amphoteric surfactants** is the term used to describe the non-ionic nature of a class of surfactants, consisting mostly of block copolymers containing chains of polydimethylsiloxane and a poly(alkylene oxide), such as those derived from ethylene or propylene oxide.

73. Surging

A term used to describe a pulsating output of an extruder, resulting from the break-up of the solid bed within the transition zone. This creates a temperature perturbation and local flow instability, which gives rise to a fluctuating pressure at the die entry and a pulsating flow rate through the die. The phenomenon can be prevented by using screws with a very long metering zone to ensure that the flow instability vanishes by the time the melt reaches the die, or by using a 'barrier' screw. (See Extrusion theory and Barrier flights screw.)

74. Suspension

Particles of a liquid or solid dispersed in a liquid medium, usually water. Sedimentation of the particles does not take place if the density of the two components is not too different. The stability of a suspension can be increased by means of surface-active agents, also known as protective colloids. These have two different types of chemical groups, each one of which is capable of forming strong bonds, or associations, with the other component of the suspension. It is also possible to increase the stability of suspensions by allowing charges to be accumulated on the surface of the droplets or particles so that their agglomeration or coalescence is prevented through inter-particle repulsions. This is sometimes referred to as 'electrostatic stabilization'. (See Suspension polymerization.)

75. Suspension Polymerization

A type of polymerization whereby a monomer is dispersed in water to form stable droplets in the region of 10–1000 μm, depending on the nature of the monomer and the type of surface-active agent, that is, the suspension stabilizer. Suspension polymerization is an effective way of maintaining constant the temperature of the polymerizing species, as it allows the heat developed to be dissipated into the surrounding water, which can be effectively controlled through appropriate heat exchangers in the form of cooling jackets or other devices. The initiator for the polymerization is dissolved in the monomer particles; alternatively, polymerization is induced by photocatalysis with the use of UV light. In this respect, therefore, suspension polymerization is similar to bulk or mass polymerization. For most vinyl monomers, the stability of the monomer suspension is achieved with the addition of poly(vinyl alcohol), containing a certain amounts of acetate groups. (See Vinyl polymers.) The stability is achieved by virtue of the $-OH$ groups forming strong associations with water, while the more hydrophobic acetate groups are solubilized into the outer surface layers of the monomer droplets, as shown.

Schematic illustration of the spatial organization of poly(vinyl alcohol) in a suspension of monomer in water. Source: Denkinger (1996a).

In most cases the polymer formed remains dissolved in the monomer until the required, or maximum, degree of conversion is achieved. The free monomer is then expelled through suitable drying procedures in the final stage of the polymerization when the polymer particles are separated from the water. In some cases, however, the polymer formed precipitates out of the monomer, while still remaining within the suspended particles. For the case of PVC, this situation is exploited for the production of porous polymer particles, used for the production of the so-called 'dry blend' grades. (See PVC.)

76. Syndiotactic

(See Tacticity.)

77. Synergism

The combined response or effect produced by two additives or components of a polymer formulation that is substantially greater than the weighted sum of the effects of the individual components.

78. Syntactic Foam

A cellular product produced with the use of hollow Ballottini spheres. (See Foam.)

T

1. Tack

A term used in the rubber industry to describe the sticking of a rubber gum (prior to vulcanization). This is an important characteristic for fabrications of rubber articles, as it allows stacks of various cuttings to remain in position during manufacturing operations. The term is also used in relation to adhesives.

Isotactic and syndiotactic configurations. Bold (thin triangular) and dashed C–X bonds denote the spatial position of the X group, respectively, above and below the reference plane (of the paper).

2. Tackifier

An additive, usually a resin (phenolic, coumarone–indene and terpene types), used in rubber mixes or hot-melt adhesives and pressure-sensitive adhesives in order to enhance the 'tack' of a gum or adhesive tapes.

Note that the regularity of the spatial position of the CH_3 groups in isotactic and syndiotactic systems allows the chains to be packed close to each other into crystalline domains. Usually, this is not possible if the structure is atactic, owing to the steric hindrance effects of the side groups.

3. Tacticity

A term used to describe the spatial position of symmetrical side groups attached to a monomeric unit of polymer chains, such as the position of CH_3 groups in polypropylene. Accordingly, when each unit has the same spatial configuration along the chains, the polymer is said to be 'isotactic'. When the CH_3 groups are arranged in an alternating fashion, the polymer is said to be 'syndiotactic'; and if they assume a random configuration, the structure is said to 'atactic'. The spatial configuration of isotactic and syndiotactic polymers is shown.

4. Talc

A magnesium silicate filler derived from rocks known as soapstone or stealites. Talc is essentially aluminium silicate, with density in the range 2.58–2.83 g/cm^3, depending on the origin and grade. There are a variety of particles in talc fillers, ranging from platelets, through rods to irregularly shaped particles, which provides a certain level of reinforcement capability. The typical geometric features of talc particles is shown in the micrograph.

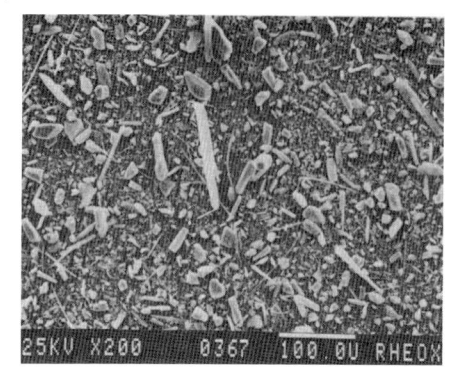

Micrograph of a talc filler, from Rheox Inc. Source: Wypych (1993).

The average particle size and surface area of talc fillers are in the range 1–8 µm and 3–20 m^2/g. (See Filler.)

5. Tan δ

Also known as tangent delta. (See Loss factor, Dynamic mechanical thermal analysis and Dielectric thermal analysis.)

6. Tangent Modulus

The modulus calculated by drawing a tangent on the stress–strain curve through the origin. The modulus is calculated as the gradient of the resulting straight line. The tangent modulus is sometimes known as the 'initial modulus'. (See Modulus.)

7. Tear Strength

A term used to denote the resistance of a flexible material to tearing. Tests are normally carried out to measure the resistance to propagation of an existing crack introduced into a rectangular specimen, which is stretched in a tearing mode, as shown in the diagram.

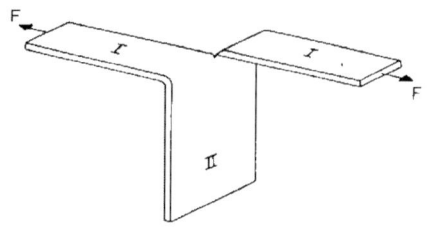

Tear test for flexible materials.

The case where the force remains constant during fracture (smooth tearing) is an example of fracture taking place without a change in compliance of the specimen during crack propagation, as the load is supported only in section I of the specimen, while section II is not subjected to stresses. This type of fracture takes place when the rate at which the specimen is stretched is higher than the rate at which the crack would propagate naturally, so that the crack has to be 'driven' in order to advance. This situation arises when fracture takes place with the formation of a fairly large yield zone at the crack tip (e.g. polyethylene film) and in the case of rubber, provided that the crack advances along a linear path in the direction of the applied load. For these tearing conditions, the strain energy release rate, G, is given by

$$G = -2(\delta W/\delta A),$$

where W is the total strain energy in the specimen and A is the surface area of the crack formed. Under constant-compliance conditions this equation becomes

$$G = 2F/B,$$

where F is the force associated with the propagation of the crack and B is the thickness of the specimen. Consequently, the tear strength, expressed as force per unit thickness, corresponds to the fundamental fracture toughness parameter G, (strain energy release rate), expressed as energy per unit area. (See Fracture mechanics.)

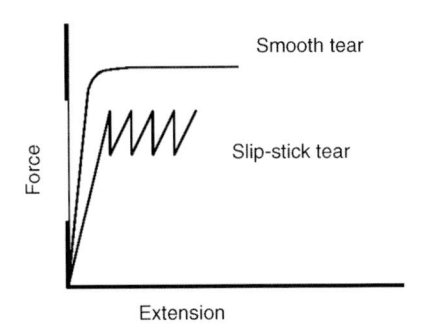

Force–extension curves recorded in tear tests.

When the natural rate of crack propagation is higher than the rate at which the specimen is stretched, fracture takes place via the so-called slip–stick mechanism, manifested in the form of a zig-zag force–deformation trace during fracture propagation. In other words the force drops when the crack starts to propagate, owing to the recovery of the deformation at the crack tip, and rises again when the extension in the specimen again reaches the condition required for the propagation of the crack. In the case of a slip–stick tear, the value of the force to be used in the equation for the calculation of the tear strength is taken as the average between the oscillating values. (See Adhesive test.)

8. Teflon

Tradename for polytetrafluoroethylene (PTFE).

9. Telechelic Oligomer or Polymer

Systems containing functional groups at the chain ends that can be used to extend the length of the chain through further reactions.

10. Tenacity

A term used to denote the tensile strength of fibres or filaments, in grams per denier (g/denier) or newtons per tex (N/tex). (See Denier and Tex.)

11. Tensile Modulus

Modulus measured in tension. Corresponds to the Young's modulus if measurements are made at small extensions. Under these conditions, the value of the modulus measured in tension is equal to the value obtained in compression. (See Modulus.)

12. Tensile Test

A test carried out in tension.

13. Tensor

A quantity (excitation or response) that has a magnitude, a direction and a position relative to a surface. An example of a tensor quantity is 'stress', as it has a magnitude (say, n pascals), a direction (x, y or z in Cartesian coordinates) and a position relative to a reference plane (xy, xz or yz). For this reason, two letters or digits are required to describe its spatial position. For example, σ_{xx} denotes a normal stress (tension or compression) that acts in the x direction and is located in the plane perpendicular to the x direction. On the other hand, σ_{xz} is a shear stress that acts in direction z and lies in the plane perpendicular to direction x. (See Stress.)

14. Termination Reaction

A reaction that stops the growth of polymer chains during polymerization. (See Polymerization.)

15. Tex

A unit used to describe the dimension of fibres (cross-sectional area), filaments and weaving tapes. The 'tex' is defined as the number of grams of the product per 1000 metre length. Other units also used are 'decitex' as the weight in grams per 10 000 metres and 'denier' as the weight in grams per 9000 metres. (See Denier.)

16. Thermal Conductivity

The coefficient that relates the heat flux (H) through a slab of material to the imposed temperature difference (ΔT) between the two opposing surfaces, that is,

$$H/A = K\Delta T/X,$$

where A is the surface area and X is the thickness. The thermal conductivity (K) of polymers is much lower than that of metals owing to the absence of free-moving electrons to transfer the thermal energy in the direction of the temperature gradient.

17. Thermal Degradation

A term that denotes the breakdown of the molecular structure of a polymer by the action of heat. In most cases this results from oxidative reactions with atmospheric oxygen. In other cases, such as vinyl polymers, different reactions may take place either prior to, or simultaneous with, thermal oxidation reactions.

18. Thermal Expansion Coefficient

A coefficient that relates the linear or volumetric expansion (ΔX) to changes in temperature (ΔT), that is, $\Delta X = \alpha \Delta T$, where α is the linear or volumetric expansion coefficient.

19. Thermal Gravimetric Analysis (TGA)

A technique that measures the change in weight of a sample with changes in temperature, under chosen environmental conditions, such as nitrogen, air or oxygen. This technique is particularly useful to measure the loss of volatiles resulting from degradation reactions. Normally measurements are made in a temperature ramp mode, but it can also be used isothermally for kinetic studies.

20. Thermal Oxidation

(See Thermal degradation.)

21. Thermochromic Pigment

Type of pigment that changes colour reversibly at specific temperatures. There are two common types.

a) **Liquid crystals** with a twisted nematic mesophase. The change in colour when a temperature is reached arises from the increase in crystal spacing, producing reflections at different wavelengths, which are perceived as a change in colour.

b) **Leuco dyes** used in the form of microencapsulations 3–5 μm in diameter. These have a less accurate temperature response than liquid crystals.

22. Thermoforming

A generic name for processes involving the heating of a sheet, or a tubular product, to the rubbery state of the polymer and drawing it by the application of vacuum to produce the desired shape, often assisted by mechanical devices. The product is cooled while still under a state of external stresses in order to prevent recovery of the imposed deformation through relaxation of the acquired molecular orientation. A characteristic feature of thermoforming is that the deformations produce molecular orientation, which imparts dimensional instability if the product were to be used at high temperatures. On the other hand, the orientation can be controlled as a means of increasing the mechanical strength and toughness of the article. Examples of sheet thermoforming processes are briefly discussed and shown schematically.

In negative (female) vacuum forming, the required shape is acquired through the simple application of vacuum to draw the sheet into a cavity. The formed article is removed after the vacuum is released.

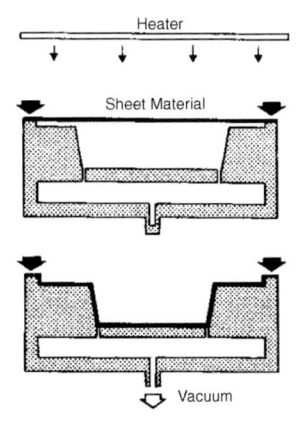

Negative (female) vacuum forming. Source: Unidentified original source.

In air-slip vacuum (male) forming, after heating the sheet to the required temperature, air is used to inflate and pre-stretch the sheet. Vacuum is then applied to draw the sheet onto a male mould, where is cooled before the vacuum is released and the moulded part removed.

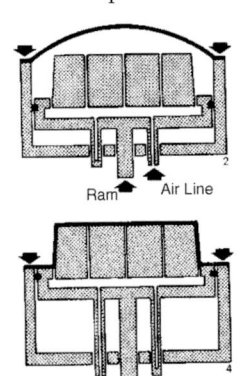

Air-slip vacuum (male) forming. Source: Unidentified original source.

In plug-assisted thermoforming, the preheated sheet is drawn into a cavity through the combined use of a 'plug' and vacuum to achieve larger draw ratios than is possible by other methods.

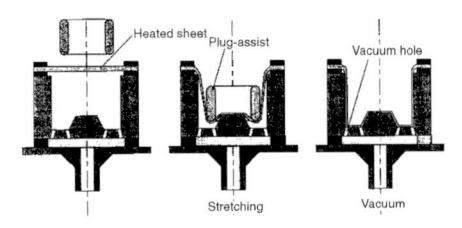

Plug-assisted forming. Source: Osswald (1998).

Note that 'stretch blow moulding' methods can be described as thermoforming of a tubular product. (See Injection blow moulding.)

23. Thermofusion Process

This can be described as a manufacturing process by which articles are produced via the interfacial fusion of smaller components. These processes comprise powder fusion operations, such as powder sintering, powder coating and rotational moulding, as well as welding processes. (See Powder sintering, Powder coating, Rotational moulding and Welding.) 'Perfect fusion' of the components brought into contact with one another is achieved if the overall adhesive forces acting across the interface become equal to the cohesive forces acting within the bulk of the components. If the adhering components are chemically similar, perfect fusion would be achieved simply by removing all surface irregularities and heterogeneities at the interface, so that the material at the interface cannot be distinguished from the material in the bulk. In order to achieve this structural state, there has to be molecular diffusion across the interface of the individual components. Chemical similarity of the fusion components can be interpreted also in terms in terms of mutual miscibility, determined by similarity in the values of the solubility parameters.

This may not be a sufficient requirement in the case of crystalline polymers insofar as the two components, although miscible in the melt state, may segregate at the interface through differential crystallization instead of co-crystallization. An example of the latter situation is the fusion of high-density polyethylene (HDPE) to polypropylene (PP), which exhibit similar solubility parameters but do not crystallize into common crystal cells on cooling. This results in a differential surface energy of the components at the interface, which prevents the achievement of 'perfect fusion', as a state of the material where the interface cannot be distinguished from the bulk. For the combination HDPE/PP, for instance, the difference in solubility parameters is only in the region of 5%, whereas the ratio of surface energy for HDPE and PP is greater than 2.

24. Thermogram

A term used to describe the plot of an event as a function of temperature, such as the weight of the sample in TGA or the heat flow in DSC experiments.

25. Thermomechanical Analysis (TMA)

A technique that measures the changes in dimensions of a sample as result of changes in temperature, using a dilatometer. The analysis is particularly useful for measuring the glass transition temperature (T_g) and the linear expansion coefficient of polymers below the T_g.

26. Thermoplastic

A polymer that will melt and flow at high temperatures, evidenced by the net displacement of entire molecules relative to each other through a reptation mechanism. To be able to fulfil these conditions, the polymer must not contain chemical cross-links. Hence, for a cross-linked polyethylene product, while it is crystalline and will melt at high temperatures, the molecules exist in the form of a low-density network and cannot be made to flow. The reason why some thermoplastic products are cross-linked after processing, therefore, is to achieve this particular state so that they can be used at temperatures above their melting point. (See Heat-shrinkable product.)

27. Thermoplastic Elastomer

A synthetic rubber that has the processing characteristics of a rigid thermoplastic polymer. There are several types of thermoplastic elastomers.

27.1 Block Copolymer Elastomer

These constitute the main class of thermoplastic elastomers, consisting of ABA-type block copolymers, where the A blocks consist of rigid units (glassy amorphous domains, usually polystyrene, $T_g = 95\,°C$, or crystalline rigid domains), while the B blocks are rubbery linear polymers of various compositions. The B units of the main types of thermoplastic elastomers are polybutadiene ($T_g = -90\,°C$), polyisoprene and isobutylene (T_g in the region of $-60\,°C$). (See Block copolymer.)

Other systems are obtained by the hydrogenation of the central units consisting of copolymers of butadiene and isoprenes to produce systems known as styrene–ethylene/propylene–styrene (S–EP–S) and styrene–ethylene/butylene–styrene (S–EB–S). The removal of the unsaturation in the central rubbery blocks provides a greater resistance to thermal and UV-induced oxidative degradation. The polystyrene content of these block-copolymer elastomers is usually in the region of 30–40%. The molecular weight of the central rubbery units is usually greater than 100 000, whereas the molecular weight of the end polystyrene units is in the region of 7000–10 000. Higher molecular weights of the PS blocks would produce large rigid domains, which will have an adverse effect on many properties, particularly transparency. The manner in which the rigid blocks of polystyrene produce physical cross-links between the high-molecular-weight (diene) segments of the block copolymer chains is illustrated.

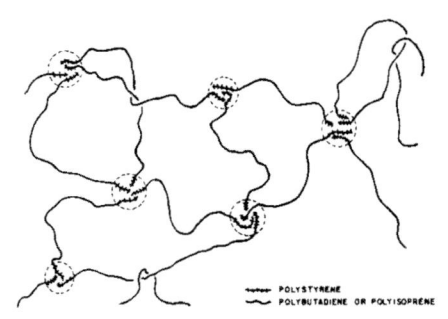

Physical cross-links provided by the strong attractive forces between the segregated segments of the rigid domains.

TEM micrographs of typical ABA block copolymers obtained on commercial thermoplastic elastomers are shown. Note that the black-stained domains correspond to the rubbery phase. Those with a lower extent of phase interpenetration ((a) and (b)) are softer elastomers owing to the more prominent effect of the co-continuous rubbery domains.

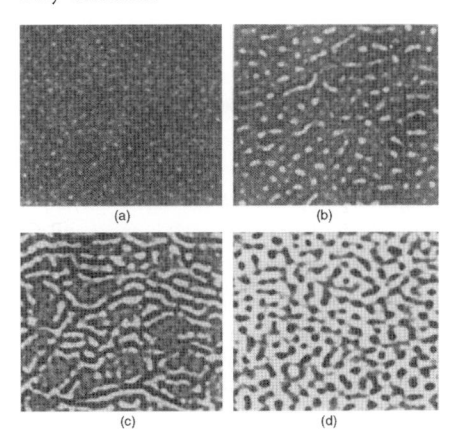

TEM micrographs of different types of thermoplastic elastomers produced from ABA block copolymers. Source: Unidentified original source.

27.2 Multi-Block Copolymer

These elastomers are based on a multi-block copolymer with an $(H-E)_n$ structure, where H represents the hard segments, which are often crystalline thermoplastics, and E represents the rubbery (soft) segments, with an amorphous structure. The H segments are formed from polymers with a regular repeating unit structure, that is, ...AAAAAAAAAAA..., capable of forming crystalline domains with melting point well above room temperature. The soft E segments are formed from polymers with an irregular (random) arrangement of monomeric units, that is, ...ABBAAABAABBBBAAB..., which produce domains with a T_g well below room temperature. The composition and size of each block are distributed at random within macromolecular linear chains with different molecular weights. The resulting block copolymers have a complex morphological structure consisting of ill-defined lamellae interconnected with chains containing both soft and hard segments. The presence of crystals immersed in the amorphous domain not only provides a strain-hardening behaviour exhibited by conventional cross-linked rubber, but will also act as reinforcing filler.

The most important types of elastomers based on multi-block copolymers usually contain hard blocks consisting of either polyurethanes, polyesters or polyamides. The soft segments are either aliphatic polyethers, such as polyoxytetramethylene, polyoxypropylene and polyoxyethylene, or polymers produced from the condensation reactions of adipic acid or sebacic acid with different types of long-chain glycols. The hard blocks of polyester elastomers are usually polybutylene terephthalate; those used for polyamide are the same as those used for conventional nylons, that is, PA 6, PA 11, PA 12 or PA 6,6. The hard segments of polyurethane elastomers are urethane blocks obtained from the reaction of diphenylmethane-4,4′-diisocyanate (MDI) and 1,4-butanediol. (See Polyester, Polyamide, Nylon and Urethane polymer and resin.) In the case of polyolefin-based thermoplastic elastomers, the hard crystalline blocks are either isotactic polypropylene or linear polyethylene (HDPE type), while the soft segments are made of random copolymers of ethylene and propylene (EPR) with a glass transition around $-60\,°C$.

27.3 Dynamic Vulcanizate

These have a structure, both chemical and morphological, quite different from that of block copolymers. The morphological structure consists of a fine continuous phase of a cross-linked elastomer (usually EPDM) in a matrix and dispersed particles of a hard thermoplastic polymer component (usually polypropylene). The thermoplastic-like processing characteristics are likely to be derived from phase inversion when the crystalline component melts. This not only results in a large volumetric expansion of the linear polymer domain but also produces a mechanism for the dissipation of the imposed mechanical energy through molecular movements. The reptation movements of the linear polymer chains will 'drag' along the cross-linked elastomeric domains by imposing on them sequential chain extension and recovery motions. These dynamic vulcanizates are obtained by mixing the non-cross-linked (or partially cross-linked) elastomer with the thermoplastic polyolefin at high shear rates in order to induce grafting reactions between the two components and to induce the required degree of cross-linking in the elastomer phase. The choice of vulcanizing agent is very important as a means of controlling both the cross-linking reactions and the resulting morphology.

27.4 Plasticized Glassy Polymer

These are thermoplastic elastomers, typified by plasticized PVC, derived from the concept that the addition of sufficient amounts of plasticizer can reduce the glass transition temperature (T_g) to values well below room temperature. In order to achieve strain-hardening characteristics of typical elastomers, the glassy polymer must contain a certain degree of crystallinity, so that the swollen crystals can act as reinforcing domains. For the case of PVC, these characteristics are often enhanced by incorporating into the formulation another polymer, usually a specially designed nitrile rubber (NBR) or a multi-block polyurethane or a chlorinated polyethylene. These polymers are intrinsically rubbery in nature and are 'compatible' with PVC. At the same time, they have the capacity to introduce more efficient reinforcing domains than is achievable with swollen PVC crystals alone.

28. Thermoset Polymer

A glassy polymer based on cross-linked macromolecular organic networks, which prevents flow occurring when heated to high temperatures.

29. Thermosetting Resin

A term used to describe a certain type of multifunctional reactive organic component, usually oligomeric compounds, which can be made to flow and be shaped at temperatures above their 'softening point'. Subsequent cross-linking reactions produce networks that make up the glassy characteristics of 'thermoset' polymers.

30. Thickener

An additive used to increase the viscosity of a solution through chain extension reactions. An example is the use of magnesium hydroxide to chain-extend an unsaturated polyester resin through the formation of magnesium carboxylate salts, which increase the molecular weight of the original resin. The term 'thickener' is often used synonymously with thixotropic agent. For instance, various cellulose-derived thickeners, such as hydroxyethyl cellulose and carboxymethyl cellulose, are often added to water solutions to increase the viscosity. This takes place by the formation of intermolecular hydrogen bonds between polymer molecules through bridges of aggregated water clusters. Other thickeners include polyurethane systems obtained by chain extension of hydroxyl-terminated block copolymers with diisocyanates. These produce hydrophilic groups within the polymer chains (e.g. polyethers or polyesters) and hydrophobic groups at the chain ends (e.g. oleyl or stearyl). The resulting structures develop surface activity characteristics so that they will produce micelles when dispersed in water in concentrations above a critical value.

In contrast to monomeric surfactants, a polymeric thickener can form more than one micelle, joined by polyurethane segments. The viscosity increase arises, therefore, through a reduction in the mobility of water molecules due to the association that they form with the micelles and the repulsions from the hydrophobic components. These types of thickeners are widely used to increase the viscosity of emulsions, where the hydrophobic component forms associations with the outer layers of polymer particles, while the hydrophilic part of the molecules becomes swollen with water. This structure can 'hold' different polymer particles together through polymeric chain bridges. Such associations can be formed also with other components of a polymer formulation, such as filler or pigment particles, as shown schematically.

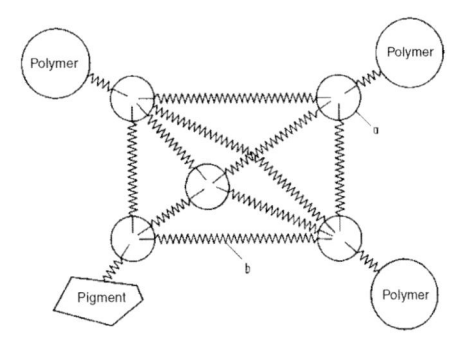

Immobilization of particles within an emulsion or dispersion through many micelles formed by the same thickener molecule. Source: Bieleman (1996).

31. Thickening

A term used to describe the increase in viscosity of a liquid resin or paste, achieved through chain extension reactions or by means of additives.

32. Thiouram

A vulcanizing agent for rubber. (See Vulcanization.)

33. Thixotropic Agent

An additive that increases the viscosity at low shear rates of a resin as a result of the formation of hydrogen bonds with polar groups within the resin. These interactions, however, break down at higher shear rates, causing the viscosity to decrease. Thixotropic agents are mixtures, usually based on high-surface-area silica mixed with glycols or glycerols as a means of enhancing their dispersion in the resin.

34. Thixotropy

A phenomenon related to the increase in viscosity of a fluid or suspension resulting from continual shearing.

35. Tie Molecule

(See Crystalline polymer.)

36. Time-Dependent Modulus

A term used within the context of viscoelasticity to describe the decrease in modulus with time, resulting from molecular relaxations.

37. Time Lag (Diffusion)

(See Diffusion and Induction time.)

38. Time–Temperature Superposition

A principle originating from the theory of linear viscoelasticity, which describes the equivalence of the effects of increasing the duration of the time under load, at a given temperature, to that of increasing the temperature at constant duration of loads. This concept arises from the verification that the curves representing the variation of the relaxation modulus, or creep compliance, with time at different temperatures are similar to those expressed as a function of temperature, measured over a constant duration of the applied strain or stress. The rationale for this principle derives from the realization that the time and temperature dependences of the modulus and compliance functions are primarily determined by the ratio of the duration of the load (t) to the characteristic time (λ), which is a parameter for the material. This ratio, sometimes referred to as the Deborah number, corresponds to the t/λ concept in the Maxwell, Kelvin–Voigt and standard linear solid models. Examining the modulus and compliance curves (see diagram) at temperatures T_R and T_1, T_2 and T_3 (where $T_R < T_1 < T_2 < T_3$), it is clear that, by sliding the curves for temperatures T_1, T_2 and T_3 along the time axis, they can be made to overlap the curve at temperature T_R. In doing so one has, in effect, kept the values of the compliance constant and changed the time t until the value of λ is reached for which the t/λ ratio has become the same as for the curve at temperature T_R. The movement (shift) that has been made along the time axis to make the curves overlap is known as the time–temperature shift factor a_T.

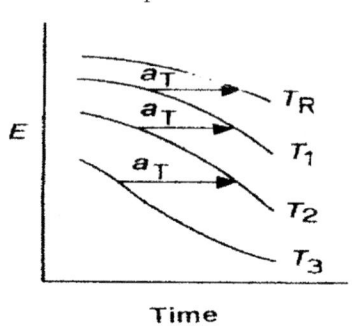

Horizontal shifting of modulus–log(time) curves at high temperatures (T_3, T_2, T_1) towards that obtained at room temperature, the reference temperature, T_R.

A more precise procedure would also make a vertical shift to account for the small changes that the temperature would have on the time-independent components of the modulus. However, this effect is very small and can be neglected. The time–temperature superposition principle can be used, therefore, to develop an extrapolation procedure for the estimation of the modulus and compliance values at long times, based on experiments carried out over short periods of time at higher temperatures. The argument for the stress relaxation modulus or compliance as a function of time applies equally well for plots of the complex modulus, or complex compliance, with the reciprocal of the frequency (or angular velocity) at different temperatures. A similar shift factor a_T would apply to make the curves overlap into a master curve.

In this case the amount of shifting, that is, the a_T value, required to obtain the overlap from one temperature to another can be calculated from knowledge of the activation energy, ΔH. With the use of the Arrhenius equation, one can derive the expression for a_T over a given temperature interval, that is,

$\log a_T = (\Delta H/2.303R)[1/T_1 - 1/T_2]$.

The activation energy ΔH can be obtained from a very rapid test under dynamic loading conditions, carried out over a wide range of frequencies and temperatures. An empirical approach to obtain the values of a_T was used by Williams, Landel and Ferry, who derived an expression known as the WLF equation, that is,

$\log a_T = [-C_1(T-T_g)]/(C_2 + T - T_g)$,

where C_1 and C_2 are constants for the material and loading conditions (i.e. creep or stress relaxation).

39. Titanate

(See Coupling agent.)

40. Titanium Dioxide (Titania, TiO$_2$)

A white pigment in two forms, 'rutile' and 'anatase', which are crystallographically similar (tetragonal structure) but with a rather different density, respectively 4.2–5.5 and 3.9 g/cm^3. The particles have a complex structure, with a predominance of Al$_2$O$_3$, SiO$_2$ and ZnO in the outer layers to block the diffusion of Ti ions, which would have a strong catalytic effect on the degradation of polymers, irrespective of whether it is thermally or UV-induced. The average particle size is in the region of 0.2–0.3 μm in order to achieve the optimum light scattering characteristics, which are largely derived from the very high refractive index, equal to 2.6.

41. Torque

The product of a shear force and the distance over which the force is applied. A concept widely used in rotation and torsion situations.

42. Torque Rheometer

An apparatus consisting of a small internal mixer, comprising kneading rotors within a chamber, which contains a specific amount of polymer and other ingredients. The apparatus records the torque developed by the mixer rotors and the temperature of the melt during mixing and continual shearing of the melt. In this respect, it can be considered as a tool complementary to a cure-meter for the characterization of rubber mixes, as well as a valuable apparatus for studying the mixing characteristics of polymer blends and particulate composites. Although the term 'rheometer' is a misnomer, insofar as it does not involve measurements of flow rates or shear rates, the variation of torque and temperature with time provide valuable information about the fusion (melting) characteristics of a mixture as well as other events that may take place during mixing, such as cross-linking and degradation reactions.

43. Torsion Pendulum

An apparatus used to characterize the dynamic mechanical properties of polymers by subjecting a specimen to free damping torsional oscillations.

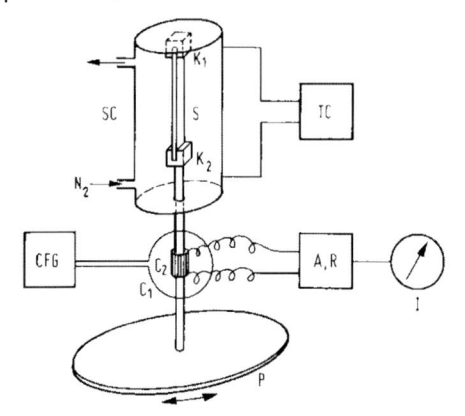

A schematic illustration of the components of a damping torsion oscillation apparatus: S is the sample; P is the inertia disc; TC is the temperature controller; other parts are electronic devices. Source: Hoffmann et al. (1977).

From the decay of the amplitude (A) of the oscillations, it is possible to obtain the logarithmic decrement (Δ), from which the loss factor ($\tan \delta$) can be calculated, that is, $A = A_0 \exp(-pt)$ and $p = i\psi - a$ (where $i = \sqrt{-1}$, ψ is the angular frequency of the oscillation and a is a constant). Thus

$$\Delta = \ln[A_n/(A_n+1)] = \ln[(A_n+1)/(A_n+2)]$$
$$= \pi \tan \delta = \pi G''/G'.$$

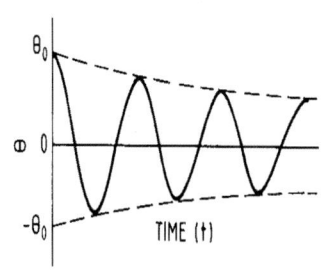

Damping of oscillations of the pendulum disc due to the viscoelastic behaviour of the polymer sample.

The elastic (G') and loss (G'') components of the complex shear modulus, on the other hand, are calculated from the applied moment of inertia (I) and the angular frequency of the torsion oscillations (ψ):

$$G' = \frac{I\psi}{\theta} \frac{\lambda_2}{(1-4\pi^2)} \text{ and } G'' = \frac{i\lambda\psi^2}{\pi\theta},$$

where $\lambda = 2\pi a/\psi$ and θ is a geometric (shape) factor for the specimen.

44. Toughness

The resistance of a material to fracture expressed in energy terms. (See Fracture mechanics and Impact strength.)

45. Tow

A term (jargon) widely used in the context of carbon fibre composites to describe a collection of fibres for filament winding or for the production of woven fabrics.

46. Tracking

A type of failure that occurs in dielectrics, which occurs via the formation of conductive (carbonaceous) surface channels, usually brought about by the presence of surface contaminants, such as salts. A 'dry band' is formed on the surface of high-voltage insulators when exposed to surface contamination, usually salt solutions, which represent the type of atmospheric conditions prevailing in areas near the sea. The local heat generated by the high voltage causes the evaporation of the water over a small area, thereby producing a drastic reduction of the surface conduction across a narrow nonconductive zone, known as a 'dry band'. These conditions result in the formation of arcs across the dry bands, which bring about rapid thermal degradation reactions, with the formation of carbonaceous conductive surface tracks.

The latter is obviously related to the thermal decomposition mechanism of the polymer. Aromatic units within a polymer chain are particularly prone to form carbonaceous tracks. Polymers such as polytetrafluoroethylene (PTFE), poly(methylene oxide) (PMO) and poly(methyl methacrylate) (PMMA), on the other hand, depolymerize completely under the influence of the arc formed across the dry bands. While insulators made of PTFE would simply create eroded paths and holes as a result of the depolymerization, those made of PMO and PMMA would catch fire because of the high flammability of the products, consisting predominantly of monomers.

The resistance to tracking of polymers is usually assessed by tests measuring the comparative tracking index (CTI). The CTI value corresponds to the voltage required to cause failure by surface tracking with the application of 50 drops of a 0.1% NH_4Cl solution onto the surface of a specimen subjected to a voltage gradient between two chisel-shaped electrodes. A test that is considered to produce results that approach more closely the type of failures experienced in service is the tracking erosion test (TERT). The test is carried out on an inclined specimen where a 1% NH_4Cl electrolyte solution is run in small drips over the lower surface to ensure that it forms only a thin surface layer.

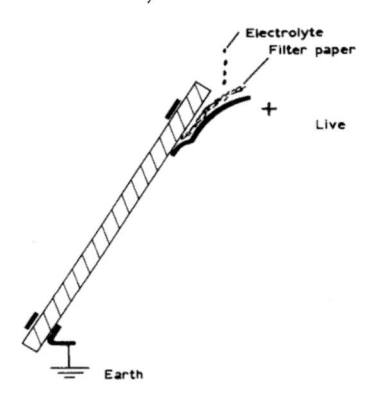

Inclined-plane method for measuring the tracking and electrical erosion resistance of polymers. Source: Mascia (1989).

The voltage is increased in steps until failure takes place due to the formation of surface tracks that leak the current to earth. Alternatively, the extent of erosion produced through a series of localized surface failures is monitored at different time intervals by measuring the weight loss. (See Antitracking additive.)

47. Transesterification

A reaction that takes place between esters from two different compounds, causing an exchange of carboxylate substituents. This can happen during melt mixing and processing of polymer blends, where the transesterification mechanism may be exploited to enhance the compatibility of the components through the deliberate production of 'compatibilizing' block copolymers.

48. Transfer Moulding

A process used for moulding thermosetting moulding powders and bulk moulding compounds (BMC), as well as vulcanizable elastomers. The latter uses a different system for removing the moulded part. The principle consists in feeding the right amount of pre-heated pellets or powder into the 'pot' of a hot mould, from which it is quickly 'transferred' into the cavities of the mould via the sprue, runners and gates. A typical set-up is shown in the diagram.

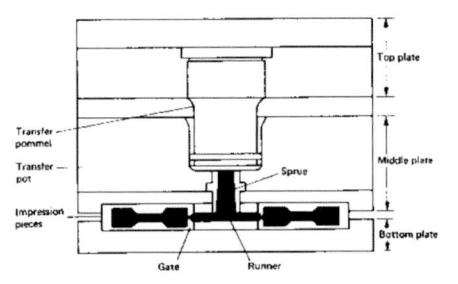

Schematic diagram of a transfer mould with an integral pot contained in the plate between the plunger carrying plate and the cavity plate. Source: Unidentified original source.

One of the main advantages of transfer moulding over compression moulding is the much lower cycle times achieved through shear heating of the melt during the flow through the various channels. The process is still slow and less versatile than screw injection moulding, which has gradually replaced the traditional transfer moulding process.

49. Transition

The change from one state to another, usually brought about by a change in temperature. There are two major transitions encountered in polymers, respectively, the glass–rubber transition (secondary transition) and the melting transition (primary transition). (See Glass transition temperature and Melting point.)

50. Transmissibility

A term used within the context of damping of oscillations in situations such as machine mounts used to isolate a structure from vibrations. The effectiveness of the isolation is expressed in terms of the transmissibility, T, which is the ratio of the transmitted force to the applied force. This is related to the dynamic mechanical property of the rubber mount by the expression

$$T^2 = \frac{1 + [\tan\delta(\omega)]^2}{[1-(\omega/\omega_0)^2 G'(\omega_0)/G'(\omega)]^2 + [\tan\delta(\omega)]^2},$$

where ω is the actual frequency of the vibrating system, ω_0 is the resonance frequency, $G'(\omega_0)$ is the storage shear modulus at the resonance frequency, and $G'(\omega)$ and $\tan\delta(\omega)$ are the shear modulus and loss tangent of the rubber at frequency ω. From this it can be deduced that machine mounts have to be designed to have a resonance frequency considerably lower than the frequency of the vibrating member from which isolation is required.

51. Transmission Electron Microscopy (TEM)

An electron microscopy technique using a high-energy electron beam to produce secondary electrons and/or backscattered electrons, as well as X-rays, when focused on a sample. (See Scanning electron microscopy.) Using samples less than a few micrometres thick, the scattered electrons produced are partially absorbed and partially transmitted through the object, forming an image of the physical heterogeneity of the sample down to about 1 nm, depending on the nature of the sample. In the case of polymers, the samples have to be extremely thin because of their low 'mass thickness' resulting from their low density (i.e. mass thickness = density × thickness of sample). The technique is more applicable, therefore, for polymers containing inorganic heterogeneities, as in nanocomposites and organic–inorganic hybrids.

52. Transparency

The ability of a material to transmit visible light. (see Optical properties.)

53. Tresca Criterion

(See Yield criteria.)

54. Trouton Viscosity

Corresponds to the elongational, or extensional, viscosity. For incompressible Newtonian fluids, the Trouton viscosity is three times the value of the shear viscosity.

55. True Shear Rate

The value of the shear rate for the flow of polymer melts, which takes into account their non-Newtonian behaviour. Assuming that the behaviour of the melt follows a power law, one can write $\tau = K\dot{\gamma}^n$, where n is the power-law index, which is a parameter that describes the deviation from Newtonian behaviour. For Newtonian liquids $n = 1$, while for a polymer melt the value of n is usually in the range 0.3–0.6. (See Apparent shear rate and Non-Newtonian behaviour.) The true shear rate ($\dot{\gamma}_T$) can be calculated from the value obtained assuming Newtonian behaviour, known as the apparent shear rate ($\dot{\gamma}_a$)

- circular channel $\dot{\gamma}_T = [(3n+1)/4n]\,\dot{\gamma}_a$
- rectangular channel $\dot{\gamma}_T = [(2n+1)/3n]\,\dot{\gamma}_a$

56. True Viscosity

The value of the viscosity of polymer melts, at a given shear rate, which takes into account their non-Newtonian behaviour, that is, $\eta_T = \tau/\dot{\gamma}_T$, where τ is the shear stress and $\dot{\gamma}_T$ is the value of the 'true' shear rate. (See True shear rate and Non-Newtonian behaviour.)

57. Tubular Film

(See Blown film.)

58. Twin-Screw Extruder

Extruder with two parallel screws mounted in a barrel with connected double bores. Twin-screw extruders are classified according to the degree by which the screws intermesh and the relative direction of their rotation. Accordingly, they are known as (a) intermeshing counter-rotating, (b) intermeshing co-rotating and (c) non-intermeshing counter-rotating types, respectively, as illustrated.

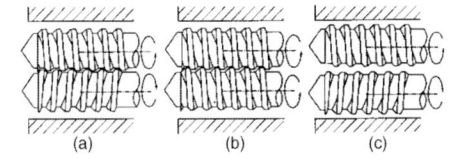

Illustration of rotational directions and flight intermeshing in twin-screw extrusion. Source: Baird and Collias (1998).

The sweeping of the melt through the C-shaped cavities of an intermeshing counter-rotating type twin-screw extruder is illustrated.

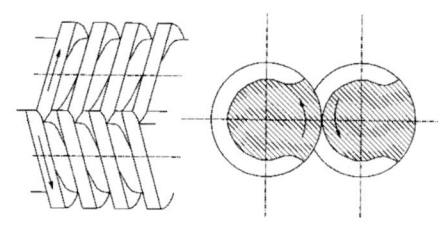

Melt sweeping action of the screw flights of an intermeshing counter-rotating twin-screw extruder. Source: Baird and Collias (1998).

While in single-screw extruders the transport of both solid feed and melt takes place by the drag action of the screw, the conveying mechanism in intermeshing twin-screw extruders is mostly by positive displacement. The maximum flow rate, Q, can be estimated, therefore, from the equation $Q = 2pNV$, where p is the number of parallel flights, N is the screw speed (revolutions per unit time) and V is the volume of the closed C-shaped channel carrying the material. The actual geometry of the C-shaped chamber for each of the two intermeshing screws is shown.

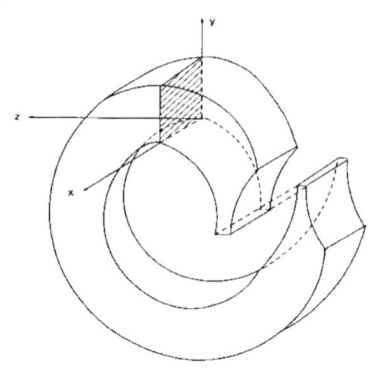

Geometry of a single C-shaped chamber used to calculate the flow rate of an intermeshing counter-rotating twin-screw extruder.

The diagram shows that there are several leakage flows taking place through the intermeshes of the screws, which take place in the opposite direction to the displacement flow. Thus the flow rate equation for the intermeshing counter-rotating twin-screw extruder can be written as

$$Q = 2p\,NV - 2[Q_f + Q_t + p(Q_c + Q_s)].$$

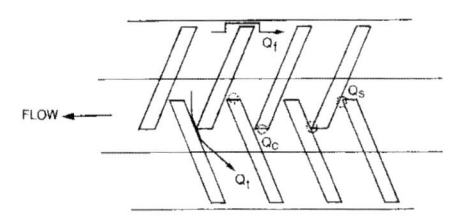

Leakage flows in intermeshing counter-rotating twin-screw extruder.

59. Two-Roll Mill

(See Mixer.)

60. Tyre Construction

The fabrication of the pneumatic tyre is one of the most complex operations in the manufacture of polymer products. The typical structural components are shown in the diagram.

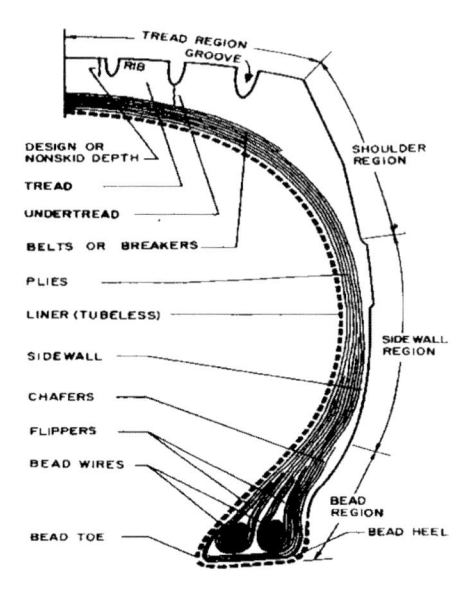

Typical structural components of a tyre. Source: Novac (1978).

The diagram shows that a tyre has the following basic components:

a) **Tread** representing the wear-resistant component that provides traction, silent running and low heat build-up. The composition of the tread component is different from the rest of the tyre and usually consists of a blend of oil-extended SBR and polybutadiene elastomer and natural rubber, compounded with carbon black, curatives, oils and other auxiliary additives. The geometrical features of the tread consist of circumferential ribs and grooves, specially designed for optimum traction and direction control with minimum heat build-up.

b) **Sidewalls** corresponding to the structural components between the tread and the beads, provide support for the weight of the vehicle. The rubber is compounded to provide a high fatigue resistance (flex life) and weather resistance.

c) **Shoulder** comprising the upper portion of the sidewall below the edge of the tread. It is the component that controls the cornering characteristics of the tyre.

d) **Bead** consisting of high-strength steel wire formed into hoops functioning as anchors for the plies and holding the assembly onto the rim of the wheel. The cross-section of the bead conforms to the flange of the wheel to prevent the tyre from rocking or slipping on the rim.

e) **Plies** consisting of layers of rubber-impregnated fabric cord, which extends from bead to bead and provides mechanical reinforcement for the tyre.

f) **Belts or breakers** consisting of narrow layers of tyre cord under the tread of the crown of the tyre. These have the function of resisting deformations in the footprint (i.e. the contact with the road).

g) **Liner** consisting of a thin layer of rubber inside the tyre to provide a seal against the escape of compressed air. This is very important in tubeless tyre construction.

h) **Chafer** consisting of narrow strips of material around the outside of the bead to protect the cord against wear and cutting by the rim, and to distribute the flex above the rim, as well as preventing moisture and dirt getting into the tyre.

U

1. Ubbelohde Viscometer

Used to measure the solution viscosity of a polymer for measuring its molecular weight. (See Molecular weight.)

2. Ultrasonic Welding

A welding technique that uses ultrasonic pulsations. (See Welding.)

3. Ultraviolet Light (UV Light)

(See Electromagnetic radiation, Radiation, Solar radiation, UV degradation, UV spectroscopy and UV stabilizer.)

4. Uniaxial Orientation

The alignment of the constituents of the structure of a material in one or two directions, usually the molecular chains of a polymer or the reinforcing fibres of a composite. A typical example is the molecular orientation of polymers in fibres or fibrillated tapes. (See Orientation function.)

5. Unit Cell

A concept used in crystallography to describe a regularly repeating element from which a crystal is formed through parallel displacements in three dimensions. The relative positions of atoms within a unit cell are constant from cell to cell.

6. Unplasticized PVC

Refers to PVC formulations that do not contain a plasticizer and are often referred to as rigid PVC or PVC-U.

7. Unsaturated Polyester Resin (UP Resin)

A resin obtained from polycondensation reactions between glycols and acid anhydride monomers, which contains double bonds at regular intervals along the backbone chains. The resin is usually dissolved in a liquid unsaturated monomer that acts as a solvent–hardener for the curing reactions by a free-radical mechanism. These resins are used primarily as matrices for glass fibre composites and for coatings. In the latter case they are often referred to as alkyds. Some use of UP resins is also made for castings and as binders for artificial stone from inorganic aggregates. The most widely used UP resins are produced as alternating copolyesters of propylene glycol phthalate–fumarate, as depicted in the chemical formula.

$$HO-\left[CH-CH_2-O-\overset{O}{\underset{\|}{C}}-CH=CH-\overset{O}{\underset{\|}{C}}-O-\overset{CH_3}{\underset{|}{CH}}-CH_2-O-\overset{O}{\underset{\|}{C}}-\underset{}{\bigcirc}\right]_n-COOH$$

Alternating copolyester structure of UP resins.

The solvent–hardener is usually styrene, at around 40 wt%. Other monomers are sometimes used as hardeners. A schematic diagram that illustrates the structure of the solvated resin before curing and that of cross-linked products is shown.

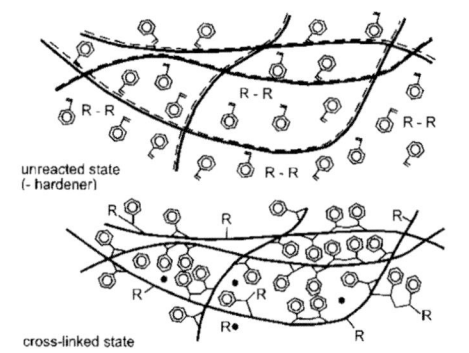

Schematic structure of a UP resin solvated with styrene monomer (top) and the resulting network after curing (bottom). Source: Ehrenstein (2001).

Variants of standard UP resins are obtained by replacing, partially or totally, the *ortho*-phthalate units in the chain with other saturated diesters. The following are typical replacements for o-phthalic anhydride:

a) adipic or succinic anhydride to increase the flexibility;
b) isophthalic acid to reduce water absorption and increase the resistance to hydrolysis; and
c) tetrabromophthalic anhydride or hexachloro-*endo*-methylene tetrahydrophthalic anhydride (known as chlorendic anhydride) to produce fire retardant resins.

Alternatives to styrene as the solvent–hardener are: (i) methyl methacrylate for better weathering resistance; and (ii) divinylbenzene or diallyl phthalate to increase the T_g through an increase in the cross-linking density. Curing reactions are induced by peroxide initiators with varying decomposition temperature, depending on the application. Room-temperature cure resins for composites usually use methyl ethyl ketone (MEK) peroxide activated with a cobalt organic salt or complex. Higher-temperature peroxides, such as benzoyl peroxide, are used in moulding compounds, such as bulk moulding compound (BMC) and sheet moulding compound (SMC), as these are processed under pressure at higher temperatures. Sometimes an inhibitor (alkylated phenols, cresols or quinones) is used to increase the storage stability of the resin. Magnesium oxide is also used in these compounds to increase the viscosity of the resin at ambient temperature to make them easier to handle.

8. Upper Critical Solution Temperature

(See Miscibility.)

9. Upper Limit

(See Law of mixtures.)

10. Urea Formaldehyde Resin (UF Resin)

(See Amino resin.)

11. Urethane Polymer and Resin

These are systems that contain urethane groups in the polymer chains or networks. These systems are often abbreviated to PU (for polyurethane). Urethane groups are formed from the reaction of an isocyanate and a hydroxyl group, that is,

$$R-N=C=O + H-O-R' \rightleftharpoons R-N(H)-C(=O)-O-R'$$

Isocyanate   *Urethane*

A linear polymer is obtained, therefore, if a difunctional isocyanate and a glycol are reacted together. A cross-linked polymer is formed if one or both reactants are multifunctional.

11.1 Polyisocyanate

The majority of isocyanates used in industrial products are aromatic. The more widely used systems are toluene diisocyanate (TDI), respectively, 2,4-TDI and 2,6-TDI, and diphenylmethane-4,4′-diisocyanate (4,4′-MDI). The structure of these isocyanates is shown.

[Structure: 2,4-TDI — benzene ring with CH_3, NCO, NCO substituents]

[Structure: 2,6-TDI — benzene ring with OCN, CH_3, NCO substituents]

[Structure: 4,4′-MDI — OCN–C$_6$H$_4$–CH$_2$–C$_6$H$_4$–NCO]

Apart from the linear structure, the 4,4′-MDI has a lower volatility than both TDIs, which is advantageous in many applications owing to the toxicity of isocyanates. Even lower volatility can be achieved with the polymeric diisocyanate poly(diphenylmethane-4,4′-diisocyanate) (PMDI), represented by the following chemical structure:

[Structure: PMDI]

Although in the majority of cases cross-links are produced using multifunctional hydroxyl compounds, known as polyols, higher-functionality polyisocyanates are sometimes used for the same purpose. A typical multifunctional diisocyanate is the adduct of TDI to trimethylolpropane, having the following structure:

[Structure: $H_5C_2-C(-CH_2-O-CO-NH-C_6H_3(CH_3)-NCO)_3$ (idealized)]

Aliphatic polyisocyanates are primarily used for light-fast products, such as coatings or fibres, owing the low UV stability of aromatic isocyanates, which give rise to degradation and discolorations through the formation of quinoids and conjugated double bonds, as shown.

$$R-O-\overset{O}{\underset{\|}{C}}-NH-\langle\rangle-CH_2-\langle\rangle-NH-\overset{O}{\underset{\|}{C}}-O-R$$

$$\downarrow [O] \; UV$$

$$R-O-\overset{O}{\underset{\|}{C}}-NH-\langle\rangle-CH=\langle\rangle=N-\overset{O}{\underset{\|}{C}}-O-R$$

$$\downarrow [O] \; UV$$

$$R-O-\overset{O}{\underset{\|}{C}}-N=\langle\rangle=C=\langle\rangle-N-\overset{O}{\underset{\|}{C}}-O-R$$

The most widely used aliphatic diisocyanate is hexamethylene diisocyanate (1,6-HDI), $OCN-(CH_2)_6-NCO$.

In general, aliphatic isocyanates are less reactive than aromatic isocyanates. For this reason, their reactions are usually catalysed with either organo-tin compounds, such as dibutyl tin dilaurate (DBTL), and tertiary amines, such as 1,4-diazabicyclo[2.2.2] octane (DABCO):

[Structure: DBTL — H_9C_4, H_9C_4 on Sn with two $O-CO-C_{11}H_{23}$ groups]

[Structure: DABCO]

The ability of isocyanates to react with any hydrogen atom that can be ionically abstracted from a chemical compound provides the opportunity to synthesize a wide range of products. There are also side reactions that can occur, such as the reaction with water:

$$R-N=C=O + H-O-H \longrightarrow \left[R-\underset{|}{\overset{H}{N}}-\underset{}{\overset{O}{\underset{||}{C}}}-O-H \right] \longrightarrow R-NH_2 + CO_2$$

This reaction is utilized in the production of flexible foams, owing to the formation of CO_2, which acts as an 'internal' chemical blowing agent and assists in the formation of cells deriving from an external physical blowing agent.

At the same time, the resulting amine reacts very rapidly with other isocyanate groups to produce urea groups, that is,

$$R-N=C=O + H_2N-R' \longrightarrow R-\overset{H}{\underset{|}{N}}-\overset{O}{\underset{||}{C}}-\overset{H}{\underset{|}{N}}-R'$$

The H atoms in the urea groups can also react with isocyanates to produce biuret groups. Other reactions of isocyanates include those leading to the formation of allophonate groups through reactions with the hydrogen atom in the urethane group and those with itself leading to the formation of cyclic dimers and trimers or to carbodiimides with the elimination of CO_2. The trimerization reaction is often exploited for the production of cyanate ester resins used in formulations for high-temperature adhesives, that is,

Isocyanates can also react with phenols, ε-caprolactam, oximes and secondary amines. The reversible nature of these reactions is exploited for the production of the so-called 'blocked isocyanate' prepolymers to protect the isocyanate groups against reactions with water during storage. When they are subsequently heated, the isocyanate groups are formed again, so that they can react with other components. Blocked isocyanates are particularly useful for the production of one-pack systems for coatings or adhesives.

11.2 Polyol

The most widely polyols are aliphatic polyether and polyester types. Polyethers are mostly obtained by the addition of cyclic ethers, such as ethylene oxide or propylene oxide, to bifunctional starter molecules either alone or mixed with multifunctional starter molecules to produce either linear or branched systems. Typical starters are ethylene glycol, 1,2-propanediol, trimethylolpropane, glycerol and sugar. Polyester-type polyols are produced from combinations of

Trimerization of diisocynates.

various diacids and glycols with different amounts of glycerol, neopentyl glycol or trimethylolpropane to produce systems with different degrees of functionality (usually 2–4) and molecular weight (normally within the range 500–5000). For coating applications, hydroxyl-functionalized polyacrylates are widely used in combination with aliphatic polyisocyanates. Typical acrylic polyols are produced from hydroxyethyl acrylate and methacrylate and also from hydroxypropyl methacrylate.

11.3 Polyurethane Adhesive and Coating

Both systems are available as one- or two-component systems. The one-component systems require blocking of the isocyanate groups to overcome the possibility of reactions taking place during storage. The disadvantage of these systems lies in the regeneration of the blocking agent, such as phenol or caprolactam, when the products are cured. The released blocking agents remain dispersed in the system as diluents.

Some systems make use of moisture infusion for curing at ambient temperatures. These are based on isocyanate-terminated prepolymer, which will react readily with water to produce CO_2 and a diamine (see the reaction schemes above). Waterborne systems, on the other hand, are specifically designed to allow them to be dispersed in water without undergoing curing. (See Water-borne coating.) The prepolymer may contain ionic groups in the structure to allow the use of ionic surfactants for the storage stabilization of the dispersion, or alternatively the polyol component can be specially designed to make it compatible with non-ionic surfactants.

11.4 Polyurethane Elastomer

The original polyurethane (PU) elastomers were designed to match the technology of conventional rubbers so that they could be vulcanized by sulfur curatives or peroxides. These were referred to as 'millable PU elastomer' and were produced as systems with superior solvent resistance and higher thermal stability than NBR or chloroprenes. Subsequently liquid thermosetting cast systems with high tensile strength and high tear strength were introduced, which were then followed by thermoplastic polyurethane elastomers (TPUs), processed by conventional extrusion and injection moulding techniques. More recently, high-pressure impingement mixing machines have become available for the production of cast thermoset elastomers by reaction injection moulding (RIM).

The chemical structures of the main units of cured PU elastomers are quite similar, differing in details regarding either the specific groups for cross-linking or the control of the rate of curing to suit the particular process. Millable PU elastomers are essentially ABA-type block copolymers produced from a polyol, containing a few double bonds for free-radical cross-linking reactions, and an aromatic isocyanate, such as TDI or MDI. TPUs with similar properties are produced from the reaction of saturated linear polyols, polyether or polyester type (soft blocks), with an isocyanate-terminated prepolymer (hard blocks), obtained from the reaction of MDI with a glycol extender, such as 1,4-butanediol or 1,6-hexanediol, ethylene glycol or diethylene glycol. TPU made from polyethers have a high hydrolytic stability, while those containing polyester blocks exhibit a higher resistance to mineral oils.

The RIM products are produced primarily from aromatic diamine extenders, instead of glycol extenders, in order to achieve faster reaction rates, so that the cycle time for curing can be reduced. These are often referred to as polyurethane–urea RIM systems. In order to increase further the rates of reactions for RIM systems and to obtain an even higher thermal stability, the OH terminal groups in the polyol can be replaced with amines, and are referred to as polyurea RIM systems.

11.5 Polyurethane Fibre

(See Spandex fibre.)

11.6 Polyurethane Foam

PU foams are divided into flexible and rigid foams.

Flexible foams are produced using TDI (or MDI/PMDI mixtures), trifunctional polyether polyols (MW range 3000–6500) and water. The blowing action takes place through the formation of CO_2 gas derived from the reaction of isocyanates with water, producing a coarse open-cell foam structure. A finer cell structure is obtained with the addition of physical blowing agents, such as methylene chloride and fluorocarbons, although these are nowadays replaced with less toxic systems.

Rigid foams are produced from MDI or PMDI and either polyether or polyester polyols, using traditional chlorofluorocarbon physical blowing agents, which are increasingly being replaced by non-fluorinated systems, such as dimethyl ether.

12. UV Degradation

This term denotes oxidative decomposition of the molecular structure of a polymer caused by the absorption of UV light. The most harmful radiation in the solar spectrum is that within the wavelength range 290–400 nm, where there is sufficient energy to break most chemical bonds of aliphatic polymer chains. The degradation reactions take place in several steps.

a) **Initiation step**: production of free radicals as a result of the absorption of UV light, that is,

$$PH(\text{polymer chain}) + UV \rightarrow P^\bullet + H^\bullet$$

b) **Propagation step**: reaction of radicals with other polymer chains, that is,

$$P^\bullet + O_2 \rightarrow POO^\bullet$$

and

$$POO^\bullet + PH \rightarrow POOH + P^\bullet$$

c) **Termination step**: deactivation of free radicals, that is,

$$P^\bullet + P^\bullet \rightarrow P$$
$$P^\bullet + POO^\bullet \rightarrow POOP$$

and

$$POO^\bullet + POO^\bullet \rightarrow POOP + O_2$$

These degradation reactions can be exemplified by reference to the UV degradation of polystyrene, as shown. (See Norrish I and Norrish II and Environmental ageing.)

Formation of radicals resulting from the absorption of UV light by polystyrene.

Reaction of free radicals with atmospheric oxygen.

Formation of hydroperoxide groups by the reaction of peroxy radicals with other polymer chains (PH), propagating the formation of free radicals.

Decomposition of hydroperoxide groups as a first step in the formation of carbonyl groups and molecular scission.

13. UV Spectroscopy

A spectroscopic analysis using radiation in the UV region, 200–600 nm, and sometimes up to the visible range. In these regions there is a strong radiation absorption by double bonds (e.g. the C=O group at 280 nm and the conjugated double bonds of aromatic rings and in polyacetylene segments of aliphatic chains, often produced by degradation reactions in vinyl polymers).

14. UV Stabilizer

An additive used to slow down the degradation of polymers induced by UV light. UV stabilizers are classified into two main groups.

14.1 UV Absorber or Screening Agent

These can absorb UV light more easily than polymers, thereby reducing the amount of energy available to cause the decomposition of polymer chains. Absorption of the damaging UV light takes the UV absorber into its excited state (higher energy level) and the absorbed energy is then released as less harmful radiation within the infrared region. The most effective UV absorbers are carbon black followed by metal oxide pigments.

Pigments act also as screeners of UV light by internal scattering at the interface with the polymer. For this mechanism to be effective, the diameter of the particles must be smaller than about 1 μm. Some pigments, such as titanium oxide, are coated to prevent the diffusion of metal ions into the polymer as a means of alleviating their catalytic effect on the propagation reactions for the degradation.

In general, UV stabilizers are used in combination with antioxidants such as phenolic and hindered amine light stabilizer (HALS) types to obtain a synergistic effect. The structure of a typical HALS is shown below:

The more widely used organic UV stabilizers are derivatives of 2-hydroxybenzophenone and hydroxybenzotriazole. The chemical structures of these compounds are shown. Additional groups, particularly chlorine, could be attached to the other benzene ring as a means of enhancing the UV absorption efficiency of the compound.

(a) Hydroxybenzophenone and (b) hydroxyphenyl-benzotriazole. Note that R and R' are long-chain aliphatic segments to make them more easily dispersed in the predominantly aliphatic polymers.

Other compounds with strong UV absorption characteristics are derivatives of hydroxyphenyl-S-hydrazine. The common feature of these additives is their ability to form internal hydrogen bonds and the formation of a conjugated structure as a light-absorbing mechanism, as shown.

Mechanism for UV absorption and dissipation of energy as IR radiation (reverse transformation).

14.2 Excited-State Quencher

These additives exert a UV stabilization function by deactivating the photoexcited groups of a polymer chain (chromophores) through the dissipation of the absorbed energy as harmless infrared radiation. The more widely known excited-state quenchers used for UV stabilization are nickel complexes such as those shown.

Typical excited-state quencher.

V

1. **Vacuum Forming**

A shaping process involving the application of vacuum to draw a heated sheet against the contours of a cavity or male part of a mould. (See Thermoforming.)

2. **Van der Waals Force**

A force acting between molecules through interactions between the dipoles within the molecular structure. These are weaker forces than those arising through hydrogen bonds or by ionic attractions.

3. **Vector**

An excitation or response that can take place in three directions without a specified position or location. A typical example of a vector is a force, which can act in any one or all three directions but does not have a starting position. An applied force, however, can be decomposed into three spatial components acting at right angles to each other.

4. **Velocity Gradient**

The change of velocity with respect to position within the channel in which the flow takes place. In shear flow the velocity gradient occurs perpendicular to the direction of flow, while for elongational flow the velocity gradient takes place along the flow direction. (See Viscosity.)

5. **Velocity Profile**

The profile that the velocity vector of a liquid assumes within the flow channels. (See Non-Newtonian behaviour.)

6. **Vent**

Small bore holes in moulds to allow the escape of air when the melt enters the cavity.

7. **Vicat Softening Point**

An empirical parameter used as a measure of the resistance of a polymer to the penetration of sharp hard objects at high temperatures. It is measured by monitoring the penetration of a loaded needle into a disc-shaped specimen, supported on a rigid plate and fully immersed into an inert liquid. The temperature is increased at a constant rate by heating the liquid, and the penetration of the needle is measured and then plotted against temperature. The Vicat softening point is defined as the temperature at which a specified penetration of the needle (usually 1 mm) is recorded. This value is often very close to the glass transition temperature of the polymer (T_g). (See Heat distortion temperature.)

8. **Vickers Hardness**

(See Hardness.)

9. **Vinyl Ester**

Originally used to describe products obtained from the reaction of an epoxy resin with acrylic acid or methacrylic acid as a means of converting the epoxy groups into unsaturated (vinyl) groups:

$$CH_2=\overset{R}{\underset{CH_3}{C}}-\overset{O}{\underset{}{C}}-O-CH_2-\overset{}{\underset{OH}{CH}}-CH_2-O-\underset{CH_3}{\overset{CH_3}{C}}-\underset{CH_3}{\overset{}{\underset{}{\bigcirc}}}-O-CH_2-\overset{}{\underset{OH}{CH}}-CH_2-O-\overset{O}{\underset{}{C}}-\underset{CH_3}{\overset{}{C}}=CH_2$$

The introduction of vinyl groups allows the resin to be cross-linked by a free-radical mechanism using either peroxide initiators or radiation sources, such as UV light or electron beaming. Often a vinyl ester resin is used as a mixture with another multifunctional monomer in order to increase the cross-linking density and/or to enhance the rate of cure. The term is nowadays used in a more general way to describe a wide range of oligomers containing unsaturated end groups, notable among which are the polyurethane types. Other vinyl esters from cycloaliphatic epoxy resins are used to obtain products that can be cured with cationic photoinitiators.

10. Vinyl Polymer

These are polymers produced from vinyl monomers, covering a very wide range of products. The polymers are produced mostly by free-radical polymerization and are predominantly atactic. The polarity, size and flexibility of the vinyl substituent group have a dominant effect on the capability of the polymer chains to form a crystal lattice, and on the glass transition temperature for the case of amorphous polymers. The most common types of vinyl monomers are briefly discussed below.

10.1 Poly(Vinyl Acetate) (PVAc)

An amorphous polymer with a T_g in the region of 28 °C, represented by the chemical formula:

$$\left[CH_2-\underset{OCOCH_3}{CH} \right]_n$$

PVAc is usually available in the form of water emulsions for use in formulations for adhesives and coatings, particularly for porous materials, such as wood, paper and fabrics. PVAc is also the starting material for the production of poly(vinyl alcohol).

10.2 Poly(Vinyl Alcohol) (PVA)

A water-soluble polymer with the following chemical structure:

$$\left[CH_2-\underset{OH}{CH} \right]_n$$

PVA is produced by the hydrolysis of poly(vinyl acetate) (PVAc), $-(CH_2CHCOOCH_3)_n-$. The preparation of the polymer by this indirect route is due to the instability of vinyl alcohol, which converts readily to acetaldehyde. PVA has biocompatibility and biodegradation characteristics. Commercial products are available with varying degrees of hydrolysis, with the vinyl alcohol component dominating over the residual vinyl acetate groups, that is,

$$(CH_2CHCOOCH_3)_n + xH_2O$$
$$\rightarrow (CH_2CHCOOCH_3)_{n-x}(CH_2CHOH)_x$$
$$+ xHOCOCH_3$$

where $x \gg n-x$.

Biodegradable grades are often available in the form of blends with starch and, possibly, a plasticizer. The fully hydrolysed grades of PVA are crystalline polymers with a T_m in the region of 240 °C and a T_g around 90 °C.

10.3 Poly(Vinyl Butyral) and Poly(Vinyl Formal)

These two polymers are produced by reacting poly(vinyl alcohol) with the respective aldehyde, and they are represented by the chemical formulae:

$$\left[-CH_2-CH\underset{O}{\overset{CH_2}{\diagdown}}CH-\right]_n \quad \left[-CH_2-CH\underset{O}{\overset{CH_2}{\diagdown}}CH-\right]_n$$
$$\qquad\qquad |$$
$$\qquad\quad (CH_2)_2CH_3$$

Poly(vinyl butyral)   *Poly(vinyl formal)*

The main application of poly(vinyl butyral) is for the production of laminated safety glass for automotive windshields. Poly(vinyl formal), on the other hand, is used primarily in lacquers.

10.4 Polyvinylcarbazole

A glassy polymer with a T_g greater than 200 °C, represented by the chemical formula:

$$\left[-CH_2-CH-\right]_n$$
$$\qquad\qquad|$$
$$\qquad\quad N\text{(carbazole)}$$

It is mainly used for paper impregnation for application in capacitors.

10.5 Poly(Vinyl Alkyl Ether)

Various crystalline polymers with an atactic molecular structure, represented by the formula:

$$\left[-CH_2-CH-\right]_n$$
$$\qquad\qquad|$$
$$\qquad\quad OR$$

The T_m and T_g values are respectively 145 and −13 °C for poly(vinyl methyl ether) (PVME) and 86 and −19 °C for poly(vinyl ethyl ether) (PVEE).

10.6 Poly(Vinyl Chloride) (PVC)

A glassy polymer produced by free-radical polymerization in emulsions (e.g. grades for pastes) and suspensions (e.g. grades for dry blends). (See PVC.) The chemical structure is represented by the formula:

$$\left[-CH_2CH-\right]_n$$
$$\qquad\quad|$$
$$\qquad\;Cl$$

The Cl atom take up random spatial positions (atactic configuration) except for some sequences in which chlorine takes up a syndiotactic configuration. Therefore, although the polymer is predominantly amorphous and glassy with a T_g in the region of 80 °C, there are a small amount of crystalline domains (around 5–10%) with a broad melting transition occurring at temperatures higher than the range normally used for processing. The quoted T_m value is 212 °C. The crystals have a density similar to the surrounding amorphous regions (around 1.4 g/cm^3), as the molecules are not tightly packed owing to the syndiotactic configuration of the segments responsible for the formation of crystals. This does not cause any internal light scattering and therefore the products are intrinsically transparent.

Commercially the molecular weight of poly(vinyl chloride) is specified in terms of the K value (named after Fikentscher) obtained from measurements of the solution viscosity. Typical values range from about 40 to 85, corresponding to number-average MW = 40 000–90 000 and weight-average MW = 70 000–500 000. A 'viscosity number' is also used as an indication of the molecular weight of PVC.

10.7 Poly(Vinyl Fluoride) (PVF)

A crystalline polymer with melting point in the region of 200 °C and a T_g around 20 °C. PVF is represented by the formula:

$$\left[CH_2-CH(F) \right]_n$$

Although essentially an atactic polymer, the small size of the fluorine atom makes it possible for the polymer chains to pack into a crystal lattice despite the atactic configuration. PVF is available mainly in the form of films, requiring a solution casting technique owing to excessive thermal decomposition at high temperature, resulting in the loss of HF (a very toxic gas). PVF is transparent to both visible and UV light and has a high resistance to outdoor weathering conditions.

10.8 Polyvinylpyridine

There are two types, respectively poly(4-vinylpyridine) and poly(2-vinylpyridine), depending on the position of the nitrogen of the pyridine ring relative to the vinyl group of the monomer. Both polymers are amorphous, owing to the atactic structure, with different T_g values, respectively 153 °C for poly(4-vinylpyridine) and 102 °C poly(2-vinylpyridine). The formula for the latter is shown below:

$$\left[CH_2-CH(\text{—Py}) \right]_n$$

A major attraction of these polymers is their ability to form quaternary ammonium salts with carboxylic acid compounds, thereby providing a mechanism for producing water-soluble products, which become infusible under dry conditions. This has created a wide interest in the production of microspheres for encapsulation, owing to the possibility of polymerizing the water-soluble monomers by gamma radiation, also in combination with water-soluble acrylic monomers.

10.9 Polyvinylpyrrolidone

A rigid brittle polymer soluble in water, with a T_g in the region of 160 °C, represented by the chemical formula:

$$\left[CH_2-CH(\text{—N-pyrrolidone}) \right]_n$$

It is used primarily as a binder for medical and cosmetic products.

11. Virial Coefficient

A coefficient that is sometimes added to an equation to account for the deviation from ideal behaviour. An example is the variation of the osmotic pressure (π) of a solution with temperature, that is,

$$\pi/c = RT[A_1 + A_2c + A_3c^2 + \ldots],$$

where c is the concentration of solute, R is the universal gas constant, T is absolute temperature, and A_1, A_2, A_3 and higher terms are the virial coefficients. The values of A_1, A_2 and A_3 are equal to zero for an ideal solution.

12. Viscoelastic Behaviour

A term used to describe the time-dependent mechanical properties of polymers. Whereas for elastic materials the Young's modulus and shear modulus are constant, the related values for polymers vary with the duration of the applied stress. The terms 'viscoelastic behaviour' and 'viscoelasticity' derive from combining the two words 'viscous' and 'elastic', which are related to the models normally used to describe the time-dependent behaviour of polymers. These models use a dashpot for the viscous deformations and a spring for the elastic response (see the subsection on modelling that follows).

The viscoelastic behaviour of polymers arises from the molecular structure (which allows the applied forces to be transmitted internally through rotations of segments of the polymer chains) and to the predominance of weak Van der Waals intermolecular forces. The rotation of polymer chains and their uncoiling in the direction of the acting forces take place through a sequence of events that require time to reach a new equilibrium position. This is a characteristic time that depends on the intrinsic nature of the polymer molecules, such as rigidity (i.e. energy required to rotate the chains) and the strength of the intermolecular forces. Restricting molecular rotations, with the introduction of cross-links or by the incorporation of bulky groups into the polymer chains, is a method used in polymer synthesis to attenuate the viscoelastic characteristics of polymers. The various movements of individual polymer chains in response to external forces are illustrated in the diagram.

Schematic illustration of the movements of polymer chains in the direction of the applied stress. Source: Mascia (1974).

12.1 Modelling the Viscoelastic Behaviour of Polymers

Such models consist of combinations of a spring and a dashpot according to the loading conditions. The extension of the spring (hence the strain) is directly proportional to the applied load (hence the stress) and, therefore, represents the 'elastic component'. The proportionality constant is Young's modulus, that is,

$$E = \sigma/\varepsilon.$$

A dashpot is a cylinder with a frictionless piston capable of moving in either direction by allowing the liquid to flow from one part to the other through an orifice in the piston. An applied force will cause the piston to move at a rate that is directly proportional to the viscosity of the liquid within the cylinder. When the force is removed, the piston stops instantaneously and the extension produced by the applied load does not recover at any time thereafter. Again, one can replace the force with stress and the deformation rate with strain rate, so that the characteristic constant for the dashpot is the viscosity, η. This can be used to represent the 'viscous component' of the viscoelastic behaviour, that is,

$$\eta = \frac{\sigma}{d\varepsilon/dt}.$$

The combination of a spring and a dashpot in series (known as the Maxwell model) gives isostress conditions for the response of the two elements and is used to model the time-dependent stress relaxation behaviour of polymers. (See Maxwell model.) The combination of the elastic and viscous elements in parallel, known as the Kelvin–Voigt model, is used to describe the time-dependent creep and recovery behaviour of polymers. (See Kelvin–Voigt model.) A hybrid combination, known as the standard linear solid, is used to model both the stress relaxation and creep behaviour of polymers. (See Standard linear solid.) The individual models mentioned here contain details of the physical interpretation and underlying equations.

13. Viscoelasticity

A theory used to analyse and characterize the time-dependent deformational behaviour of polymers, under different loading conditions. Static loading conditions are used to describe the time-dependent evolution of the strain resulting from the application of a constant stress, and the time-dependent relaxation (decay) of the stress under conditions in which the strain is kept constant. Cyclic loading conditions are used for the dynamic mechanical analysis of the viscoelastic behaviour of polymers. (See Viscoelastic behaviour.)

14. Viscometer

An apparatus used to measure the viscosity of liquids or low-viscosity emulsions, suspensions or pastes. The apparatus used for measurements made on high-viscosity liquids is generally known as a rheometer. The more widely used viscometers for polymer systems are the Ubbelohde viscometer for solutions, such as those for measurements of molecular weights, and the Brookfield viscometer for emulsions and pastes.

15. Viscosity

A property that denotes the resistance of a liquid to flow. Viscosity, η, is defined as the ratio of the stress, σ, acting on the liquid to the related strain rate, $d\varepsilon/dt$ (or velocity gradient, dV/dy), that is,

$$\eta = \frac{\sigma}{d\varepsilon/dt} = \frac{\sigma}{dV/dy}.$$

For shear flow situations, the viscosity is known as the shear viscosity, or simply viscosity, and the strain rate corresponds to the shear rate, while the velocity gradient corresponds to the gradient perpendicular to the flow direction, as shown.

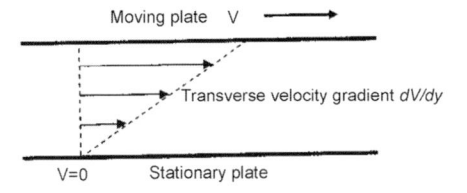

Velocity gradient in shear flow.

For elongational flow conditions, the viscosity is known as the elongational viscosity, or extensional viscosity, and the strain rate corresponds to the elongational rate, where the velocity gradient is experienced along the direction of flow, as shown. (See Elongational flow.)

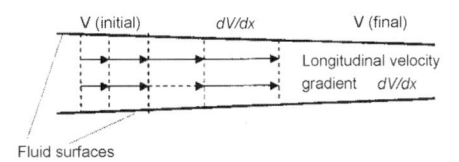

Velocity gradient in elongational flow.

For Newtonian liquids, there is an exact relationship between the elongational viscosity (η_e) and the shear viscosity (η_s), that is, $\eta_e = 3\eta_s$. For polymer melts, on the other hand, the elongational viscosity is many times greater than the shear viscosity, depending on the temperature and molecular weight of the polymer. Under dynamic or cyclic flow conditions, the non-Newtonian nature of the flow behaviour of polymers can be expressed in terms of a complex viscosity, η^*, comprising a real component, η', and an imaginary component, η'', that is, $\eta^* = \eta' - i\eta''$, where $i = \sqrt{-1}$ (the imaginary number). The imaginary term, η'', arises from the melt elasticity characteristics and is directly related to G', the real component of the shear modulus when the behaviour is expressed in complex solid-like characteristics, that is, $\eta'' = G'/\omega$, where ω is the angular frequency of the oscillatory motion.

The viscosity of liquids in general decreases exponentially with temperature and follows very closely the Arrhenius equation, as shown by the plots of the zero-shear viscosity against the reciprocal of the absolute temperature in the diagram. The data show that the viscosity of poly(vinyl butyral) (PVB) is much more sensitive to changes in temperature than is polyethylene, that is, PVB has a much higher activation energy. (See Arrhenius equation.)

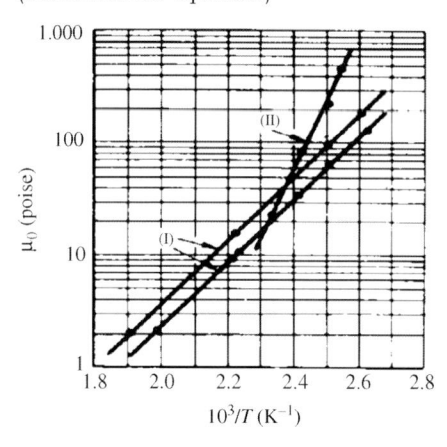

Change of viscosity of polymer melts with absolute temperature for two grades of polyethylene (I) and poly(vinyl butyral) (II). Source: McKelvey (1957).

16. Viscous Flow

Flow is a term generally used to denote irreversible deformations in polymers. Flow occurs by the displacement of entire molecules, which takes place through reptation movements, involving segments of the polymer chains. These molecular movements are feasible only when the polymer is in the melt state and, therefore, viscous flow has to be differentiated from what is sometimes known as 'plastic flow', which takes place at low temperatures through yielding and cold-drawing deformations. In the latter case, the molecules stretch out in the direction of the applied forces and assume an oriented configuration. (See Cold drawing and Yield failure.) Contrary to viscous flow, plastic flow deformations of polymers are recoverable owing to the instability of the orientation of the polymer chains, which will recoil back to their original random position in order to regain their stable (high-entropy) configuration when the temperature is increased above the glass transition temperature or melting point of the polymer.

17. Viscous State

A state of matter in which flow can take place. Within the context of the deformational behaviour of polymers, the viscous state can be considered to correspond to the state in which melt elasticity vanishes and the flow assumes a Newtonian behaviour.

18. Visible Light

(See Radiation.)

19. Vitrification

The state that a thermosetting resin system develops subsequent to 'gelation', when the system reaches the glassy state, that is, the T_g assumes a value greater than ambient temperature.

20. Void

These are extensively found in composites due to the difficulty of expelling completely the air entrapped between the fibres by the resin during manufacture. Voids are also found sometimes in thick sections of injection-moulded thermoplastics articles. (See Moulding defect.)

21. Voigt Model

(See Kelvin–Voigt model.)

22. Volume Resistivity

A property that describes the resistance of a dielectric material to the flow of an electric current through the bulk. From the general definition of resistivity, that is, $\rho = $ electrical stress/current density, a formula can be derived for the volume resistivity (ρ_V) from measurements made on the electrical circuit shown, that is,

$$\rho = (V/d)(A/I) = (A/d)R_V,$$

where A is the average area of the top and bottom electrodes and d is the thickness of the plaque used as specimen. Note that the dimensions of volume resistivity are ohm metre (Ω m). (See Surface resistivity.)

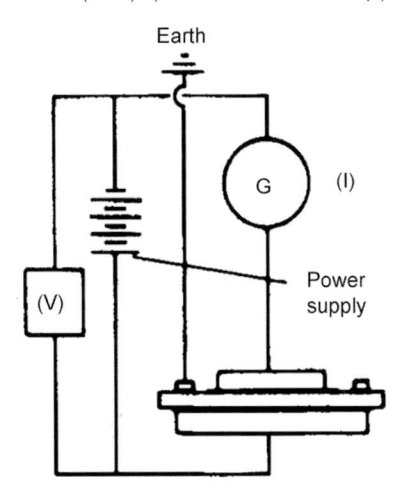

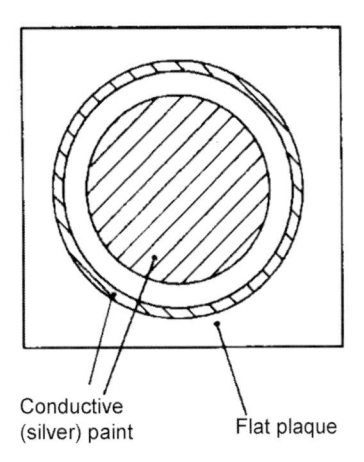

Top electrode and guard ring attached to the plaque used as specimen for measuring volume resistivity.

Circuit for measuring the current flowing between the top inner electrode and the bottom electrode (i.e. through the thickness of the plaque).

The geometric set-up for the electrodes (top and bottom) and the guard ring is shown.

23. Von Mises Criterion

(see Yield criteria.)

24. Vulcanizate

A rubber sample or article that has gone through a vulcanization process.

25. Vulcanization

A term used in the rubber industry to denote an operation that produces cross-links through the use of sulfur and sulfur-containing curatives (vulcanizing agents). The term derives from *Vulcan*, the Greek god of fire, to denote the use of heat to 'cure' the rubber. Originally, vulcanization was carried out by the addition of sulfur. It was later discovered that the cure rate could be increased considerably by incorporating zinc oxide in the formulation and increased further again by the addition of a fatty acid, such as stearic acid, and the use of an 'accelerator'. The fatty acid is absorbed on the surface of the zinc oxide particles and produces the desirable quantities of zinc ions, which activate the accelerator through the formation of complexes. The various type of cross-links between polymer chains obtained in the vulcanization of rubber with sulfur are shown.

A typical reaction scheme showing the participation of the accelerator in the formation of sulfur cross-links between elastomer chain is shown.

Reaction scheme for the formation of sulfur cross-links involving the action of the accelerator.

Sulfur cross-links in vulcanized rubber. Ac stands for accelerator molecule.

Diene rubbers can also be cross-linked with phenolic compounds, particularly diphenol type, such as resorcinol formaldehyde resins. These isomerize to quinones with the formation of allylic double bonds that can react readily with the unsaturation of the diene elastomer chains, as shown.

Reaction between resorcinol formaldehyde resin and the chain of a diene rubber.

W

1. Wall Slip

A term used to describe the non-zero velocity of polymer melts at the surface of the flow channel. (See Rheology and Mooney equation.)

2. Warping

(See Distortion and warping.)

3. Water-Borne Coating

A coating deposited from a micro-suspension of a polymer in water. The polymer is specially designed through synthesis or modification of existing polymers to produce ammonium salts, which provides the desired level of 'affinity' for water. An ionic surfactant is used to stabilize the dispersion against segregation and agglomeration. After removing the water by drying or through an electrolytic separation process, the coating is 'stoved' or cured at high temperatures to induce cross-linking reactions with other components. (See Electrolytic deposition.) This operation also drives off the ammonia from the carboxylate salt. In other cases, such as epoxides, the required ammonium salts are produced from amine-extended epoxy resins with the addition of a weak monomeric carboxylic acid, such as lactic acid. For electrolytic coating deposition, the choice of type of ion bound to the polymer particle depends on whether a cathodic or anodic deposition process in used. Water-borne polyurethane systems are widely used in view of the excellent mechanical properties, such as abrasion resistance, provided by urethane blocks within the final polymer network. Both one-pack and two-pack systems are used. In two-pack systems, one component contains a blocked isocyanate, and the other contains the polyol, so that, after drying out the water, reactions can take place between the polyol of one pack and the unblocked isocyanate groups in the other pack. (See Urethane polymer and resin.)

4. Weak Boundary Layer

An interlayer between an adhesive and the adherend, which prevents a joint from reaching the maximum achievable bond strength. Weak boundary layers are usually formed from contaminants that are either present in the atmosphere (typically water) or exude out from the bulk of the adherend. In the case of polymer adherends, the formation of a weak boundary layer is often caused by the migration of external lubricants to the surface during processing. In moulded products, the mould release agent used to ease the ejection of a moulded part can give rise to the formation of a weak boundary layer if it is not soluble in the adhesive.

5. Weathering

A term used to describe the deterioration of the properties of materials or the performance of products brought about by adverse climatic conditions. The damaging atmospheric agents in the weather include radiation (mainly the UV range), temperature, oxygen, moisture, contaminants and rainfall. The deterioration of materials by exposure to weather conditions is due almost entirely to chemical reactions, particularly oxidation through reactions of atomic oxygen with tertiary $H-C$ bonds, $CH=CH$ bonds, and the hydrolysis of groups within the backbone chain of a polymer, such as esters, amides, imides and urethanes. Oxidation reactions are accelerated by the action of UV light (wavelength in the range 280–400 nm) and metallic contaminants present in other components of a formulation, such as fillers. The hydrolytic degradation reactions, on the other hand, are highly sensitive to the presence of acidic or basic contaminants in the environment. (See Hydrolysis and Degradation.) The extent of degradation due to weathering varies widely in different parts of the globe and also between summer and winter months. The cause of the seasonal variations is rooted in the difference in the intensity of light falling on Earth, as illustrated in terms of the variation of the 'irradiance' (radiation power) as a function of the wavelength within the UV region in the diagram.

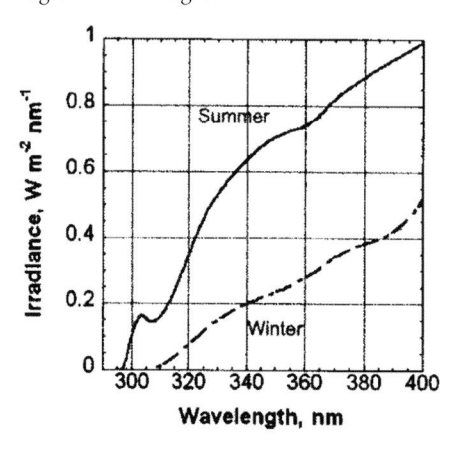

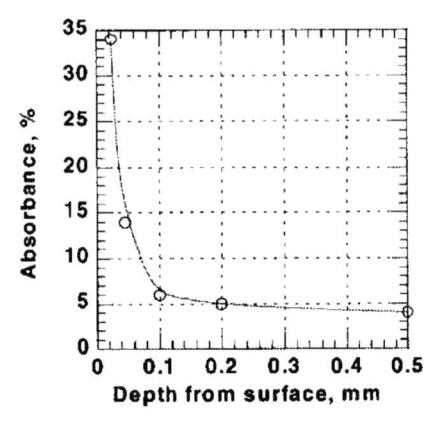

Carbonyl absorbance of samples of a polyethylene pipe, taken at different depths from the surface, after one year's exposure to natural weather. Source: Allen et al. (1997).

The degradation reactions cause severe embrittlement of the outer surface layers, with the formation of small craters and cracks. This is illustrated for a sample of polypropylene examined at different times of exposure to tropical weathering conditions. In this case a thin skin of polymer, possibly a highly oriented layer, was found to be highly susceptible to degradation.

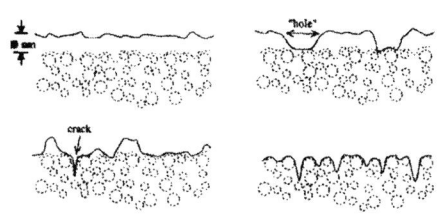

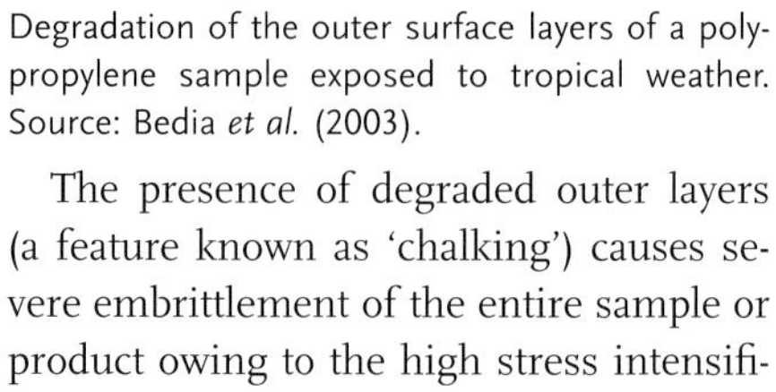

Typical solar spectrum for summer and winter seasons. Source: Brennan and Fedor (1988).

Degradation of the outer surface layers of a polypropylene sample exposed to tropical weather. Source: Bedia et al. (2003).

Weathering affects the outer surface of a product and gradually progresses to a limiting depth depending on the material and environmental conditions. It takes place through a combination of chemical reactions and erosion or loss of products of the degradation process as volatiles or leachable compounds. The diagram shows a plot of the absorbance of carbonyl groups, measured on a polyethylene pipe, as a function of the distance from the surface (depth) after one year's exposure to natural weather.

The presence of degraded outer layers (a feature known as 'chalking') causes severe embrittlement of the entire sample or product owing to the high stress intensification caused by the crevices and cracks. This is illustrated in the diagram in the form of plots of the flexural strength measured on weathered ABS sheets after removing layers of different thickness from the exposed surface, relative to the flexural strength of the original sample.

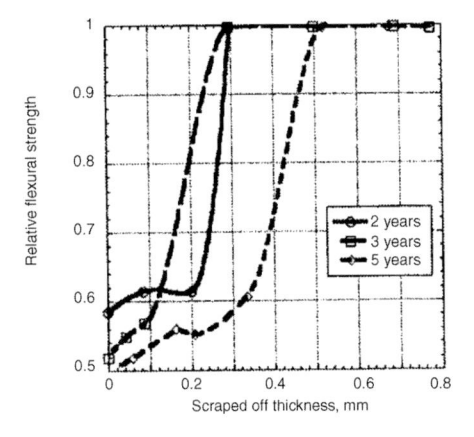

Change in flexural strength of ABS sheets, weathered for two, three and five years, as a function of the thickness of the layer removed from the surface. Source: Watanabe et al. (1981).

Natural weathering evaluations are prone to large variations in climatic conditions and may take too long for the evaluation of new formulations or for comparison of different systems. For this reason, laboratory tests frequently use sources that emit light that fairly closely matches the solar spectrum. In some systems there is also the possibility of introducing other climatic variants, such as spraying water or salt solution, as well as being able to operate at higher temperatures to accelerate the degradation reactions. The main sources of UV light for laboratory weathering tests are: carbon arc, xenon arc, fluorescent UV lamps, metal halides and mercury lamps. Filters are frequently used to fine-tune the wavelength of the radiation used for the tests relative to some local natural light conditions.

6. **Weight-Average Molecular Weight**

The average molecular weight calculated on the basis of the weight fraction of polymer chains of specific size. (See Molecular weight.)

7. **Weissenberg Rheogoniometer**

This is also known as the cone-and-plate rheometer. (See Cone-and-plate rheometer.)

8. **Weld Line**

The line that identifies the location of two melt fronts that have come into contact to form a weld. Apart from welding processes, weld lines are observed in extrusion blow-moulded containers at the 'pinch' line, and in injection-moulded products at those points where two melt fronts meet within the cavities of the mould, as shown.

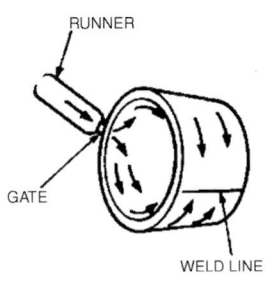

Typical weld line in injection-moulded products. Source: Unidentified original source.

Weld lines are also found in extruded tubular products at the points within the die where the melt fronts are reunited after separation by the 'spider legs' before reaching the die lips. The typical geometry of the spider legs used for fixing the inner core of the die to the outer body is shown. (See Extrusion die.)

Cross section of spider leg

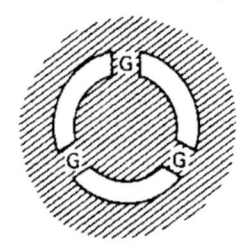

Spider legs joining the central core of a tubular die to the main body. Source: Rosato (1998).

9. **Welding**

A manufacturing or fabrication process by which two or more, usually large, components are brought into contact and joined together by promoting molecular diffusion across the interface through a rapid localized increase in the temperature. This implies, therefore, that only thermoplastic products or components can be welded. (See Thermofusion process.) The main welding processes used for plastics are briefly described.

9.1 Hot Gas Welding

A technique used for large-area fabrications adopted from the oxy-acetylene welding of metals, using hot gas (usually heated air or nitrogen) to raise the temperature of the welding rod and the V-groove formed by adjacent edges of the components of the jointed parts. The diagram shows an example of how the welding rod can be guided into the welding groove through a nozzle.

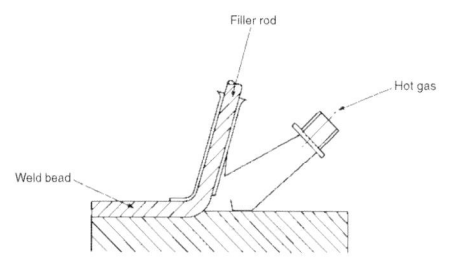

Schematic illustration of hot gas welding of plastics. Source: Bernam (1988).

The welding rod is chemically similar to the parts used for fabrication, but with a slightly lower viscosity in order to ensure that the groove is filled completely.

9.2 Hot Plate Welding

A simple technique that uses a hot metal plate to heat the parts to be welded, then forcing the two components against each other for a sufficient length of time to allow the weld to be formed. This technique is also known as butt fusion, as illustrated.

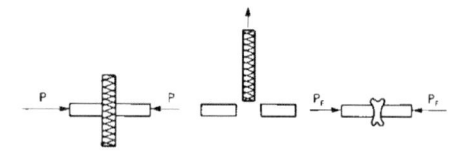

Illustration of the principle of hot plate welding.

The principle of hot plate welding has been widely exploited for the joining of pipes and has been elaborated through the use of sockets to enhance the performance of the joint, a technique known as electrofusion. The socket is often heated electrically by inserting a coiled wire element into the actual socket so that it can be connected to a power source to produce the required heat for melting the inner surface of the sleeve, as well as the outer surface of the pipe fronts, as illustrated.

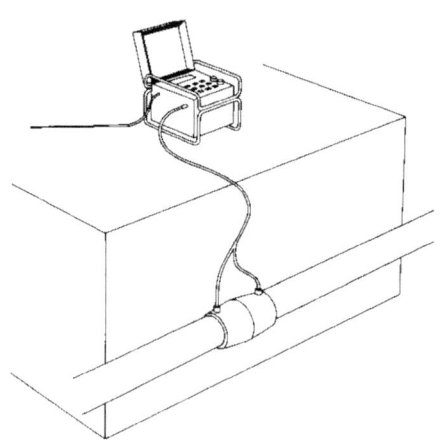

Typical set-up for electrofusion welding of plastic pipes. Source: Stafford (1988).

9.3 Friction Welding

The principle consists of heating the joining fronts through mechanical friction produced by vibrating or spinning one of the parts against the other. The principle is illustrated.

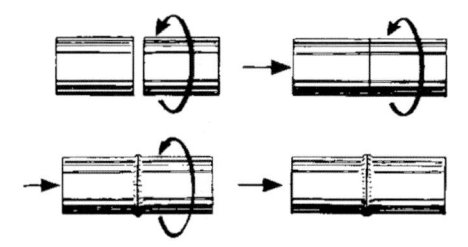

Principle of friction or spin welding. Source: Dunkerton (1988).

9.4 High-Frequency (RF) Welding

The principle is based on the ability of polymers with a high dielectric loss factor (tan δ) to generate heat under the influence of an AC field produced by a generator with a 2–5 kW output. The electrical energy loss W (W/cm³) that is converted to thermal energy for heating the welding area can be calculated from

$$W = 0.555 \times 10^{-12} \times f \times \tan \delta \times F^2,$$

where f is the frequency (standard value of 27 MHz) and F is the electric field or electrical stress (V/cm). This technique is widely used for the welding of flexible PVC sheets owing to the very high loss factor and the low temperature required to induce the required molecular diffusion through the welding surfaces. The electrodes used to apply the RF field to the welding area are also used to exert the pressure for the thermofusion.

9.5 Ultrasonic Welding

A technique that uses ultrasonic pulsations (frequencies between 20 and 50 kHz) generated by a piezoelectric crystal, amplified and transmitted via a metal sonotrone known as the 'horn', to the welding area of the polymer component supported on a rigid metal base. The set-up is illustrated.

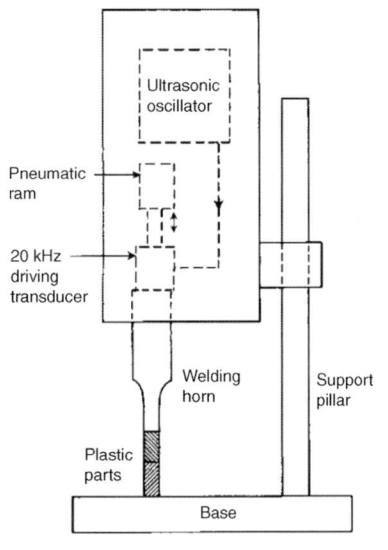

Set-up of an ultrasonic welding device. Source: McCrum et al. (1988).

Welding takes place as a result of the transformation of the mechanical energy delivered by the horn to thermal energy, which heats up the polymer in the welding area. This technique takes advantage of the low thermal conductivity of polymers, which ensures that only the contact regions reach high temperatures. For polymers with a high damping factor (tan δ value), the horn has to be close to the welding area (near-field mode). For glassy polymers with low attenuation or damping characteristics at ambient temperatures, the horn can be applied at fairly high distances from the welding area (far-field mode), which is particularly attractive for welding hollow components, as shown.

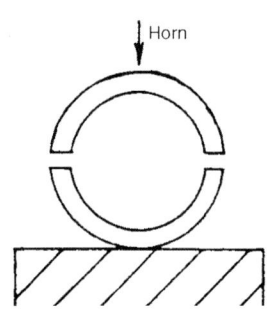

Example of 'far-field' ultrasonic welding. Source: Mascia (1989).

10. Wide-Angle X-Ray Diffraction (Wide-Angle X-Ray Scattering, WAXS)

A technique used to measure the degree of crystallinity and orientation in polymers. The measurements are based on the principle that X-rays are diffracted from a system of parallel equidistant lattice planes of crystals if the Bragg equation is satisfied, that is,

$$n + \lambda = 2d \sin \vartheta,$$

where n is an integer that defines the order of diffraction, λ is the wavelength of the incident X-rays, τ is the angle of diffraction, and d is the spacing between lattice planes, as shown in diagram.

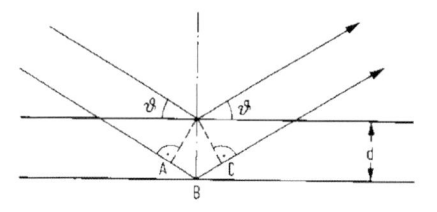

Diffraction of X-rays at a set of lattice planes in a crystalline substance.

The measured diffracted X-ray intensity F from the sample analysed is plotted as a function of 2τ. An amorphous material exhibits a broad spectrum of diffraction, whereas a crystalline substance is identified by sharp diffraction peaks. Since crystalline polymers consist of mixtures of amorphous and crystalline domains, the resulting spectrum consists of a superimposed combination of both types of spectra, as shown.

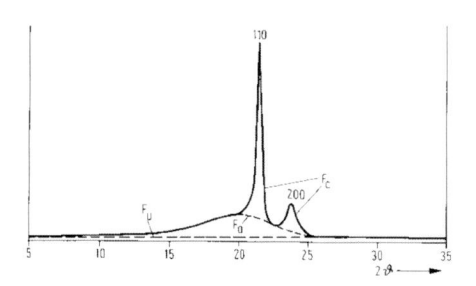

X-ray diffraction of a low-density polyethylene sample. Source: Unidentified original source.

From measurements of the overall diffracted intensity of the two components (F_c and F_a), through integration over appropriate intervals, it is possible to calculate the relative proportions of crystalline (X_c) and amorphous domains (X_a), that is, $X_c = aF_c$ and $X_a = 1 - F_c = bF_a$, where a and b are constants determined by suitable calibration measurements. With the goniometric techniques normally used for probing the crystallinity in polymers, it is possible to identify also whether the constituent domains are oriented, and to calculate the values for the respective orientation functions for the crystalline and amorphous regions. The photographs below show the typical X-ray patterns for an isotropic crystalline polymer and for the same polymer after being stretched longitudinally to introduce orientation.

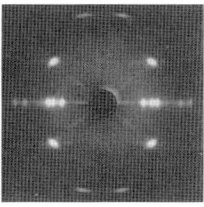

X-ray flat chamber photographs on polypropylene: (left) isotropic sample; (right) stretched sample at 12 : 1 draw ratio. Source: Kampf (1986).

11. WLF Equation

The abbreviation stands for Williams–Landel–Ferry, the three authors who derived the equation. They used an empirical approach to obtain values of the shift factor a_T for the production of master curves from isochronous creep or stress relaxation experiments carried out at various temperatures. The procedure is based on the time–temperature equivalence of the viscoelastic behaviour of polymers. The WLF equation is written as

$$\log a_T = [-C_1(T-T_g)]/(C_2 + T-T_g),$$

where C_1 and C_2 are constants for the material and loading conditions (i.e. whether creep or stress relaxation). (See Time–temperature superposition.)

12. Wood Flour

A filler obtained by the comminution of wood. Mainly used as a reinforcing filler for phenolic moulding powders.

13. Work of Adhesion

(See Adhesive wetting.)

X

1. X-Ray Diffraction (XRD)

(See Wide-angle X-ray diffraction and Small-angle X-ray scattering.)

2. X-Ray Photoelectron Microscopy (X-Ray Photoelectron Spectroscopy, XPS)

An analytical technique used for surface analysis, based on the principle that an atom excited by photons with energy level $h\nu$ from monochromatic X-rays will knock an electron out of the L shell. This (photo) electron possesses a kinetic energy equal to $h\nu - E_L$, where E_L is the characteristic energy of the L shell of a specific atom. XPS analysis makes it possible to determine the chemical nature of the atom from experimental measurements, from which the energy E_L can be calculated from the overall kinetic energy.

3. X-Rays

Rays produced by the acceleration of electrons emitted from a heated cathode in a high vacuum using a high voltage (3–100 kV) and allowing them to impinge on an anode (usually water-cooled). The kinetic energy of the electrons is mainly converted into heat, but part of it is emitted as X-rays. These X-rays possess a characteristic energy distribution (spectrum) that is dependent on the excitation voltage used and on the nature of the anodic material. (See Radiation.)

4. Xenon Arc Lamp

A source of light that simulates natural daylight through the use of appropriate filters. Xenon arc lamps are widely used for accelerated weathering tests.

Y

1. Y Calibration Factor

A geometrical parameter that takes into account the geometry of the specimen, the crack length, a, and the width of the specimen, W, in the calculation of the stress intensity factor (K), that is, $K = Y\sigma\sqrt{a}$, where σ is the applied stress. (See Fracture mechanics.)

2. Yellowness Index

An empirical parameter used to describe the extent of degradation of a polymer that takes place through weathering and is manifested by the development of a yellow tint. It is determined using UV spectroscopy by measuring the absorption in the wavelength range 570–590 nm. Some standard tests compare the absorbance of the polymer to that of magnesium oxide, as a reference point.

3. Yield Criteria

Criteria to determine the conditions that bring about yield failure in multiaxial stress situations. (See Yield failure.) The most commonly used criteria are the Tresca criterion and the Von Mises criterion. Although these were originally used for metals, they have been found to apply equally well to rigid polymers. Small modifications have been suggested to improve the accuracy of the predictions of these criteria for polymers.

3.1 Tresca Criterion

According to this criterion, yielding takes place when the maximum shear stress (τ_{max}) reaches a critical value (τ_{crit}), which is determined by the nature of the material. (See Lüder lines.) The Tresca criterion is usually written as

$$\tau_{max} = (\sigma_{max} - \sigma_{min})/2 = \tau_{crit},$$

where σ_{max} and σ_{min} are the maximum and minimum principal stresses acting along the main axes. Since the yield strength (σ_Y) of a material is measured under uniaxial stress conditions, that is, tension or compression, the application of this criterion gives $\tau_{max} = \sigma_{max}/2 = \tau_{crit}$ because σ_{min} (stress in other principal directions) is zero, while σ_{max} corresponds to the yield strength of the material, σ_Y. For situations where the yield strengths in tension and compression are equal, as in the case of most metals, the Tresca criterion can be represented graphically with the identification of contours for the magnitude of the principal stresses that have to be reached for yielding failures. The diagram shows the graphical description of the yielding conditions by the Tresca criterion for plane stress conditions, that is, when one of the principal stresses is zero (as is often the case when the thickness of the structure or product is very small).

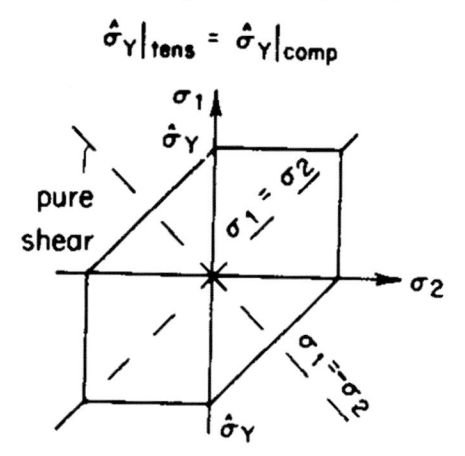

Tresca criterion for plane stress conditions for materials where the yield strength in tension is equal to the yield strength in compression. Source: Adapted from Mascia (1989).

For materials, such as polymers, where the yield strength in compression is greater than the yield strength in tension, the graphical representation is modified in the manner shown in the diagram.

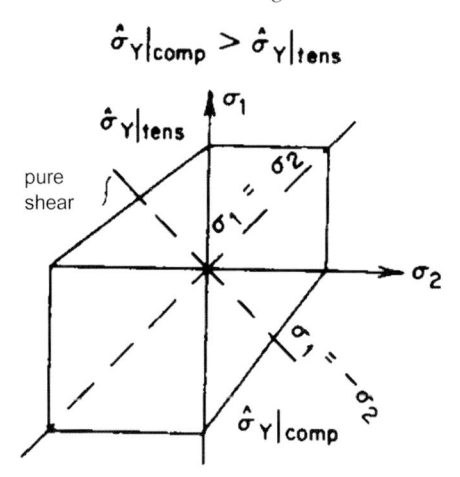

Tresca criterion for plane stress conditions for polymers, where the yield strength in tension is lower than the yield strength in compression. Source: Mascia (1989).

3.2 Von Mises Criterion

This criterion states that yielding conditions are reached when the maximum distortional strain energy reaches a critical value determined by the nature of the material. For materials where the values of the yield strength in tension and compression are equal, the Von Mises criterion can be written in terms of the corresponding principal stresses as

$$(\sigma_1-\sigma_2)^2 + (\sigma_1-\sigma_3)^2 + (\sigma_2-\sigma_3)^2 = 2\sigma_Y^2,$$

where σ_1, σ_2 and σ_3 are the three principal stresses. For plane stress conditions, that is, when σ_3 is equal to zero, the Von Mises criterion becomes

$$\sigma_1^2 + \sigma_2^2 - \sigma_1\sigma_2 = \sigma_Y^2.$$

The graphical representation of this equation is an inclined ellipse at 45° with respect to the two axes of principal stresses σ_1 and σ_2.

The diagram shows a graphical comparison of the Tresca and Von Mises criteria, which indicates that they predict very similar conditions for yielding, with a maximum discrepancy around 14% occurring in the tension and compression quadrants.

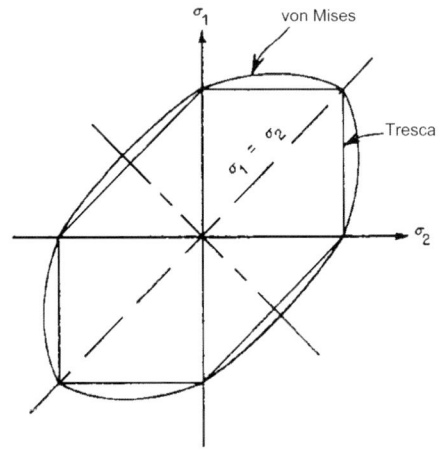

Comparison of Tresca and Von Mises criteria for materials where the yield strength in tension is equal to the yield strength in compression.

4. Yield Failure

A failure arising from large unrecovered deformations when the stress reaches a critical value, known as the yield strength. This is exemplified in the diagram, where the yield failure conditions are identified as the yield point.

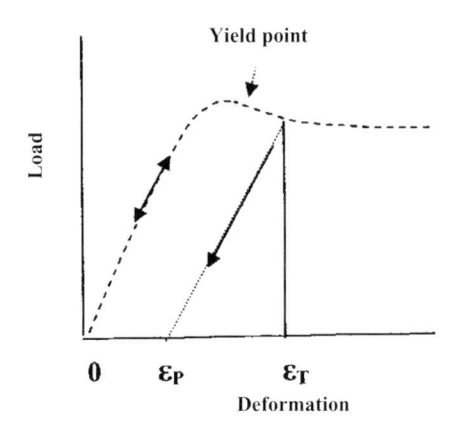

Graphical representation of conditions leading to yield failure and identification of the yield point.

When the applied load is lower than the value required for reaching the yield point, the deformations are reversible (i.e. the strain recovers when the stress is removed). After exceeding the yield point, on the other hand, only a small part of the deformation recovers. The largest part of the deformation, corresponding to the plastic strain, ε_P, remains after the load is removed. This is identified in the diagram by the arrow after the yield point. Yield failures take place through the sliding of planes of crystal lamellae with the uncoiling of random tie molecules, followed by the alignment of the crystals in the direction of the applied stress, as depicted in the diagram.

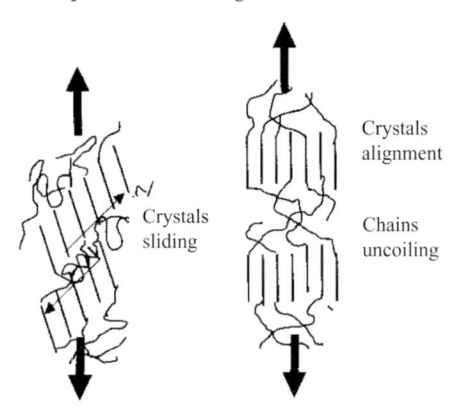

Yield failure mechanism for crystalline polymers.

For the case of amorphous (glassy) polymers, yielding also takes place through shear deformations along the planes where the shear stresses are highest, as exemplified by the formation of Lüder lines shown in the diagram for a polystyrene sample in a test carried out under plane stress conditions.

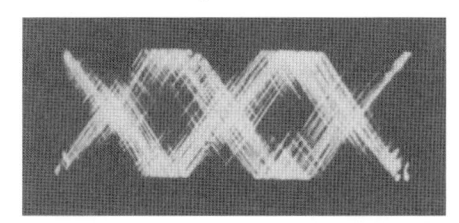

Lüder lines as evidence for the shear deformations in yield failure of glassy polymers. Source: Bucknall (1977).

5. Yield Point

The point that identifies the failure by yielding in load–deformation curves, normally obtained by tensile tests. In some cases the yielding conditions cannot be easily identified as a maximum or a clear discontinuity on the load–deformation curve. A technique, known as the Considere construction, is sometimes used to assign a precise values of the stress and the strain on a stress–strain curve at which yielding occurs. This is rarely done in the case of polymers.

6. Yield Strength

The value of the stress at which yielding takes place, calculated as the load/original cross-sectional area of the specimen, normally from tensile tests. As there can be a substantial amount of deformation before the yield point is reached, the actual value of the stress, based on the actual cross-sectional area of the specimen, can be appreciably higher. Although the value of the true stress can be calculated from the value of the nominal stress, this is not normally done.

7. Young's Modulus

A material property defined as the proportionality coefficient between an applied stress and resulting strain in either tension or compression, for conditions below those that cause failure either by yielding or brittle fracture, that is,

Young's modulus = stress/strain.

The Young's modulus of polymers decreases with the duration of the applied load, owing to their viscoelastic behaviour. Even for a typical engineering polymer, such as poly(ether sulfone) (PES), the creep modulus decreases from about 3 GPa for very short durations of applied load (say, about one minute) to less than 1 GPa when the load is maintained for a very long time (say, about two years).

The value of the Young's modulus allows a design engineer to estimate the deflection resulting from an applied load. For instance, in the case of a cantilever beam, the deflection Δ can be calculated using the formula

$$\Delta = (P/E)[L^3/6bd^3],$$

where P is the applied load, E is the Young's modulus, L is the length of the beam, b is the width and d is the thickness. With this formula, the designer can choose the most appropriate material, knowing the modulus values. To contain the amount of deflection resulting from the applied load, the designer can either choose a material with higher modulus or increase the width or thickness of the beam. The latter is usually the preferred choice, because of the cubic relationship instead of a linear one with respect to width or modulus. The use of typical design equations for rigid polymers implies that one cannot assign a specific value to these materials for the Young's modulus because the value would depend on the loading history, which has to be known in order to use an appropriate value to perform the calculations. (See Modulus.)

Z

1. Z-Blade Mixer

A typical design for mixing doughs and pastes. (See Mixer.)

2. Zero-Shear Viscosity

The value of melt viscosity of polymers within the Newtonian plateau at low shear rates, that is, shear rates tending to zero. (See Non-Newtonian behaviour.)

3. Zeta Potential (or ζ Potential)

A parameter used to characterize the level of charge on a surface, particularly objects with a high surface-to-volume ratio, such as particles and fibres. The ζ potential is a useful parameter in flocculation, as it makes it possible to determine the optimum pH of the suspension medium to achieve the highest stability. These conditions are determined by the chemical nature of the surface of the particle, which can vary considerably from one filler to another, as illustrated.

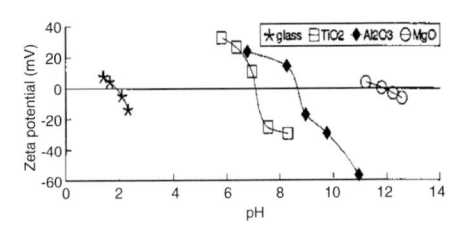

Zeta potential of water suspension of various fillers as a function of pH of the suspending medium. Source: Schroder (1991).

From an examination of the isoelectric point (at which the ζ potential is equal to zero) of the different fillers, one can deduce that the glass surface is strongly acidic, while that of MgO is very basic. The surface of TiO_2, on the other hand, is practically neutral. The ζ potential of the filler, therefore, determines the nature of chemical species that can be adsorbed on the surface, which is particularly useful for the surface treatment of fillers as a way of improving their dispersion characteristics in polymer compounds and preventing agglomerations in liquid suspensions.

4. Ziegler Catalyst

A coordination-type catalyst (known also as a Ziegler–Natta catalyst) used for the polymerization of linear polyethylene (LLDPE and HDPE) and for stereoregular polymers, such as isotactic PP. The most common type is a coordination complex produced from the interaction of aluminium trialkyl compounds and titanium tetrachloride to form a complex of the type:

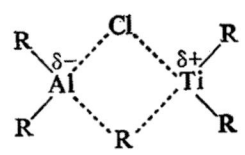

Polymerization of an olefin takes place through the break-up of the coordination complex by the polarization effect of the π-electrons in the double bond of the monomer, as shown.

As a consequence, this causes the splitting of the double bond, with the formation of a positive charge:

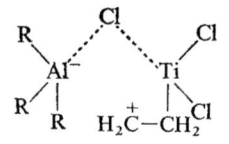

$$\begin{array}{c} \text{R} \quad\quad \overset{Cl}{\underset{\cdot\cdot\cdot}{}}\quad\quad Cl \\ \diagdown\!\text{Al}^{-}\!\cdots\cdots\text{Ti}\diagup \\ \text{R}\diagup\phantom{\text{Al}}\big|\phantom{\cdots\text{Ti}}\big|\diagdown \\ \text{R}\quad \overset{+}{\text{H}_2\text{C}}\!-\!\text{CH}_2\ Cl \end{array}$$

This is then followed by the lengthening of the chain (propagation reaction) through the addition of successive monomer units at the Ti side of the complex. The isotactic configuration of the side groups in the polymerization of an α-olefin arises from the spatial restrictions imposed by the catalyst on the addition of successive monomer units at the coordination complex side of the growing chain.

5. Zinc Oxide

An additive used in combination with a fatty acid, usually stearic acid, as an accelerator for the sulfur vulcanization of rubber. (See Accelerator.) It has a high level of purity (99.99% ZnO) and is also used in combination with magnesium oxide to cross-link chloroprene rubbers. It has a density of 5.6 g/cm^3 and particle size around 0.1–0.4 µm, with a specific surface area of about 10–20 m^2/g. The grade used as an accelerator for the vulcanization of rubber is often coated with stearic acid or propionic acid to assist its dispersion.

6. Zisman Plot

A plot of the contact angle ($\cos\theta$) formed by a series of liquids on a solid surface against the surface energy of each liquid. This produces an approximately straight line that can be extrapolated to $\cos\theta = 1$ (i.e. $\theta = 0$), which corresponds to the conditions for spontaneous wetting of the surface of the solid. The value of the surface energy for which $\theta = 0$ is known as the 'critical surface energy' or 'critical wetting tension'. The plot is based on an empirical equation put forward by Zisman in the form

$$\cos\theta = 1 + b(\gamma_C - \gamma_{LV}),$$

where γ_C is the critical surface energy and γ_{LV} is the interfacial energy between the liquid and air saturated with liquid vapour. (See Surface energy.)

References

Consulted works with permission to reproduce copyright images

Allen, N.S., Palmer, S.J., and Gardette, J.L. (1997) *Polym. Degrad. Stab.*, **56** (3), 265.

Ashton, H.C. (2010) in *Functional Fillers for Plastics* (ed. M. Xanthos), Wiley-VCH, Ch. 17.

Baird, D.G. and Collias, D.I. (1998) *Polymer Processing: Principles and Design*, John Wiley & Sons.

Bedia, E.L., Pablicawan, M.A., Bermas, C.V., Tosaka, S.T., and Kohjiya, M.(2003) *J. Appl. Polym. Sci.*, **87** (6), 931.

Bell, M.S., Lacerda, R.G., Teo, K.B.K., and Milne, W.I. (2006) in *Carbon: The Future Material for Advanced Technology Applications* (eds G. Messina and S. Santangelo), Topics in Applied Physics, vol. 100, Springer, pp. 77–93.

Bernam, T.R. (1988) Science & Business Media in *Joining Plastics in Production* (ed. M.N. Watson), The Welding Institute.

Bieleman, J. (1996) in *Resins for Coatings: Chemistry, Properties, and Applications* (eds D. Stoye and W. Freitag), Hanser, Ch. 10.

Birley, A.W., Haworth, B., and Batchelor, J. (1991) *Physics of Plastics*, Hanser.

Brennan, P. and Fedor, G. (1988) *Kunststoffe*, **78**, 323.

Bucknall, C.B. (1977) *Toughened Plastics*, Applied Science.

Chua, K., Leong, K.F., and Lim, C.S. (2003) *Rapid Prototyping: Principles and Applications*, World Scientific.

Demirer, H. (2000) PhD thesis, Loughborough University.

Denkinger, P. (1996) in *Resins for Coatings: Chemistry, Properties, and Applications* (eds D. Stoye and W. Freitag), Hanser, Ch. 4.

Denkinger, P. (1996a) in *Resins for Coatings: Chemistry, Properties, and Applications* (eds D. Stoye and W. Freitag), Hanser, Ch. 8.

Domininghaus, H. (1992) *Plastics for Engineers: Materials, Properties, Applications*, Hanser.

Dunkerton, S.B. (1988) in *Joining Plastics in Production* (ed. M.N. Watson), The Welding Institute.

Echte, A., Haaf, F., and Hambrecht, J. (1981) *Angew. Chem., Int. Edn.*, **20**, 344.

Ehrenstein, G.W. (2001) *Polymeric Materials: Structure – Properties – Applications*, Hanser.

Eirich, F.R. (ed.) (1978) *Science and Technology of Rubber*, Academic Press.

Elias, H.G. (1993) *An Introduction to Plastics*, Wiley-VCH.

Flory, P.J. (1953) *Principles of Polymer Chemistry*, Cornell University Press.

Gachter, R. and Muller, H. (1990) *Plastics Additives Handbook*, 3rd edn, Hanser.

Goodman, S.H. (1998) *Handbook of Thermoset Plastics*, Noyes Publications.

Herpels, J. (1989) MSc Thesis, Loughborough University.

Hoffmann, M., Krömer, H., and Kuhn, R. (1977) *Analysis of Polymers II*, Georg Thieme.

ICI (1965) *Landmarks of the Plastics Industry*, John Wiley & Sons/ICI.

Jansen, J. (1990) in *Plastics Additives Handbook* (eds R. Gachter and H. Muller), 3rd edn, Hanser, Ch. 18.

Kampf, G. (1986) *Characterization of Plastics by Physical Methods: Experimental Techniques and Practical Applications*, Hanser.

Koppelmann, J. (1979) *Progr. Colloid Polym. Sci.*, **66**, 235.

Kraus, G. (1978) in *Science and Technology of Rubber* (ed. F.R. Eirich), Academic Press, Ch. 8.

Lavorgna, M. (2009) PhD thesis, Loughborough University.

Lee, N.C. (1998) *Blow Molding Design Guide*, Hanser Gardner.

Lee, S.-T. (2009) in *Polymeric Foams: Technology and Developments in Regulation, Process, and Products* (eds S.-T. Lee and D. Scholtz), CRC Press, Ch. 1.

Lee, S.-T. and Scholtz, D. (eds) (2009) *Polymeric Foams: Technology and Developments in Regulation, Process, and Products*, CRC Press.

Maddock, B.H. (1959) *Soc. Plastics Engrs J.*, **15**, 38.

Margolis, J.M. (ed.) (1986) *Decorating Plastics*, Hanser.

Mascia, L. (1974) *The Role of Additives in Plastics*, Edward Arnold.

Mascia, L. (1978) *Polymer*, **19**, 325.

Mascia, L. (1989) *Thermoplastics: Materials Engineering*, 2nd edn, Elsevier Applied Science.

Mascia, L. and Margetts, G. (1987) *J. Macromol. Sci. – Phys. B*, **26**, 237.

Mascia, L., Wooldridge, P.G., and Stockwell, M.J. (1989) *J. Mater. Sci.*, **24**, 2775.

Matthews, G. (1982) *Polymer Mixing Technology*, Applied Science.

McCrum, N.G., Buckley, C.P., and Bucknall, C.B. (1988) *Principles of Polymer Engineering*, Oxford University Press.

McKelvey, J. (1957) *Polymer Processing*, John Wiley & Sons.

Messina, G. and Santangelo, S. (eds) (2006) *Carbon: The Future Material for Advanced Technology Applications*, Topics in Applied Physics, vol. 100, Springer Science & Business Media.

Michaeli, W. (1992) *Extrusion Dies for Plastics and Rubber: Design and Engineering Computations*, Hanser.

Michaeli, W., Cramer, A., and Florez, L. (2009) in *Polymeric Foams: Technology and Developments in Regulation, Process, and Products* (eds S.-T. Lee and D. Scholtz), CRC Press, Ch. 4.

Minoshima, W., White, J.L., and Spruiell, J.E. (1980) *J. Appl. Polym. Sci.*, **25**, 287.

Monk, J.E. (1997) *Thermoset Plastics: Moulding Materials and Processes*, Longman.

Muccio, E.A. (1994) *Plastics Processing Technology*, ASM International.

Murphy, B.M. (1969) *Chem. Ind.*, **8**, 290.

Novac, K.F. (1978) in *Science and Technology of Rubber* (ed. F.R. Eirich), Academic Press, Ch. 14.

Osswald, T.A. (1998) *Polymer Processing Fundamentals*, Hanser.

Plati, E. and William, J.C. (1973) *J. Mater. Sci.*, **8**, 949.

Pocius, A.A. (2002) *Adhesion and Adhesives Technology*, Hanser.

Prezzi, L. (2003) PhD thesis, Loughborough University.

Pye, R.G.W. (1989) *Injection Mould Design*, Longman.

Riew, C.K. (ed.) (1989) *Rubber-Toughened Plastics*, American Chemical Society.

Rosato, D.V. (1998) *Extruding Plastics: A Practical Processing Handbook*, Chapman & Hall.

Ross, G. and Birley, A.W. (1974) *Thermoplastics: Properties and Design*, John Wiley & Sons/ICI.

Rudnik, E. (2008) *Compostable Polymer Materials*, Elsevier.

Sato, Y. (1969) *Rep. Progr. Polym. Phys. Japan*, **9**, 369.

Schmitt, B.J. (1979) *Angew. Chem.*, **91**, 286.

Scholtz, D. (2009) in *Polymeric Foams: Technology and Developments in Regulation, Process, and Products* (eds S.-T. Lee and D. Scholtz), CRC Press, Ch. 2.

Schroder, J. (1991) *Progr. Org. Coat.*, **19**, 227.

Stafford, T.G. (1988) in *Joining Plastics in Production* (ed. M.N. Watson), The Welding Institute.

Stoye, D. and Freitag, W. (eds) (1996) *Resins for Coatings: Chemistry, Properties, and Applications*, Hanser.

Tadmor, Z. (1974) *J. Appl. Polym. Sci.*, **18**, 1753.

Teegarden, D. (2004) *Polymer Chemistry: Introduction to an Indispensable Science*, NSTA.

Tinker, A.J. and Jones, K.P. (1998) *Blends of Natural Rubber*, Chapman & Hall.

Todd, D.B. (2010) in *Functional Fillers for Plastics* (ed. M. Xanthos), Wiley-VCH, Ch. 3.

Ugbolue, S.C.O. (2009) *Polyolefin Fibres: Industrial and Medical Applications*, Woodhead.

Ver Strate, G. (1978) in *Science and Technology of Rubber* (ed. F.R. Eirich), Academic Press, Ch. 3.

Wake, W.C. (1992) *Adhesion and the Formulation of Adhesives*, Applied Science.

Watanabe, Y., Kitajima, F., and Shirota, T. (1981) *Kenkyu Hokoku-Kogyo Gijutsuin*, **2**, 155.

Watson, M.N. (ed.) (1988) *Joining Plastics in Production*, The Welding Institute.

Wilson, A.A. (1995) *Plasticizers: Principles and Practice*, Institute of Materials.

Wypych, G. (1993) *Fillers*, ChemTech Publishing.

Wypych, G. (2008) *Handbook of Materials Weathering*, ChemTech Publishing.

Xanthos, M. (ed.) (2010) *Functional Fillers for Plastics*, Wiley-VCH.

Zitouni, F. (1994) PhD thesis, Loughborough University.

General consultation

Brydson, B.J. (1981) *Flow Properties of Polymer Melts*, Goodwin.

Brydson, J.A. (1997) *Plastics Materials*, 7th edn, Butterworth-Heinemann.

Hofmann, W. (1980) *Rubber Technology Handbook*, Hanser.

Jenkins, A.D. (ed.) (1972) *Polymer Science: A Materials Science Handbook*, North-Holland.

Koo, J.H. (2006) *Polymer Nanocomposites*, McGraw-Hill.

Mark, H.F. (2004) *Encyclopedia of Polymer Science and Technology*, 3rd edn, Wiley-Interscience.

Rupprecht, L. (ed.) (1999) *Conductive Polymers and Plastics in Industrial Appplications*, SPE/Plastics Design Library.

Tadmor, Z. and Gogos, C.G. (2006) *Principles of Polymer Processing*, 2nd edn, SPE/Wiley-Interscience.

Ulrich, H. (1993) *Introduction to Industrial Polymers*, Hanser.

Veselosky, R.A. and Kestelmann, V.N. (2002) *Adhesion of Polymers*, McGraw-Hill.

White, J.L. (1996) *Rubber Processing: Technology – Materials – Principles*, Hanser.